Textile Science

Textile Science

Edited by

Deepali Rastogi
Sheetal Chopra

Orient BlackSwan

TEXTILE SCIENCE

ORIENT BLACKSWAN PRIVATE LIMITED

Registered Office
3-6-752 Himayatnagar, Hyderabad 500 029, Telangana, INDIA
e-mail: centraloffice@orientblackswan.com

Other Offices
Bengaluru, Chennai, Guwahati, Hyderabad, Kolkata,
Mumbai, New Delhi, Noida, Patna, Visakhapatnam

First published 2017
Reprinted 2023

ISBN 978-93-86392-66-4

037087

Typeset in
Adobe Garamond Pro 11.5/13.5
by Shine Graphics, Delhi 110 094

Printed in India at
Rajiv Book Binding House, Delhi

Published by
Orient Blackswan Private Limited
3-6-752, Himayatnagar,
Hyderabad 500 029, Telangana, INDIA
e-mail: info@orientblackswan.com

Contents

Tables, Boxes, Figures and Images

Tables

Boxes

Figures

Images

PREFACE

Textile Science is aimed at imparting a comprehensive knowledge of textiles, right from the fibre to the finished fabric. It provides information needed by students of textile and fashion institutes, people working in the textile and garment industry, and consumers who have to make the right decisions in selection, use and care of textile products. It will also serve as a good reference book for teachers of textile science.

This volume of fourteen chapters has been organised in three units. The first unit of five chapters discusses the basics of fibres, the production and properties of different natural and man-made fibres and yarns. Brief mention has also been made of new developments in cotton fibres—such as Bt cotton, organic cotton, coloured cotton, etc. as well as all the varieties of silk available in India—such as, Muga, Eri, Tussar, etc. The stages and methods of blending have been described in the last chapter of this unit.

The second unit describes various techniques of fabric formation—i.e. weaving, knitting and non-wovens—in three chapters. The production and properties of these fabrics have been discussed along with their use and care. Other methods of fabric construction such as braiding, knotting, laces, tufted fabrics, etc. have also been briefly discussed.

The third unit gives a detailed account of finishing, dyeing and printing of fabrics—processes that impart various desirable properties to the fabrics. While Chapter Nine discusses finishes which are carried out on fabrics as a matter of routine, Chapter Ten deals with finishes that impart special functional and aesthetic properties to fabrics. Basic concepts of colour, dyes and dyeing, along with the principles of colour fastness have been dealt with in Chapter Eleven. Chapter Twelve is a detailed study on the chemistry, mechanism and application conditions of various dyestuffs, including natural dyes. Chapter Thirteen gives an account of the methods and styles of textile printing. Last but not least, Chapter Fourteen discusses labelling, certifications, standardisations and organisations dealing with consumer protection and redressal. All the chapters in this volume have a list of questions at the end to help reinforce the concepts explained therein.

A unique feature of this book is that it has been written keeping in mind the vast textile industry and rich textile heritage of India, and at the same time without compromising on the global perspective. For instance, when discussing the production and properties of textile fibres, information on major places of production includes international and Indian sites. A strong foundation in the discipline would help students take this subject as their area of specialisation, we believe.

This book is the combined effort of the members of the faculty of the the Department of Fabric and Apparel Science, Lady Irwin College, University of Delhi. We hope that this book will serve as a valuable reference for undergraduate students of not only the home science colleges but also other institutes offering the basics of textile science in their course curriculum. Any suggestions by the readers are welcome. We will try our best to incorporate them in subsequent editions.

23 March 2017
New Delhi

Deepali Rastogi
Sheetal Chopra

UNIT I

FUNDAMENTALS OF FIBRES AND YARNS

Textiles are an integral part of our existence. They stay with us from the cradle to the grave. We are surrounded by textile materials and products in various forms. An understanding of the various textile components such as fibres, yarns, fabrics and finishes and their inter-relationship is important not only for the manufacturers of textiles products but also for all of us as consumers.

The basic building block of textiles is fibre. Fibres are joined together to form long strands which are known as yarns. These yarns are further converted into fabrics using various techniques. This first unit of five chapters is dedicated to the fundamentals of various aspects of fibres and yarns. The first chapter helps us understand the basics of fibres—their internal structure and properties which make them suitable for making textiles. The classification of textile fibres has also been included in this chapter. In the second and third chapters we learn about the production, properties, use and care of various natural and man-made fibres, respectively. The fourth chapter discusses the manufacture of yarns, their types and their characteristics. Different types of fibres can be blended so that the desirable properties of the different fibres can be incorporated in the final product. The last chapter of this unit discusses the various, methods and objectives of blending.

1

INTRODUCTION TO TEXTILE FIBRES

HIGHLIGHTS

- Definition of textile fibres
- Classification of textile fibres
- Basic units of textile fibres
- Inter-polymer bonds in textile fibres
- Fibre morphology
- Properties of textile fibres

Textile fibres have been used to service humankind since prehistoric times. Archaeological excavations support the fact that our ancestors derived fibres from plant and animal sources. These were twisted to form cords, and were converted into fabric through processes such as plaiting, weaving, felting and knitting. Through history, commonly used fibres were cotton, flax (linen), wool and silk. Cotton was used in hot climates, wool in cold conditions and silk, being an expensive fibre, was worn by nobility and richer sections of the society. Commercial-manufacturing of fibres which began in 1885 revolutionised cloth-making. The desire to produce cheaper alternatives to silk which is famed for its lustre, softness and smoothness, led to the development of the first man-made fibre—rayon (also known as artificial silk) in 1925. Many other man-made fibres such as nylon, polyester, acrylic, olefin, etc. were developed subsequently. The history and development of natural and man-made fibres have been dealt in detail in Chapters Two and Three.

Definition of Textile Fibres

A fibre is a linear, pliable material, natural or man-made, that possesses qualities required to be processed into suitable textiles. It is the fundamental unit used in production of textile yarns and fabrics.

> **Textile** is any kind of material that is woven, knitted, tuffted, knotted, non-woven or otherwise made by any other fabric construction technique to be produced into a flexible material.

CLASSIFICATION OF TEXTILE FIBRES

For thousands of years many systems have been used to classify fibres into various groups. Based on their origin, fibres were classified into vegetable, animal and mineral fibres. However, with the development of man-made fibres, this systematic arrangement has become obsolete and further subclassifications have been devised. In 1960, the Textile Fibre Products Identification Act (TFPIA) came into effect in the USA. According to this now-globally-accepted law, it became mandatory to mention the fibre content on the label of textile product being sold to the consumers. The classification and mention of the type of textile fibre helps consumers in making the right decision in terms of selection and care and maintenance. Classification of textile fibres involves the following factors:

- Origin of the fibre
- General chemical constitution—i.e. cellulose, protein, mineral, synthetic
- Generic term or family name: This factor provides scientific basis for grouping man-made fibres that belong to a common or similar chemical nature/structure. During sale of some of the textile products made from man-made fibres, the generic term is emphasised since consumers may not be sure of the trade name. For example acrylic, polyester, nylon and spandex are generic names of fibres that are known by various common names.
- Common names/trade names: Many consumers are more familiar with common names and trade names for fibres. Thus, to help consumers correctly identify fibres, examples of these are also included in classification along with their generic names. For example, orlon and cashmilon are mentioned along with the generic name acrylic and lycra is mentioned along with the generic term spandex.

On the basis of origin, the textile fibres are broadly classified as natural and man-made fibres. Natural fibres are those obtained from nature in fibrous form. There are various sources from where these natural fibres can be obtained—plants, animals and natural minerals. On the other hand, man-made fibres are manufactured under controlled conditions. The raw material for these manufactured fibres can either be obtained from nature or can be synthesised in the laboratory. Table 1.1 gives a detailed classification of textile fibres.

Natural Fibres

i. *Plant fibres*: Various parts of plants can be used to obtain fibres. For example, fibres may be obtained from seed hair (cotton), stem (linen) and leaf (banana).
ii. *Animal fibres*: Fibres from animal sources may be either obtained from their hair (wool) or may be extruded (silk). Fibres can also be obtained from bird feathers.
iii. *Mineral fibres*: Fibres can also be obtained from various mineral sources. For example, asbestos fibres are obtained from rocks containing silicates of magnesium and calcium.

Table 1.1: Classification of textile fibres

Natural fibres	*Man-made fibres*
i. Plant fibres/Cellulosic fibres • *Seed hair*: cotton, kapok • *Stem* or *bast*: flax, jute, kenaf, industrial hemp, ramie and vine fibres. • *Leaf*: pina from pineapple leaves, banana fibre, sisal from agave leaves • *Fruit*: coconut fibre (coir) • *Stalk*: straw of wheat, rice and barley, bamboo, bagasse, corn husk and grass.	i. Regenerated fibres • *Cellulosic*: rayon, lyocell, modal • *Protein*: azlon (fibres from soybean, corn)
ii. Animal fibres • *Animal Hair*: Sheep wool; speciality hair fibre such as goat hair (mohair, cashmere), alpaca hair, vicuna, camel hair, horse hair, yak hair, llama hair; animal fur such as those of rabbit, angora and mink • *Extruded fibres*: cultivated and wild silk, spider silk • *Avian fibres*: feathers and feather fibres from birds such as ducks, geese, emu and ostrich	ii. Regenerated modified fibres Acetate, triacetate
iii. Mineral fibres Asbestos	iii. Synthetic fibres Nylon, polyester, acrylic, modacrylic, olefin-polyethylene and polypropylene, aramid, spandex
iv. Natural rubber	iv. Inorganic fibres Glass, carbon, metallic (gold and silver) v. Synthetic rubber

Source: Compiled by the authors.

iv. *Natural rubber*: The major commercial source of natural rubber latex in India is the Pará rubber tree (*Hevea brasiliensis*).

Man-made Fibres

i. *Regenerated fibres*: These fibres are obtained after reforming (regenerating) natural raw material such as cotton fibres, wood pulp or natural protein. Rayon is a regenerated cellulosic fibre that was first manufactured using cotton linters and wood pulp as raw material.

ii. *Regenerated modified fibres*: These fibres are regenerated from natural cellulosic raw material and modified by partially replacing the hydroxyl groups in cellulose chains with acetyl groups. Examples are acetate and triacetate fibres.

iii. *Synthetic fibres*: Fibres that are manufactured from synthesised chemicals are known as synthetic fibres. These are manufactured under controlled conditions to form a polymer which does not exist in a natural state. Nylon was the first synthetic fibre to be developed in 1938 by DuPont in the USA. This was followed by development of other fibres such polyester, acrylic, olefin, etc.

iv. *Inorganic fibres*: Some inorganic substances such as glass, and a few metals have been explored to obtain fibres. In this case the raw material is first softened and then formed into thin, long, pliable filament strands.

v. *Synthetic rubber*: Synthetic rubber is obtained from petroleum byproducts. The most prevalent synthetic rubber is styrene-butadiene rubber (SBR) derived from styrene and 1,3-butadiene. Other synthetic rubbers are prepared from isoprene, chloroprene and isobutylene.

BASIC UNITS OF FIBRES

Fibre, the basic unit of any textile, is made up of polymers which are in turn made up of monomers.

Monomers and Polymers

The Latin word 'monomer' comes from '*mono*' means 'one' and '*mer*' which means 'unit'. The monomer is the basic repeat unit of a polymer. Monomers have low molecular weights and are chemically reactive. Generally, 100–15,000 monomer units together to form a polymer. The Latin word 'polymer' comes from '*poly*' which means 'many' and '*mer*' which means 'unit'. A polymer may be formed from one monomer unit which repeats itself several times or from different monomers that may join together in some systematic manner. Polymers have high molecular weights and are chemically stable. Polymers can be of different types—linear, branched or cross-linked. However, textile polymers are essentially linear chained.

Polymerisation

The chemical process through which simple, low-weight monomer units join end to end through covalent bonding to form a more complex, high-weight polymer is known as polymerisation. Figure 1.1 illustrates the building blocks, i.e. the monomer and polymer, of cotton fibre.

Figure 1.1: Building blocks of cotton fibre

Note: The atoms are carbon, hydrogen and oxygen; the monomer is the cellobiose unit made with two glucose units; and the polymer is cellulose.

Degree of Polymerisation

The number of monomer units that join together to form a polymer is referred to as the degree of polymerisation (*dp*). It is determined by the following equation:

$$\text{Degree of polymerisation} = \frac{\text{Average molecular weight of polymer}}{\text{Molecular weight of the repeating unit in the polymer}}$$

Table 1.2: Degree of polymerisation in various fibres

Fibre	*Degree of Polymerisation*
Cotton	5000
Flax	18,000
Rayon	175
Acetate	130
Triacetate	225
Nylon 6	200
Nylon 6,6	50–80
Polyester	115–140
Acrylic, Modacrylic	2000

Source: Compiled by the authors.

Types of Polymerisation

There are two types of chemical reactions during the polymerisation process:

i. *Addition polymerisation*: In this kind of polymerisation reaction, monomers react to form the polymer without any byproduct. Fibres such as acrylic, modacrylic, polyethylene, polypropylene, etc. are produced through addition polymerisation reaction.
ii. *Condensation polymerisation*: In this kind of polymerisation, monomers polymerise and liberate a simple compound like water, hydrogen chloride or ammonia as byproduct. In such cases, usually two different types of monomer units form the polymer. Examples of fibres produced through condensation polymerisation are nylon, polyester, elastomers, etc.

Types of Polymers

As mentioned above, during the polymerisation process, a polymer may be formed by joining together many units of one kind of monomer or two or more types of monomers. In case only one type of monomer units join together the polymer thus formed is known as a **homopolymer**. For example, polymers of nylon, polyester, polyvinyl chloride (PVC), polypropylene (PP), polyethylene, polyacrylonitrile, etc. are formed by only one type of monomer. If a polymer is formed by joining different monomers, it is referred to as a **copolymer**. An example is modacrylic, which is made up of acrylonitrile and polyvinyl chloride. In the formation of a copolymer the monomers may be arranged in four different sequences:

Figure 1.2: Sequences of monomer arrangement

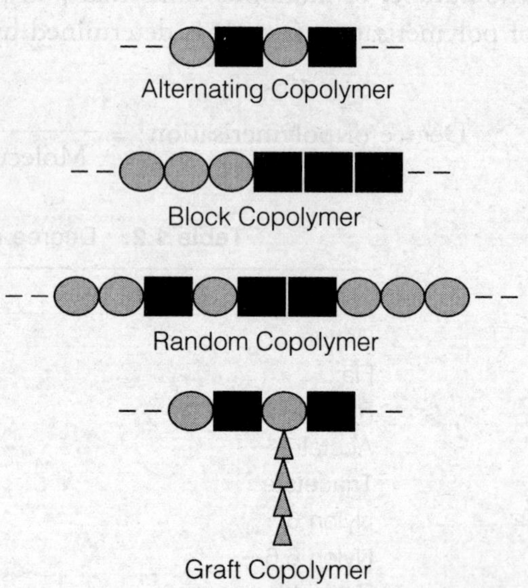

Alternating Copolymer

Block Copolymer

Random Copolymer

Graft Copolymer

- *Alternating copolymer*: In this kind of copolymer two types of monomers are arranged in an alternating sequence.
- *Block copolymer*: Here different monomers react first in blocks and then link-up to form the polymer.
- *Random copolymer*: In this type, monomers arrange themselves randomly to form the polymer.
- *Graft copolymer*: In this kind, a polymer segment attaches itself as a side chain to an already existing polymer chain to form a branch/graft.

INTER-POLYMER BONDS IN TEXTILE FIBRES

Types of Inter-polymer Bonds

1. Van der Waals forces
2. Hydrogen bonds
3. Salt linkages
4. Covalent bonds

Polymer chains in a textile fibre are held together by bonds known as inter-polymer bonds, also known as inter-polymer forces of attraction. There are four types of inter-polymer bonds:

Van der Waals Forces

These are the weakest inter-polymer forces of attraction that are present in all fibres. Van der Waals forces of attraction come into existence in the mere vicinity of molecules. Closely packed parallel polymer chains about 0.2 nm apart in a fibre system are held by these forces of attraction.

Hydrogen Bonds (H-bond)

These are weak electrostatic bonds formed between two atoms with slight opposite charges. The most common electrostatic H-bond in the polymer system exists between hydrogen and oxygen atoms. When these atoms are close to each other, at a distance of less than 0.5nm, the hydrogen atoms develop a slight positive charge and the strongly electronegative oxygen atoms assume a slight negative charge. These bonds may form between oppositely charged hydrogen and chlorine/nitrogen/flourine atoms. The strength of H-bonds depends upon the electronegativity of the oxygen/nitrogen/chlorine/fluorine atoms. The higher the H-bonds in a fibre polymer the greater the strength, durability, moisture absorbency and heat-setting properties of the fabric. H-bonds are present in protein, cellulose, nylon and polyvinyl alcohol. Figure 1.3 illustrates H-bonds in nylon fibre.

Figure 1.3: Hydrogen bonds in nylon fibre

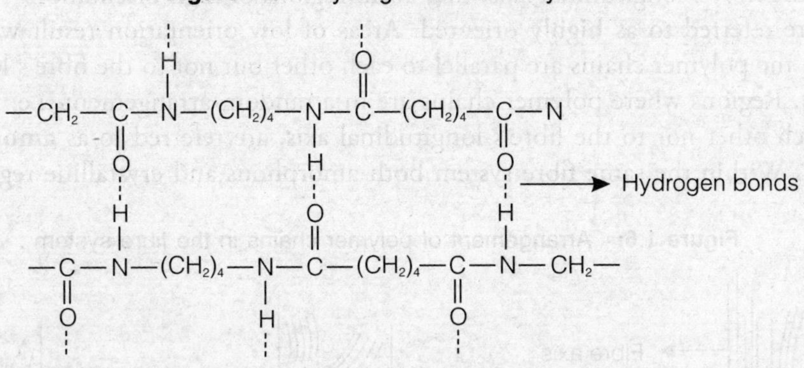

$—CH_2—C—N—(CH_2)_4—N—C—(CH_2)_4—C—N$

Hydrogen bonds

$—C—N—(CH_2)_4—N—C—(CH_2)_4—C—N—CH_2—$

Source: Drawn by the authors.

Salt Linkages

Also referred to as salt bridges, electrovalent or ionic bonds, these bonds are stronger than Van der Waals forces and H-bonds. These occur between positively and negatively charged ions that are present very closely in a fibre system. Salt linkages result in better cohesiveness, tenacity and durability of the fibre polymer. The polar nature of these linkages attracts more water molecules due to which the fibre is more absorbent, and hence easy to dye, and more comfortable to wear. Salt linkages occur between negatively-charged carboxyl groups on one polymer chain and positively charged amino groups on an adjacent polymer chain in nylon and protein fibres. This is illustrated in Figure 1.4.

Figure 1.4: Salt linkage between carboxyl group and amino group

$–COO^-$ ——Salt bridge—— $^+NH_3–$

Source: Drawn by the authors.

Covalent Bonds

Also known as cross links, these are the strongest bonds between two adjacent polymers chains of a fibre. Covalent bonds render chemical stability to a fibre. The number of covalent bonds in a fibre polymer affects its stiffness or rigidity and, consequently, its flexibility. Usually, the higher the number the lesser the flexibility. Figure 1.5 illustrates covalent bond in wool fibre in the form of disulphide linkage.

Figure 1.5: Disulphide linkage in wool fibre

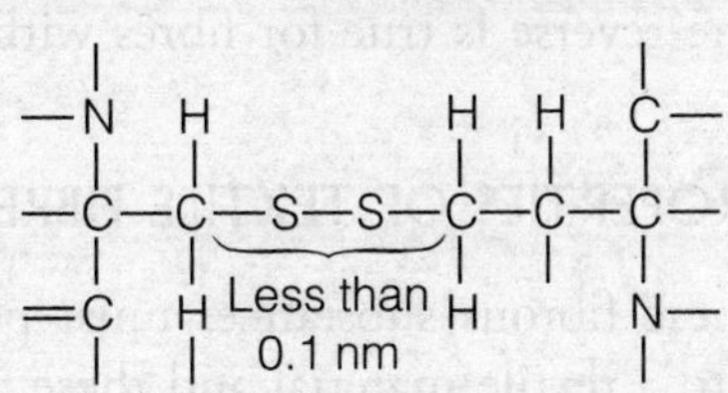

—C—C—S—S—C—C—C—

Less than 0.1 nm

FIBRE MORPHOLOGY

Polymer chains can arrange themselves in a variety of ways within the fibre system. The study of arrangement of polymer chains within a fibre system is defined as fibre morphology. The polymer chains may exhibit regions of ordered arrangement where they are parallel to each other, and these are known as **crystalline regions**. If in a crystalline region the polymers are parallel to each other as well as to the fibre's longitudinal axis, they form regions of high orientation (Figure 1.6a) and such fibres are referred to as highly oriented. Areas of low orientation result within crystalline regions when the polymer chains are parallel to each other but not to the fibre's longitudinal axis (Figure 1.6b). Regions where polymer chains are in a random arrangement, i.e. they are neither parallel to each other nor to the fibre's longitudinal axis, are referred to as **amorphous regions** (Figure 1.6c). Within the same fibre system both amorphous and crystalline regions may exist.

Figure 1.6: Arrangement of polymer chains in the fibre system

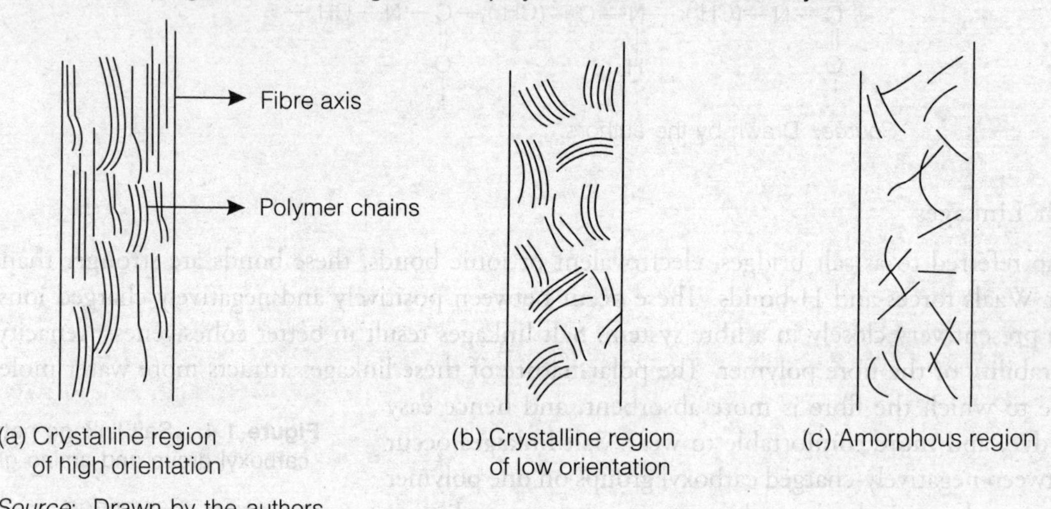

Source: Drawn by the authors.

Such crystalline and amorphous arrangements of polymer chains affect the fibre properties. When force is applied, amorphous regions allow the polymer chains to slip and get oriented, whereas crystalline regions allow reorientation to a lesser extent and highly-oriented crystalline regions even less. Crystalline fibres, therefore, exhibit higher strength, lustre, elastic recovery, chemical resistence and stiffness, but low absorbancy, dyeability, elasticity, elongation and poor flexibility. The reverse is true for fibres with a higher degree of amorphous regions.

PROPERTIES OF TEXTILE FIBRES

These fibrous substances must possess certain essential characteristics in order to be converted into a textile material and these are called as primary properties of textile fibres. There are other characteristics that are not essential but are desirable to form textile materials with better servicibility, performance and qualities and these are referred to as secondary or non-essential properties of textile fibres. Based on the presence or absence of these propeties, a fibre may be unique in its usage.

> **Primary properties of textile fibres**
> 1. Length to width ratio
> 2. Cohesiveness
> 3. Tenacity
> 4. Flexibility
> 5. Uniformity

Primary Properties

The primary properties of textile fibres are:

Length to Width Ratio

The relationship between fibre length and width is referred to as length:width (L:W) ratio. Textile fibre must have minimum L:W ratio of 100:1. This means that the fibre must be 100 times longer than its width. Fibres obtained from nature vary considerably in their length and width and thus possess different L:W ratio. For example, for a cotton fibre of 2.54 cm length and 0.0018 cm diameter, the L:W ratio is 1400. Fibres shorter than 1.3 cm are seldom used in manufacturing yarn. Fibres should be sufficiently long (over 1.5 cm) so that they can be twisted into a regular, even and cohesive mass and made into yarn.

Based on their lengths, fibres can be classified as staple fibres and filament fibres. **Staple fibres** are short fibres measured in centimeters or inches or fractions of these, whereas **filament fibres** long fibres measured in terms of kilometres or metres. By and large most natural fibres are staple fibres; the length of cotton fibre ranges from 0.6 cm to 5.1 cm and that of wool from 3 cm to 40 cm. All synthetic fibres produced by man are filament fibres and their lengths are controlled during extrusion. These long synthetic filament fibres may be cut to staple lengths if they are to be blended with other natural fibres. Silk fibre, which is several kilometres long, is the only natural fibre that is a filament fibre.

Cohesiveness

Textile fibres must possess cohesiveness so that they are capable of clinging to each other during yarn manufacture. Also referred to as spinning quality, this primary property affects yarn fineness and fabric surface texture, appearance, drapability and durability.

Factors affecting cohesiveness

- Cross-sectional shape of the fibre
- Fibre length or longitudinal contour of fibre
- Surface texture or contour of the fibre

Natural fibres such as wool and cotton have inherent irregularities in their cross-sectional as well as longitudinal contours such as convolutions or twists which allows them to adhere to each other and be easily spun into yarns. For instance, the presence of convolutions in cotton and scales in wool makes it possible to easily spin them into yarn. By virtue of their length, fibres may be twisted together to form a yarn. Silk and other man-made filament fibres, due to their long lengths, require considerably less twist to form a yarn. Cohesiveness or clinging quality may also be imparted in them through texturisation which introduces coils, crimps and other surface modifications to provide spinning quality.

Tenacity

A fibre must possess adequate strength to undergo various stages of mechanical and chemical processing during manufacturing and use as yarn and fabric. This strength is referred to as tenacity or tensile strength and is expressed as force per unit linear density of the unstrained specimen. Tensile strength is the force required to break a fibre's cross sectional mass equivalent to one unit of measure used. It is measured in pounds per square inch (psi) or newtons per square meter. In the case of fibres and yarns it is expressed as grams of force per denier (g/d) or grams of force per tex (g/t). Thus, in other words, tenacity is number of grams required to break a fibre or yarn of one denier or tex (denier and tex are measures of yarn fineness where one denier is the weight in grams of 9000 meters of yarn and tex is the weight in grams of 1000 meters of yarn).

Tenacity varies from one fibre to another. The range of tenacity of a few fibres are: 3–5 g/d for cotton, 1–1.7 g/d for wool, 2.4–5.1 g/d for silk, 2.6–7.7 g/d linen. Tenacity of a fibre may differ when it is wet or dry. For instance, cotton fibre is stronger when wet whereas wool and rayon are weaker when wet and, therefore, need to be handled very gently during washing or any wet treatment. Depending upon fibre tenacity, fabrics made from them can be put to various end uses. Fabrics meant for work and other industrial applications will have better durability if they are made from fibres of high tenacity.

Flexibility

To be processed to yarns and made into fabrics, fibres must bend or flex without rupture. Most of the uses of textile products, be it in apparel, home furnishing or industry, require bending, folding, creasing, etc. Therefore, stiff or brittle fibres cannot make useful fabrics. Pliability of fibres refers to the freedom of movement of the fibre as per body movements. It influences various other important fabric properties such as drapability, crispness and softness. It allows the fabric to be creased again and again without rupturing.

Uniformity

For all the fibres to be converted to a uniform yarn and thereafter to a uniform fabric, it is important that fibres possess limited variations—they should be fairly similar in length and width, strength, cohesiveness and flexibility. Yarns and fabrics made from uniform fibres will be regular, exhibit smoother appearance, and accept dyes and finishing chemicals uniformly. In general this property is more vital in the case of natural fibres where there is lesser control on fibre to fibre variations. During the manufacture of man-made fibres high degree of uniformity is usually achieved. To bring about this uniformity in natural fibres, fibres from different lots are often blended with each other before they are processed.

Secondary Properties

These are those properties that are not essentially required for a fibre to be used as a textile fibre. The presence or absence of these properties may impart certain unique characteristics and usages to a fibre. For instance, we have textile fibres of different levels of moisture absorbancy. Polypropylene, which has 0 per cent moisture absorption, finds extensive usage in packaging and cement bags where the

Secondary properties of textile fibres

1. Physical shape of fibre
2. Density
3. Lustre
4. Fibre colour
5. Moisture content, absorption and regain
6. Elongation and elastic recovery
7. Resiliency
8. Flammability and other thermal properties
9. Abrasion resistance
10. Chemical resistence
11. Biological resistence
12. Sensitivity to sunlight and other environmental factors

presence of moisture can spoil the product. On the other hand, any wet treatment of polypropylene is difficult. Cotton and wool with good moisture absorbancy provide comfort along with easy dyeabilty. Secondary properties also influence processing, selection, use and care of textiles. The following are secondary properties of textile fibres.

Physical Shape of Fibre

Physical shape includes fibre's longitudinal section, cross-section, surface contour, average length and irregulatities. These shape characteristics help determine both macroscopic and microscopic appearance of the fibre. It is important to understand the macroscopic and microscopic appearance of fibres because they have a direct impact on properties such as, lustre, spinnability, smoothness and flexibility.

Density

Density is the mass per unit volume of a fibre, expressed as grams per cubic centimeter. The variations in density of fibres result in variations in the weights of fabric produced. If all the other factors such as yarn construction, fabric construction and finishing operations applied are kept constant, low density fibres such as acrylic will give lighter weight fabrics as compared to fabrics made from higher density fibres such as cotton.That is why, quilts filled with synthetic fibres are lighter in weight as compared to those filled with cotton fibre. Table 1.3 enlists density of various fibres in descending order.

Table 1.3: Density of some common textile fibres

Fibre	*Density (g/cm^3)*
Cotton	1.54
Flax	1.50
Rayon	1.50
Polyester	1.38
Wool	1.30
Acetate	1.30
Silk	1.25
Acrylic	1.16
Nylon	1.14
Polypropylene	0.86–0.95

Source: Compiled by the authors.

Lustre

Lustre is the gloss, sheen or shine of a fibre. The amount of light reflected back from the surface of the fibre determines the lustre. The reflection of light by fibre surface is directly related to the shape of its cross-section. A regular fibre surface will reflect the light evenly. A fibre will be dull if the light reflected back is less even. Fibres with oval (such as wool), round (acrylic) and triangular (silk) cross-sections will be more lustrous since they evenly reflect the light falling on them. Figure 1.7a illustrates the effect of fibre cross section on reflection of light and fibre lustre. Trilobal (trilobal nylon), kidney shaped (cotton), multilobal (or dog-bone section, as in acrylic) or other irregular (viscose rayon) cross sections unevenly reflects light, and therefore those fibres have a low lusture. Figure 1.7b illustrates the effect of irregular cross-sections on light reflection.

Large variations can be seen in the lustre of natural fibres due to variations in their cross-sectional shapes. For example, silk fibre has high lusture due to its triangular cross section whereas cotton with bean shaped cross section and convolutions has low lusture. Man-made fibres, as compared to natural fibres, have very high lustre due to their very regular cross-section. Such high lustre is not always desirable. The lustre can be controlled in case of man-made fibres by adding

Figure 1.7a: Effect of regular cross-section on light reflection

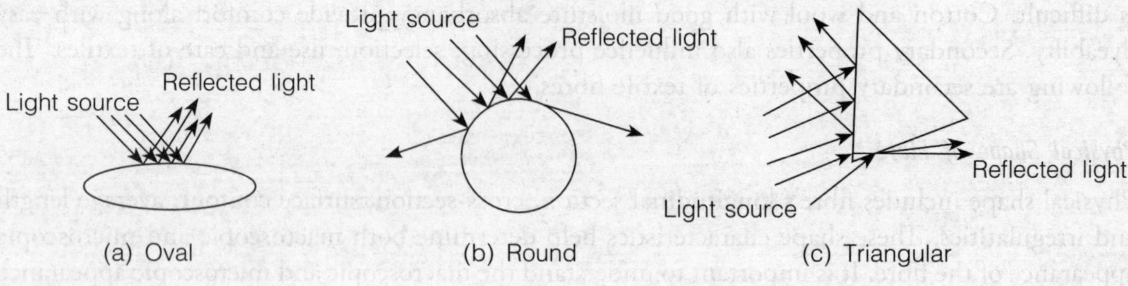

Source: Drawn by the authors.

Figure 1.7b: Effect of irregular cross-section on light reflection

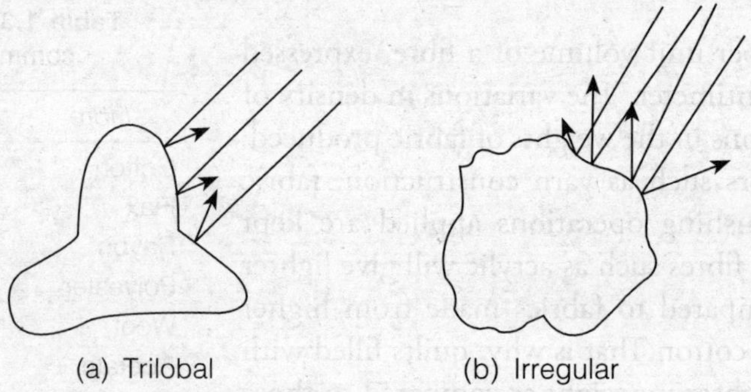

Source: Drawn by the authors.

a delustrant into the spinning solution prior to fibre formation. Titanium dioxide is generally used for this purpose. Delustrants give a speckled cross-sectional surface, thereby allowing the light falling on it to scatter resulting in low lustre.

Fibre Colour

Fibres are available in a wide range of colours. Large variations can be seen in natural fibres that range from pure white to deep grey, tan or black. On the other hand, man-made fibres are usually white or off-white in colour. White/colourless fibres are prefered over coloured fibres since these can be easily bleached, dyed in lighter shades or printed with any colour for surface ornamentation.

Moisture Absorption, Content and Regain

Moisture absorption is the ability of fibre to absorb moisture from the atmosphere. Moisture absorption depends on whether the fibre is hydrophobic (water-hating) or hydrophilic (water-loving) in nature. Hydrophilic fibres such as cotton, rayon, wool and silk exhibit good moisture absorption property. On the other hand, hydrophobic fibres such as, nylon, polyester and acrylic show low moisture absorption properties.

Absorption of moisture can be expressed as either 'moisture content' or 'moisture regain'. The **moisture content** of a textile fibre is the amount of moisture expressed as the percentage of its conditioned weight. Conditioned weight is the weight of a sample of fibre that has been kept under standard atmospheric conditions (since atmospheric humidity affects fibres, the values for moisture content and regain are measured under standard atmospheric conditions, i.e. 65 per cent (± 2) relative humidity and 20°C ± 2°C temperature). The formula used to calculate moisture content is:

$$\text{Moisture content (\%)} = \frac{\text{conditioned weight} - \text{dry weight}}{\text{conditioned weight}} \times 100$$

On the other hand, **moisture regain** is the amount of moisture expressed as the percentage of the weight of the moisture free specimen. Moisture-free or 'dry state' samples are obtained when samples are oven dried to bone-dry state. The samples are dried, weighed, redried and weighed till a constant weight is reached. The formula used to calculate moisture regain is:

$$\text{Moisture regain (\%)} = \frac{\text{conditioned weight} - \text{dry weight}}{\text{dry weight}} \times 100$$

It is generally assumed that moisture picked up by fibres is absorbed by them. But some water molecules may remain only on the surface of fibre in which case it is said to be **adsorbed**. The process of adsorption results in **wicking**, that is, the transfer of moisture along the fibre surface. Some fibres irrespective of their low moisture regain are still considered to be comfortable to wear owing to their wicking property.

Moisture absorption affects the processing, care and comfort properties of textile fibres. Good moisture absorption properties render fibres more amenable to wet processing such as dyeing and finishing. However, such fibres will lose moisture slowly and will take a lot of time to dry. On the other hand, fibres with low moisture regain will be difficult to dye, but will dry faster. Certain fibres vary in their strength under wet and dry states while others may not show any difference. For instance, rayon when wet becomes weaker in its strength while cotton increases in its strength when wet. Thus, depending upon the wet strength the fibre needs to be handled accordingly during wet processing and laundering. Fibres such as nylon, polyester, etc. with low moisture regain, will have drip-dry property, i.e. they dry easily but they will be uncomfortable to wear during hot climate. This is due to the presence of a film of sweat in contact with the wearer's skin. However, when this film evaporates it gives a cooling effect. Moisture from the skin may thus be adsorbed by the fibre and wicked away to another layer of the apparel. Latest sportswear utilise this property of synthetic fibres to achieve the dry-feel effect.

Elongation and Elastic Recovery

Textile fibres must possess elasticity so that the fabric is able to come back to its orignal shape and size after any load or force is removed from it. Any change in length that may occur in a fibre due to stretching is referred to as **extension**. **Elongation** or percentage elongation in a fibre can be found out by calculating the ratio of extension of a fibre to its actual length prior to stretching. The following equation illustrates this ratio calculation:

$$\text{Elongation (\%)} = \frac{\text{(extended length} - \text{original length)}}{\text{original length}} \times 100$$

If a fibre recovers completely from any deformation after the force or load applied is removed, i.e. if it comes back to its original length, it is said to be 100 per cent elastic. In other words, such a fibre is said to exhibit 100 per cent elastic recovery. Thus, elastic **recovery** can be defined as the ability of a fibre to recover to its original length after the removal of load. It is not always necessary that all fibres exhibit 100 per cent elastic recovery, they may show permanent extension as well. This may happen due to molecular rearrangment within the fibre and hence permanent deformation. On 2 per cent elongation cotton has 75 per cent elastic recovery while silk shows 92 per cent recovery and wool has 99 per cent recovery. On the other hand, at a higher elongation of 8 per cent, spandex (lycra) and nylon show 100 per cent elastic recovery.

Fibres that exhibit excellent elastic recovery may show low elongation and thus have limited usage. Similarly, if a fibre has high elongation but cannot come back to its original length, then usage of this fibre also becomes undesirable, since products made out of it will loose their shape and size. Both elongation and elastic recovery are, therefore, important fibre properties. Besides affecting comfort, they also influence the ability of fabrics to retain their appearance.

Resiliency

Resiliency can be defined as the ability of a fibre to return to its original shape after removal of any form of deformation such as folding, creasing, twisting, stretching, flexing or compression. Materials with good elastic recovery exhibit good resiliency. This property helps determine crease recovery of a fibre or fabric. A carpet pile or wrinkles formed in a garment will show recovery to its original position due to resiliency after being compressed. Polyester, wool and nylon show high resiliency as compared to silk, cotton and rayon. Linen has the lowest resiliency.

Flammability and Other Thermal Properties

Most textile fibres burn when exposed to flame. They exhibit different behaviours when approaching flame, in the flame and away from it; they produce specific odours and residues. While some fibres may ignite and burn at high temperatures, others may melt or soften. The manner in which a fibre will react to heat and flame determines its thermal properties. These properties are very important for the safety of the consumer. It is on the basis of the thermal properties of a fibre that the care instructions, safe ironing temperatures and optimum processing temperatures are determined.

Those fibres that melt or soften when exposed to high temperatures or heat are called **thermoplastic fibres**. Some examples are nylon, polyester and olefin fibres. Such fibres have a definite glass transition temperature (T_g) at which point the fibres soften. At T_g, the amorphous regions of the fibre flow but the material retains its basic fibre form. The fibre becomes pliable—it may be reshaped—and will maintain this new shape till it is heated again to its T_g. The melting point (T_m) is the temperature at which the fibre liquidifies. At this temperature permanent fibre damage occurs.

Abrasion Resistance

Abrasion is referred to as the wearing away of a fabric when it rubs against another surface, i.e. it is the resistance of a material against wear due to friction. It is desirable for any textile material to resist this kind of damage so that it gives better serviceability. Generally, three types of abrasions are seen in textiles and garments:

- *Flex abrasion*: It is seen when a fabric flexes/bends or folds and rubs against another surface, as in the elbow or knee positions of apparels.
- *Flat abrasion*: It is seen when a fabric's flat surface rubs against another surface, as in the thigh area in jeans and the underarm areas in blouses, and in bed sheets and pillow covers.
- *Edge abrasion*: This is seen on curved edges or folds in a garment, as in collar edges or cuff ends or trouser hems.

Natural fibres, such as cotton and wool, show low abrasion resistance whereas synthetic fibres, such as polyester, nylon and acrylic, show very high abrasion resistance.

Other Miscellaneous Properties

Other properties of textiles that are of major concern when these are put to specific end uses include resistance to chemicals, resistance to microbes or insects (biological resistance) and sensitivity to sunlight and various other environmental factors.

Resistance to chemicals plays an important role in deciding the application of finishes and the care of textiles while using acids, alkalis and other organic compounds. For example, cellulosic fibres are resistant to alkalis but are harmed by acids. Thus acids are not used in processing these fibres. On the other hand, wool and silk are damaged by alkalis but can be processed in an acidic medium.

Resistance to insects and microbes are again important in deciding the end uses and maintenance of textiles. For instance, cotton is easily damaged by mildew which aggravates under humid conditions. Similarly, wool is eaten up by carpet beetles if not stored properly. Due to this, napthalene balls are used in the storage of wool.

Sensitivity to sunlight and other harmful pollutants present in the environment also affects serviceability of fabrics. **Actinic degradation** is the weakening of fibres due to exposure to sunlight. For example, protein fibres weaken more when exposed to sunlight than glass, acrylic and polyester. Similarly, some fibres get discoloured by air pollutants such as acrylic and thus cannot be used in areas with high levels of air pollution.

SUMMARY

- Fibres are basic units of a textile material. Textile fibres have been classified into various groups based on their origin, chemical type, generic or family name and common or trade names.
- A textile fibre is made of polymers. Polymers are formed by joining hundreds of monomers through a chemical reaction known as polymerisation.
- The polymer chains in a textile fibre are held together by four kinds of inter-polymer bonds: Van der Waals forces, hydrogen bonds, salt linkages and covalent bonds.
- Fibre morphology affects the fibre properties of strength, flexibility, elasticity, elastic recovery, moisture regain and lustre.
- A fibre must possess certain essential characteristics, referred to as primary properties to be converted into a textile material, and they are: length to width ratio, cohesiveness, flexibility, tenacity and uniformity.
- Secondary properties are not truly essential but are desirable to form a textile material with better serviceability, performance and qualities, and they are: physical shape of fibre; density; lustre; fibre colour; moisture absorption, content and regain; elongation and elastic recovery; resiliency; flammability and other thermal properties; abrasion resistance; and other miscellaneous properties.

KEY WORDS

Monomer: A basic repeat unit of a polymer.

Polymer: Backbone of a textile fibre, made up of monomers.

Polymerisation: The chemical reaction in which small monomer units of low molecular weight join end to end through covalent bonds to form a polymer of high molecular weight.

Fibre morphology: The study of size, shape, form and structure of a fibre. Morphology also studies the arrangement of polymer chains within a fibre system.

Amorphous regions: Regions with random arrangement of polymer chains in which they are neither parallel to each other nor to the fibre's longitudinal axis.

Crystalline regions: Regions of ordered arrangement in which polymer chains are parallel to each other but may or may not be parallel to the fibre's longitudinal axis.

EXERCISES

1. What is the expansion of TFPIA?
2. Give the detailed classification of fibres with examples.
3. Define the following terms.
   (a) Monomer
   (b) Polymer
   (c) Polymerisation
   (d) Degree of polymerisation (dp)
   (e) Addition polymerisation
   (f) Condensation polymerisation
   (g) Homopolymer

(h) Copolymer

(i) Staple and filament fibres

4. Name the monomer and polymer present in cotton fibre.
5. What do you understand by inter-polymer forces of attraction? Enlist their various types.
6. Define fibre morphology. Explain the types of polymer chain arrangement present within the fibre structure.
7. What do you understand by primary properties of fibres? Enlist and explain these.
8. Why should wool and rayon fibres be handled gently while washing?
9. 'Reflection of light by the fibre surface is directly related to the shape of its cross-section.' Comment and support your answer with illustrations and examples.
10. Define glass transition temperature (T_g) and melting point (T_m).
11. What do you understand by the term 'abrasion'? What are its types? Give examples to explain them.

REFERENCES

Gohl, E. P. G. and L. D. Vilensky. 1983. *Textile Science*. New Delhi: CBS Publishers and Distributors.

Hollen, N. and J. Saddler. 1973. *Textiles*. Fourth edition. New York: Macmillan.

Joseph, M. 1992. *Introductory Textile Science*. Sixth edition. California: Harcourt College Publishers.

Kadolph, S. J. 2009. *Textiles*. Tenth edition. New Delhi: Dorling Kindersley (India).

2

NATURAL FIBRES

HIGHLIGHTS

- Production, processing and properties of natural cellulosic fibres
  - Cotton
  - Flax
  - Jute
  - Other cellulosic fibres
- Production, processing and properties of natural protein fibres
  - Wool
  - Specialty hair fibres and furs
  - Silk
- Other natural fibres
  - Rubber
  - Asbestos

NATURAL CELLULOSIC FIBRES

Natural cellulosic fibres, also called vegetable or plant fibres, are those which are obtained from plant sources. Since cellulose is the main building block of these fibres they are known as cellulosic fibres. These fibres are mainly obtained from four parts of the plant—seed hairs, the fruit husk, the stem and the leaf. Cotton and flax are the most commonly used natural cellulosic fibres. Cotton is gathered from the seed hairs whereas flax and jute are obtained from the stem or bast portion of a plant. Other cellulosic fibres like coir and abaca are obtained from coconut husk and leaves of the abaca plant, respectively.

Cotton

History

The exact history of the first attempts at domesticating cotton remains unclear. This is partly because cotton is easily damaged by micro-organisms, and is therefore extremely difficult to be found in archaeological excavations. The word cotton is derived from *qutun*, an Arabic word which means 'a plant found in conquered lands'. This may indicate that many countries invaded and conquered by Arabs may have already been growing and using cotton. The antiquity of cotton in India can be traced back to the fourth millennium BC. In fact there is evidence to suggest that cotton has been used in India for over five thousand years. Cotton fabrics, made with fibre from a species closely related to *Gossypium arboretum*, were found during the excavation at Mohenjo-daro in the Indus valley and have been dated back to c. 3000 BC.

The earliest literary reference to cotton is found in the *Rigveda* written around 1500 BC. It may be noted that the stages of ginning of seed cotton, spinning the lint and weaving the yarn are covered in this religious text, suggesting the use of cotton fabric in India by 1000 BC. More than a thousand years later, c. 445 BC, the Greek historian Herodotus wrote that Indians possessed 'a kind of plant, which, instead of fruit, produces wool, of a finer and better quality than that of sheep; of this the Indians make their clothes' (Dhantyagi 1996). It is not exactly known when Europe started trade with India, but the use of word *carbasina*, which means 'cotton' in Sanskrit, suggests the usage of cotton even before 200 BC. The Greeks, who arrived in India with Alexander the Great in 326 BC, were surprised to see cotton. They called it 'wool produced in nuts' and praised it saying that it surpassed sheep-wool in beauty and excellence. Indian cotton was famed even in the times of the Buddha where one finds references to fine Banaras and Dacca muslins (ibid.). Nowadays, cotton is considered as the most important textile fibre even though factors like the value of the dollar as compared to other currencies, changing weather conditions and government policies in the US may affect the production and demand of cotton everywhere. The major producers of cotton in the world are China, India, Pakistan and Turkey.

Cotton production in India during the 2015–16 cotton season was projected at 352 lakh bales of 170 kg each. The main states producing cotton are Andhra Pradesh, Gujarat, Haryana, Karnataka, Madhya Pradesh, Maharashtra, Meghalaya, Orissa, Punjab, West Bengal, Rajasthan and Sikkim. This industry provides employment to around 30 million people directly and up to 60 million if indirect employment is also included. The major Indian organisations working for the welfare of the cotton industry are Cotton Corporation of India (CCI), Cotton Export Promotion Council (TEXPROCIL), Cotton Textile Research Association (CTRA), Indian Council for Agricultural Research (ICAR), Central Institute for Cotton Research and Central Institute for Research on Cotton Technology (CIRCOT). Major international organisations for the cotton industry are International Cotton Advisory Committee (ICAC), Cotton Incorporated and Cotton Council International.

Cotton in India

India is the second largest producer of cotton in the world (18% of global production).

It has the largest area under cotton cultivation in the world.

The Indian textile industry is mainly cotton-based.

Growth and Production

Cotton belongs to the family *Malvaceae* which include plants like Okra and Hibiscus, and to the genus *Gossypium*. Asian cotton varieties usually belong to either the *Gossypium arboreum* or

Gossypium herbaceum species. Cotton requires a long growing period with a warm climate, adequate rain or other sources of irrigation and a frost-free period of 6–7 months. During the initial active growing period, the crop requires plenty of water, but for the fibre to mature the following period has to be dry. Seeds are sown in early spring, in February or March, and the plants grow to a height of 3–6 feet. The blooms appear 80–110 days after planting. On the first day of blooming, the flowers look like Hibiscus or Hollyhock and are creamy white or yellow. By the second day, the flowers turn pink, lavender or red and drop off by the end of the second or third day. This leaves behind the seed pod or the cotton boll on the stem which contains the cotton seeds that begin to grow seed hairs. After 50–80 days, the pods mature and burst open exposing the cotton fibre inside (Image 2.1). The cotton is now ready for picking or harvesting. A defoliant spray might be used to make the leaves fall-off, facilitating easier removal of cotton bolls during picking.

Image 2.1: Cotton plant

Source: 'Baumwollkapsel, cotton boll, Guangxi, China' by Azzurro, (CC by SA 3.0; Wikimedia Commons).

Picking can be done either by hand or by machines. Hand picking allows for selective picking of fully mature fibres and lesser trash. Mechanised picking is much quicker, but it picks several rows at a time regardless of the maturity of the fibre, and may pull the whole boll along with the fibre leading to more impurities later.

Processing

The harvested cotton can be processed manually or mechanically using the ginning machine. The American inventor Eli Whitney invented the first ginning machine. The harvested cotton is processed in the ginning machines (Figure 2.1) where cotton fibres are separated from the seed. The fibre that is removed from the seed is called lint cotton while the short fibres that may be left on the seed are called **linters**. The **ginning** process starts with the cotton being sent to dryers to reduce the moisture content. The cotton bolls with seed are fed into the ginning machine, where revolving circular saws pull the lint through closely placed metal mesh that prevents the seeds from passing through. The separated lint is removed from saw teeth by air blasts or rotating brushes and sent for compressing into rectangular packages called **bales** which weigh 500 lbs each. These bales are assigned grades based on the length, fineness, trash content, colour and ginning preparation of the fibres. The bales are then sent to various units for spinning yarns and then manufacturing fabric from the yarns.

Figure 2.1: Cotton gin

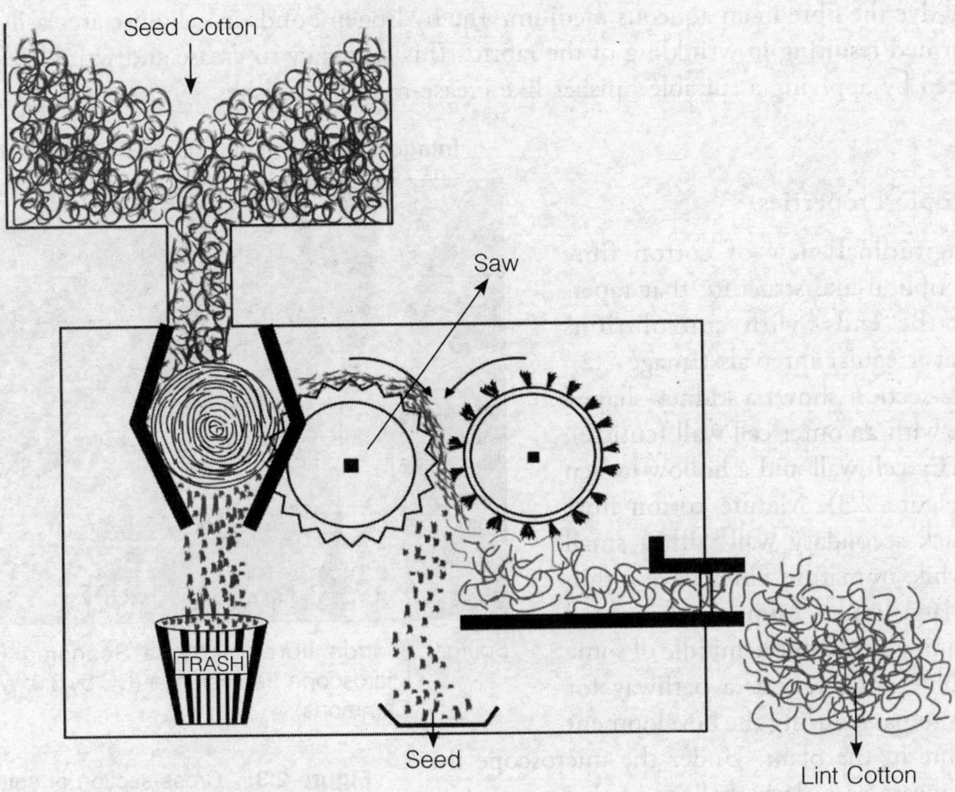

Source: Drawn by the authors.

Chemical Composition

Cellulose is a polymeric sugar or polysaccharide which is represented commonly as $(C_6H_{10}O_5)_n$. **Cellobiose**, the unit that joins together to form the polymer cellulose, is made up of two glucose units (Figure 2.2).

Figure 2.2: Structure of cellulose

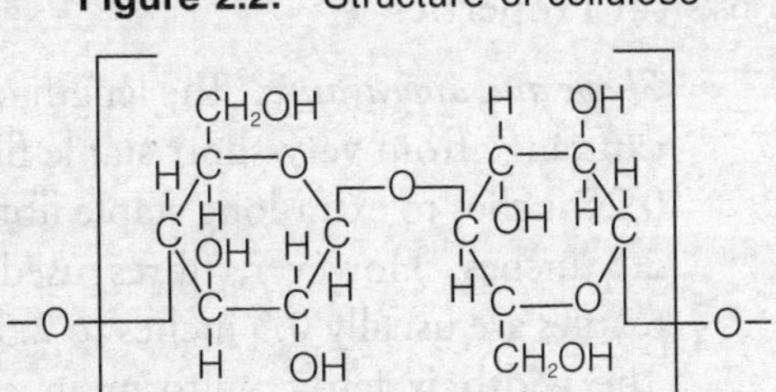

The degree of polymerisation (DP) varies with the source of the fibre and is depicted by '*n*'. Since 5000 cellobiose units form a polymer chain in cotton fibre, we can say that the DP of cotton is 5000. Cellulose molecules form orderly structures which permit the hydroxyl functional groups in adjacent chains to form hydrogen bonds. The polymer chains also associate with each other through Van der Waals forces.

The presence of hydroxyl groups influences the moisture absorption property of the fibres. The oxygen atom in water is attracted to the hydroxyl groups in cellulose, so any free hydroxyl groups will absorb perspiration. Due to this, it is comfortable to wear cotton and easier to chemically treat and dye the fibre in an aqueous medium. The hydrogen bonds in cellulose are easily broken and reformed resulting in wrinkling of the fabric. This tendency to crease and wrinkle easily can be reduced by applying a suitable finishes like crease-resistant finishes.

Properties

Microscopic Properties

The longitudinal view of cotton fibre shows a ribbon-like structure that tapers towards the ends, with convolutions (twists) at irregular intervals (Image 2.2). The cross-section shows a kidney-shaped structure with an outer cell wall (cuticle), a secondary cell wall and a hollow lumen inside (Figure 2.3). Mature cotton fibre has a thick secondary wall with a small lumen while immature fibres have a thin wall and large lumen. **Lumen** is the central canal running through the middle of some natural fibres. It serves as a pathway for nutrients to pass, during the development of the fibre in the plant. Under the microscope the lumen appears as a dark shadow in the centre of a mature fibre. It may also appear as striations on the fibre. It can vary from being elliptical to almost round in some fibres, more irregular in immature fibres.

Image 2.2: Longitudinal view of cotton fibres

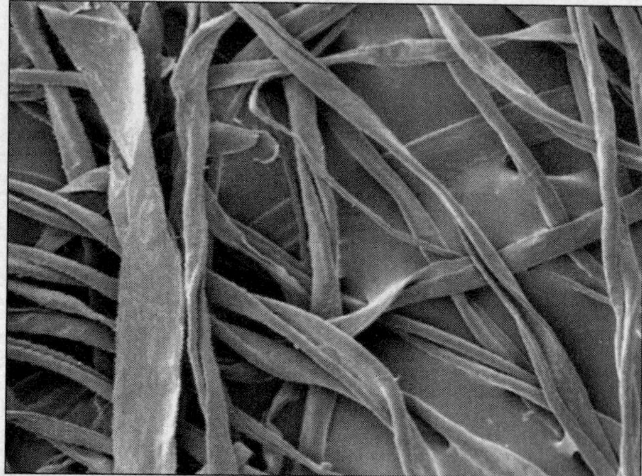

Source: 'Cotton fibres under a Scanning Electron Microscope' Featheredtar (CC by 3.0; Wikimedia Commons).

Figure 2.3: Cross-section of cotton fibre

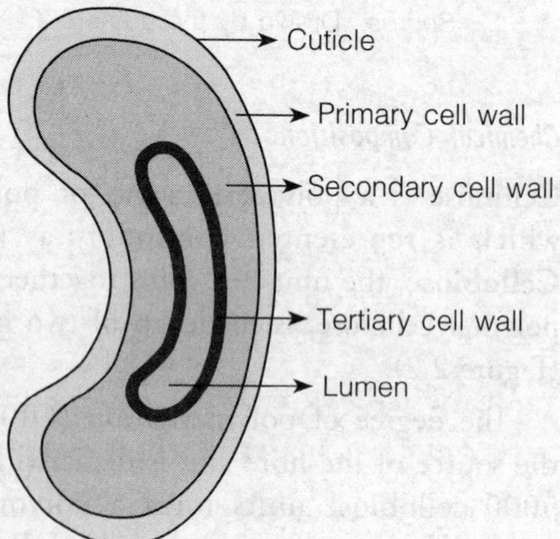

Source: Drawn by the authors.

Physical Properties

- *Shape and appearance*: The length of cotton fibre can range from very-short staple fibres of about 0.25 inches to extra-long staple fibres measuring 2.5 inches. However, fibres used for making textiles are usually 0.5 inches to 2.5 inches long. The width is fairly uniform in cotton fibres, varying from 12 to 20 microns.
- *Colour and luster*: The colour of the different varieties of cotton can vary from almost pure white to yellowish to grey. Cotton fibres usually have a low lustre unless special finishes like mercerisation are applied. On being mercerised, cotton fibres swell up and have a regular cross-section, thus increasing reflection of light from the surface of the fibre, which in turn leads to increased lustre.

- *Tenacity*: Cotton has an average strength ranging from 3–5 grams/denier. When wet, it increases in strength by 10 to 20 per cent of its dry strength.
- *Elongation and Elastic Recovery*: It has an elongation of 3–7 per cent at breaking point. Elastic recovery is poor being about 75 per cent at 2 per cent extension.
- *Resiliency*: Cotton has low resiliency due to weak hydrogen bonds, which makes the fabric prone to creasing and wrinkling.
- *Density*: Cellulosic fibres are heavier than most other commonly used fibres. Cotton fibre is fairly dense with a density of 1.54 g/cm^3.
- *Moisture regain and absorbency*: Cotton fibres have a standard moisture regain of 8.5 per cent. At 100 per cent saturation, moisture regain can increase up to 25–27 per cent. Mercerised cotton can absorb more moisture (~12 per cent) due to its enlarged/swollen structure. The high absorbency makes cotton a very comfortable choice for apparel as it can absorb sweat from the skin, especially in hot and humid weather. In the wet state, cotton also swells and becomes more flexible, making it easier to give a flat, smooth finish to cotton fabrics while pressing and finishing.
- *Dimensional stability*: Dimensional stability is the degree to which a fibre, a yarn or a fabric retains its shape and size after having been subjected to wear and maintenance. Cotton fabrics do not stretch or shrink in normal state. However, relaxation and residual shrinkages are seen in the first few washing cycles. This is due to the relaxation of stress encountered during the processes of yarn and fabric making and/or application of finishes.

Chemical Properties

Cotton is highly resistant to alkalis. Alkalis in various forms are commonly used in processing, finishing and laundering of the fibre. Mercerisation involves the usage of 22–27 per cent sodium hydroxide solution which brings about significant improvement in fibre properties such as tenacity, moisture regain and dye uptake. Highly alkaline conditions also do not adversely affect the fibre, therefore commonly used detergents and laundry aides can be safely used. However undiluted sodium hydroxide, if poured directly, can degrade the fibre.

Cotton is not damaged by brief exposure to cold, dilute acids, if rinsed and neutralised afterwards. Concentrated and hot dilute acids will cause fibre disintegration. Acids are used in some finishing processes but their application are carefully regulated, with appropriate reaction time, temperatures and proper rinsing and neutralising procedures.

Cotton is also highly resistant to organic solvents, hence household solvents and stain removal agents do not harm it. Chlorine bleaches can be safely used, if the usage instructions are properly followed.

Thermal Properties

Cotton catches fire quickly and burns easily with a smell similar to burning paper. It leaves a fluffy grey ash-like residue with an afterglow. The safe ironing temperature for cotton is 200°C.

Temperatures above 250°C rapidly deteriorate the fibre. Starched cotton should be carefully ironed as it can scorch easily.

Biological Properties

Cotton is a biodegradable fibre. It can be attacked by fungi such as mildew and bacteria. Mildew will rot and discolour the fibre, and eventually degrade it. Mildew causes more damage if cotton is stored in damp, dark and warm conditions. Moths and carpet beetles do not attack or damage cotton but silverfish may eat cotton cellulose if the fabric has been heavily starched or has food stains.

Resistance to Sunlight

Prolonged exposure to sunlight will cause cotton to turn yellow and eventually degrade it.

Use and Care

Cotton is a highly versatile fibre with many aesthetic and functional advantages. It is comfortable and pleasant to touch with a high capacity to absorb moisture, even under extremely humid and hot conditions. Cotton is a universal favourite and is mainly used in apparel, household goods, home furnishings and industrial fabrics. It can be made into low-cost garments as well as expensive high-fashion couture garments. Its comfort and absorbency properties make it highly effective for use in hosiery, towels, bed sheets and pillowcases. Many finishes can be applied and the fibre can be used in blends to enhance the fibre properties and provide various design features.

Cotton fabrics are quite easy-care with good wearability, excellent launderabilty and good colourfastness, if proper dyestuffs have been used. Cotton and cotton blends can be easily machine-washed at home but special care should be taken to avoid wrinkling and creasing. Detergents and dry-cleaning agents can be used safely on cotton. Oxidising and chlorine bleaches are safe to use on cotton if proper directions are followed. Undiluted chlorine bleach poured on cotton can cause formation of pinholes in the fabric.

Organic Cotton

Environmental concerns and increasing awareness have resulted in the rise in the demand for organic cotton. Cotton can be certified as organic, if the field producing it has been following organic practices, such as no chemical fertiliser or pesticide application, for at least three years. The cotton produced during the first three years of organic agriculture is called **transitional cotton**.

Bt Cotton

This genetically engineered cotton produces a type of protein that is toxic to pests like the cotton bollworm. The toxin is found in nature inside a bacterium called *Bacillus thuringiensis* which gives Bt cotton its name. A gene from this bacterium, '-cry1ac-b', is inserted into the cotton genetic structure and this modification makes the cotton poisonous to bollworms. Bt Cotton is produced by Monsanto Inc., an American company. USA, Argentina, Mexico, Australia, South Africa, China and India have approved the cultivation of Bt Cotton for commercial use.

The benefit of Bt Cotton includes the doubling of the yield besides requiring just half the amount of pesticides used to grow regular cotton. However, its usage has been controversial due to various disadvantages like resistance development in bollworm, high cost of seeds that cannot be reused since they lose vigour after one generation. Other concerns include possible adverse effects on non-target organisms, flow of Bt genes to other wild and cultivated relatives, resistance to pink bollworm.

Naturally-coloured Cotton

When cotton is found in a naturally-coloured state, i.e. other than white, it is called coloured cotton. The common shades are those of green and brown. Coloured cottons have been known to be used by mankind since 2500 BC. The development of lint colour depends on the soil type, water hardness and exposure to sunlight during the growth of the crop. Coloured cotton has been garnering importance due to its eco-friendly nature—it requires no dyeing or bleaching, thus dispensing the requirement of harmful chemicals besides saving time and cost. The disadvantages of raising a crop of coloured cotton include lower yields, lower fibre strength and length. However, these can be overcome by blending fibres or yarns.

Flax (Linen)

History

The word linen has been derived from the celtic word *llin*. The popular use of linen for many household items is reflected in the use of terms such as 'bed linen' and 'table linen' even today. Flax or linen is considered by many to be the oldest fibre used in the western world and also the first of plant origin to be utilised. The history of flax dates back over 8000–10,000 years. Linen fabrics found in the Swiss lake dweller civilisation showed that the Neolithic man was skilled in the production of flax. Egypt was called the 'land of linen' in ancient times. The use of very fine linen fabrics prior to 4500 BC in Egypt has been verified through fragments and pictures in the tombs of the pharaohs. Linen travelled to the Mediterranean through the trading Phoenician sailors and later to Britain. By 1685, Ireland also became a centre for the production and trade of linen. In the case of India, materials made from *kshauma* or flax have found mention in the *Arthashastra*. In the days of Lord Buddha, linen was the choice of clothing for many *bhikshus* (monks).

Globally, flax is now produced mainly in Western Europe, United Kingdom, Germany, the Netherlands, Switzerland, Russia, Belarus and New Zealand. In India, Flax is mainly grown in Madhya Pradesh, Uttar Pradesh, Bihar, Jharkhand and Orissa. Flax is cultivated in India mainly for flaxseed and oil, with about 25 per cent suitable for fibre use. India does not have a significant presence in flax production globally and therefore it has to import flax to meet the demand, especially in the defence sector.

Growth and Production

Image 2.3: Flax plant

Source: 'Flax plants in June.' by Handwerker (CC by 3.0; Wikimedia Commons).

Flax is obtained from the stem or bast of *Linum usitassimum*. The bast is the outer edge/phloem of the stem. Flax thrives in temperate climate with adequate moisture. Bright sunlight and high temperature may damage the fibres unless abundant rainfall is received intermittently. In most countries flax is planted in April or May. Plants grown for fibre are sown close together to produce closely packed long, thin stems, without branching. The plant (Image 2.3) grows to a height of 3–4 feet and bears flowers which are delicate blue, white or pink in colour. It is harvested in about 80–100 days after sowing. The plant cultivated for fibre is pulled from the roots before the seeds ripen.

Processing

Pulling and Rippling

Flax for fibres is *pulled* either by hand or machine. The entire plant is pulled keeping the roots intact as the flax fibres extend below the ground surface into the root system. If it is not pulled properly and is broken then it causes discolouration of the fibre. Harvesting is done in late August when the plant is rich brown in colour.

After pulling, a process called **rippling** is carried out by passing the stems through threshing machines to remove the seed pods. The stems are then left to dry.

Retting

The flax fibres lie in bundles just beneath the outer covering or bark of the flax stem. They are sealed together by substances like pectins, waxes and gums. To obtain the flax fibres, the outer woody covering of the stem must be removed and the fibres loosened up. This is done by a process which rots away the outer woody portions with the help of bacteria or some chemical agents. This process of rotting is known as **retting**. The various methods of retting are described below.

1. *Dew retting*: This is a traditional method of retting. It involves the spreading of flax stems on the ground and exposing them alternately to the action of dew and sunlight, causing rotting of the outer portions of the stem. The process takes a long time, around four to six weeks, but produces the most durable and strong linen. It is hardly used today as it requires a lot of time and space.
2. *Stream retting*: In this process the flax bundles are placed in a stream of running water which causes the rotting to take place. This method is also not commonly used nowadays.

3. *Pool retting*: In this traditional process, the flax is packed in sheaves (bundles) and immersed in pools of stagnant water. The bacteria in water rot away the outer woody covering. The time taken is two to four weeks. It produces a bluish-grey coloured fibre. The disadvantage of this method is that the stagnant water produces a foul smell.
4. *Tank retting*: This is a modern technique of retting similar to pool retting. In this process flax is stacked in tanks filled with warm water which increases the speed of bacterial action. The time required is just a few days and the fibre produced is light coloured with uniform strength.
5. *Chemical retting*: This process involves stacking the flax stems in tanks filled with water to which a chemical like sodium hydroxide, sodium carbonate or dilute sulphuric acid is added. This is completed in a few hours, but it must be carefully controlled to prevent any damage to the fibre.
6. *Enzymatic retting*: This type of retting uses certain pectinolytic and cellulolytic enzyme mixtures. These enzymes are mixed with water and flax stems are submerged or sprayed with this formulation. The flax is then allowed to ret for several hours at high humidity. This process is faster than water retting because it results in better yield and production of finer fibres. The disadvantages are the cost of enzymes and likelihood of lower fibre strength due to continued activity of enzymes.

Breaking and Scutching

After completion of retting, the flax fibres are rinsed and dried. The rotten outer woody covering has to be removed mechanically by breaking and scutching. The stalks are bundled together and passed between fluted rollers which break the softened outer covering into small particles. Scutching (agitation) is then done to separate the broken stem particles from the fibres.

Heckling or Combing

In this step, flax fibres are hackled or combed to separate the short fibres (tow) from the long fibres (line). This is done by passing the fibre through several sets of pins, each successive set finer than the one preceding it. This results in a combed sliver with fibres that are separate and more parallelly-arranged fibres.

Spinning

The sliver is then drawn out into a yarn and twist is inserted. Spinning can be done in wet or dry state with wet spinning often being preferred. The final yarn processing is similar to that of cotton.

Properties

Microscopic Properties

Flax fibre is actually made up of bundles of fibrils or fibre cells held together by a gummy substance. The longitudinal view shows flax stems in varying thicknesses and looking similar to a bamboo shoot due to the presence of nodes or irregular joint formations (Figure 2.4). The central lumen casts a shadow giving a darker colour at the centre with striations but no convolutions. The cross-sectional view of a fibre shows a polygonal shape with a thick outer wall and a lumen at the centre.

Figure 2.4: Longitudinal view of flax fibre

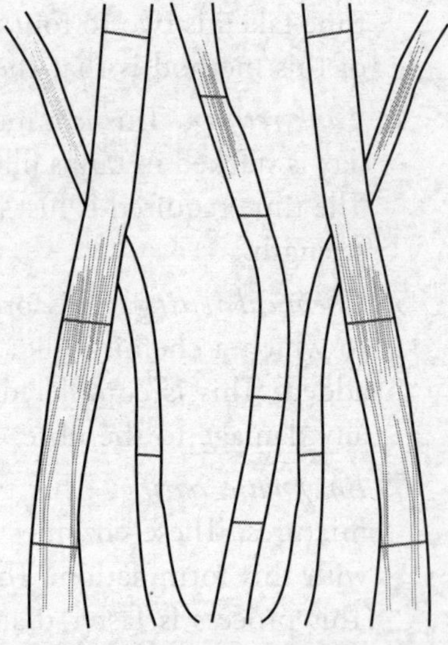

Source: Drawn by the authors.

Physical Properties

1. *Shape and appearance*: Flax fibres are longer than cotton but not as fine in width. The length of the fibre is usually 10–15 inches, though some can be as long as 40 inches. The width is usually about 15–21 microns. Longer line fibres are more than 12 inches long whereas tow fibres are shorter.
2. *Colour and luster*: Flax has a natural colour which varies from light yellow to ivory and grey. Flax has a high natural lustre which gives it an attractive sheen.
3. *Tenacity*: Flax is one of the strongest fibres with a dry tenacity usually ranging from 5.5 to 6.5 grams/denier. Like cotton, flax also shows increase in strength when wet. The wet strength can be up to 20 per cent more than the dry strength.
4. *Elongation and elastic recovery*: Flax is crystalline in nature and therefore shows very little elongation of about 2.7–3.3 per cent before breaking. It has poor elastic recovery of 65 per cent at 2 per cent extension.
5. *Resiliency*: Resiliency of flax fibres is poor and even lower than that of cotton. They are prone to wrinkling and creasing very easily unless given some special finishes.
6. *Density*: Flax is almost as dense as cotton with a density of 1.5 g/cm^3. It can therefore feel heavier than silk, polyester or nylon even if they have the same construction and weave type.
7. *Moisture regain and absorbency*: Flax has a higher standard moisture regain than cotton at about 11–12 per cent. At saturation, it has similar moisture regain as other cellulosics. Linen also has very good wickability, which is the ability of moisture to travel readily along the fibre. It also gives up moisture quickly and dries quickly. This makes it highly comfortable and apt for usage in towels and warm weather clothing.
8. *Dimensional stability*: Flax fibres do not stretch or shrink to a noticeable extent but they may exhibit relaxation shrinkage like cotton fabrics.

Chemical Properties

Flax is highly resistant to alkalis, organic solvents and other organic compounds. It is resistant to cold dilute acids but concentrated and hot dilute acids will cause damage. Most detergents are safe to use on flax. It is also safe to bleach flax, though it may be harmed more easily than cotton by chlorine bleach.

Thermal Properties

Like cotton flax burns readily producing fluffy grey ash and the smell of burning paper. It can withstand dry heat up to 150°C for long periods of time with gradual degradation above 150°C. Safe ironing temperature for linen ranges from 200°C up to 260°C with limited exposure.

Biological Properties

Dry linen has excellent resistance to mildew, but if it is stored in damp and warm conditions it will be quickly attacked and damaged by mildew. Most household insects do not harm linen but silverfish may damage linen if it is starched.

Resistance to Light

Flax has good resistance to sunlight but prolonged exposure will gradually degrade it.

Use and Care

Flax possesses high strength, good moisture absorbency and high lustre. It can be made into a range of fabrics from light weight to heavy weight. This makes it suitable for usage in apparel, for both men's and women's wear. Its ability to wear well, look elegant and lie flat when finished well, also makes it an ideal choice for household items like sheets, tablecloths and napkins. It is also used for a variety of other items like canvas for painting, draperies, upholstery, slipcovers, terry towels, awnings and car seat covers. It is even used to make water bottles (*chaggal*) and storage tanks which can be folded up and carried along when needed (used in defence).

Linen is being used by many designers to create high-fashion garments due to its unique texture, cool feel in summer and ability to enhance with age. Even the wrinkles occurring in linen are considered to be a status symbol by some.

Flax is easy to maintain due to its resistance to hot water, detergents, bleaches, laundry aids and dry-cleaning solvents. Care labels must be followed for any additional precautions. However, it is difficult to iron, and should be ironed while still damp to achieve the desirable crisp look. The problem of wrinkling can also be avoided by imparting wrinkle-resistant finishes. Due to its brittle nature linen should be stored flat or rolled up as it can break when repeatedly folded in the same place.

Jute

History

Like flax, jute is also a bast fibre. It is commonly called burlap or hessian. The name jute is said to be derived from the Bengali word *jhuto*, which means 'to be entangled', referring to the ability of these irregular fibres to readily mesh together. The fibre has been used to make many handicraft items in India since ancient times. Early Sanskrit literature mention jute as a useful household plant used both as herb and as a fibre. Several historical documents like the *Ain-e-Akbari* state that poor villagers in India used to wear clothes made of jute. Jute was also said to be in use during the biblical era, probably in making sackcloth (Image 2.4).

Image 2.4: Close-up of jute cloth

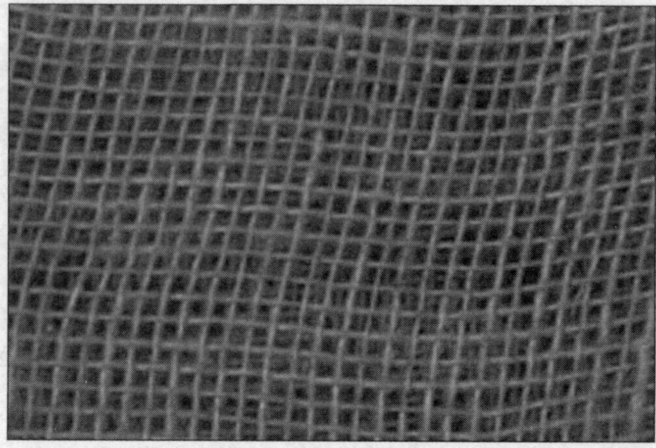

Source: 'Yute–Sacos' by Tamorlan (CC by 3.0; Wikimedia Commons).

Jute is mostly grown in India and Bangladesh, but other minor producers are China, Thailand and other South-Asian countries. India is the largest producer of jute in the world. The Indian jute industry dates back to 1855 with the setting up of the first jute mill. Jute is produced mainly in West Bengal, Bihar, Assam, Uttar Pradesh, Tripura and Meghalaya. The industry provides employment to 2.6 lakh people directly and employs another 1.4 lakh through indirect means. India produces on an average 10,843 thousand bales of jute each year. The main organisations working in the jute sector in India are the Jute Corporation of India Limited (JCI), Indian Jute Industries Research Association (IJIRA), Jute Manufacturers Development Council (JDMC), Birds Jute Exports Limited (BJEL), National Council for Jute Development (NCJD) and National Jute Manufacturers Corporation (NJMC).

Growth and Production

Jute is obtained from plants of the *Corchorus* family, specifically from *Corchorus capsularis* and *Corchorus olitorius*. It is an annual crop that grows to a height of 15–20 feet. It is cultivated in a manner similar to flax. It has a cylindrical stalk with no branches except at the top. It requires alluvial soil with standing water and temperatures ranging from 20 to 40°C. The stalks are ready for cutting just after the flowers fade. The best fibre is obtained by hand-stripping where each stalk is peeled separately. The processing of jute is similar to flax fibres, beginning with retting followed by breaking, skutching, heckling and spinning.

Properties

Microscopic Properties

Jute is similar in microscopic structure to flax—it consists of bundles of fibrils held together by a gummy substance. The fibres also show nodes in the longitudinal view. However, the lumen is smaller as compared to flax.

Physical Properties

- *Shape and appearance*: The fibres are long and may vary in length from 1.5 to 3 meters and a diameter of 17–20 microns.

- *Colour and lustre*: Jute is naturally beige to brown in colour. It is difficult to bleach jute to a pure white colour; thus many jute articles retain jute's natural colour. It has a silky lustre which determines the quality of jute—higher the lustre, better the quality.
- *Tenacity*: Jute is one of the weakest cellulosic fibres which has an average strength of about 3.5 grams/denier. Wet jute kept moist for long periods of time will deteriorate and lose strength.
- *Elongation and elastic recovery*: It has a low elongation of less than 2 per cent before breaking and a low elastic recovery of 74 per cent at 1 per cent extension. This means that flax is a brittle fibre with a low capacity for stretching, and can split and snag easily.
- *Resiliency*: Jute has low resiliency resulting in easy crushing and creasing.
- *Density*: Similar to flax, jute has a density of 1.5 g/cm^3.
- *Moisture regain and absorbency*: Jute has a standard moisture regain of 13.7 per cent, therefore it can absorb moisture easily like flax. It will deteriorate when kept under moist conditions.
- *Dimensional stability*: It has good dimensional stability.

Chemical Properties

The chemical properties of jute are similar to other cellulosic fibres. It is highly resistant to alkalis and exhibits good resistance to dilute acids. Concentrated acids will cause fibre deterioration.

Thermal Properties

Reactions to heat and flame are much like those of other cellulosic fibres.

Other Properties

Jute shows good resistance to micro-organisms and insects. It is also highly resistant to sunlight but turns brittle on prolonged exposure to air.

Use and Care

Jute is widely used for household and industrial fabrics. Its ability to stack without slippage is highly useful in sacking, bagging, carpet-backing and floor-mats. It has other varied uses in the form of ropes, geotextiles, home furnishings, food grade jute products, gifts and novelty items. Upholstery fabrics made of polyester and jute blends are very popular. In apparels, it is mostly used in the blended form with other natural and synthetic fibres such as cotton, wool and polyester.

> **Geotextiles** are permeable fabrics, generally used in association with soil, that have the ability to separate, filter, reinforce, protect or drain.

Care of jute is similar to flax but coloured jute may require dry-cleaning as jute is difficult to dye with good colour fastness.

OTHER CELLULOSIC FIBRES

Coir

Coir is obtained from the husk of the coconut found between the outer husk and the nut (mesocarp tissue). It is called the 'golden fibre' as it is often gold in colour after cleaning. India is the largest producer of coir in the world, accounting for about 80 per cent of the world's production. Sri Lanka is the other major producer of coir.

Coir is a stiff, strong fibre, which has good resistance to abrasion. It is tough, durable and resistant to dampness and moisture. It is the only natural fibre which is resistant to damage by salt-water. It is also highly resilient and springs back to shape quickly even after constant use. It is flame-retardant and provides good insulation against temperature and sound. It is also completely static-free besides being moth-proof and resistant to bacteria and rot. The mature, longer bristle-like fibre husk is stronger, less flexible and naturally rich brown in colour but it is difficult to bleach. Therefore, it is generally used in naturally coloured state or dyed in darker colours only.

Coir is used to make outdoor mats, aquarium filters, cordage, rope, garden mulch and peat moss. Brown coir is generally used in brushes, doormats, sacking and mattresses. Needle-felted brown coir is used as a geotextile to prevent soil-erosion. Rubberised coir is used in automobile upholstery padding. It is also used for insulation and packaging. White coir is mostly used for making rope.

Bamboo

Bamboo fibre is also called 'green gold'. It is a fast growing plant which can reach a height of up to 30 meters. It can be grown in many climates with little irrigation. However, much of commercial production is done in China.

The processing of bamboo stalk to obtain most fibres is similar to viscose rayon, thus it can also be called a regenerated fibre. The hard, woody stems are steamed and then crushed to obtain the pith, which is further treated with sodium hydroxide solution to get a starchy pulp. This pulp can then be extruded through a spinneret into a dilute acid bath to obtain fine bamboo filaments. However, some bamboo stalks can be processed like flax, undergoing retting to produce litrax bamboo or bamboo linen. This type of bamboo fibre is less soft than regenerated bamboo.

Bamboo is composed of 61 per cent cellulose and 32 per cent lignin. The fibres are smooth and round and can vary in length from 38 to 76 mm. Bamboo fibres have a strength of 2.3 grams/denier which decreases when it is wet. It has high absorbency which keeps it cool, with moisture regain of 13 per cent. It also has a soft handle, is breathable, with a specific gravity of 0.8. It has good elongation of 23.8 per cent at break. It is also said to be anti-bacterial and deodorising in nature. Bamboo fibres are also claimed to protect against UV rays with a score of 50 on UV protection scale. This also means that the fibre is not affected by exposure to sunlight.

Bamboo fibres are used to make bathrobes, towels, innerwear, medical textiles and food-packing bags.

Ramie

Ramie is a bast fibre obtained from the stalk of the ramie plant, *Boehmeria nivea*. It is also known as rhea, grasscloth, china grass and army/navy cloth. Ramie is a fast growing, tall perennial shrub from the nettle family, requiring hot and humid climate. It is ready for harvesting after 60 days of sowing. It is cut and not pulled for obtaining the fibre. Ramie fibres are separated from the outer bark and woody portion of the plant by a process called **decortication**. The manual process is labour-extensive and not commercially viable, therefore mechanical methods to decorticate ramie have been developed. Ramie is mainly produced in China, the Philippines and Brazil.

Ramie is a white, long, fine, lustrous, brittle and stiff fibre. It has absorbency, density and microscopic properties similar to flax. It has a high molecular crystallinity resulting in the tendency to be brittle and break when folded repeatedly. Ramie is one of the strongest natural fibres, with a tenacity of 5.3–7.4 grams/denier which increases in strength when wet. It has low resiliency, elasticity and elongation. However, special treatments can be given to make it wrinkle-resistant. It is resistant to insects, rot and mildew, and is dimensionally stable with low shrinkage.

Ramie is used in apparel like sweaters, shirts, blouses and suits. It can also be blended with cotton and other natural fibres for many apparel and furnishings uses. Ramie is also used in making window treatments, pillows, table linen besides being used in ropes, twines, nets, cigarette paper, banknotes and geotextiles for erosion-control.

Hemp

Hemp is obtained from the stem of a plant *Cannabis sativa*, which is a tall herb belonging to the Mulberry family. It is an annual plant grown in warm and hot climate. It is a fast-growing crop and is mainly cultivated in China and the Philippines. Other countries that produce hemp are Italy, France, Chile, Russia, Poland, India and Canada. However, hemp is not legally allowed to be commercially grown as it can also be used to produce the drug marijuana.

The processing of hemp is also similar to flax—retting, scutching and heckling to obtain the fibre. The plant can produce three types of fibres depending on the region of the stalk from which they are obtained. The outer region of the stalk produces the longest and finest fibres, the inner region produces fibres which are shorter, the innermost core has woody fibres.

Hemp is similar to flax in macroscopic and microscopic appearance, though hemp is somewhat coarser and stiffer than flax. Hemp fibres are very long, ranging from 3 to 15 feet in length. The colour of the fibre can vary from creamy white, brown, grey, green or almost black depending on the processing method used to remove the fibres from the stem. It has a tenacity of 5.8–6.8 grams/denier but has low elongation and elasticity. It has good absorbency and comfort properties. Hemp is resistant to insects but can be damaged by mildew.

Hemp fibres from the outer stalk region are mostly used in apparel and furnishing either as 100 per cent hemp or in blends with fibres like cotton, linen or silk. Shirts, t-shirts, hats, backpacks, jeans, hats and shoes are being made from hemp. The two inner fibres are used for non-ovens, industrial applications, mulch and animal bedding. Besides this, hemp is also used in making cordage, twine, rope, thread and sacking material.

Kenaf

Kenaf is a bast fibre obtained from the kenaf plant. It is a fibre similar to jute and is mostly grown in India and Pakistan besides Africa, Central Asia and some Central American countries. The cultivation and processing of kenaf is similar to jute. It can grow to a height of 10 feet but the fibres are shorter as they break during processing. The fibre obtained is long, light yellow to grey in colour. It is harder and more lustrous than jute, with excellent resistance to water. However, it has poor strength and low elasticity. Like jute kenaf is mostly used for ropes, cordages, twine, canvas and other industrial usages. It is also being researched upon as a fibre for making paper and as a blend with cotton.

Pina

Pina fibres are obtained from the leaves of the pineapple plant which is processed mainly in the Philippines. Pina fibre is fine, soft, lustrous, flexible and strong. It is white or ivory in colour with a length of about 2–4 inches. It is highly resistant to water but easily damaged by acids and enzymes. Therefore, pina fabrics should not be pre-soaked in detergent or enzymes and any fruit stains should be promptly removed as pectin and acids in some fruits can breakdown the fibre. The pina fabric is soft, delicate, lightweight, sheer and crisp. These fabrics are often embroidered and used for formal and wedding wear in the Philippines. It is used for making clothes, accessories, bags, table linen, mats and rope, etc.

Abaca

It is obtained from the leaves of *Musa textilis,* which is a member of the Banana family, grown mostly in the Philippines and Mexico. It is also known as Manila hemp though actually it is not a variety of hemp. Abaca fibres are long (up to 15 feet) and coarse. The fibre is off-white to dark grey or brown in colour. The good grade fibre is mostly light in colour. Abaca fibres are strong, durable and flexible which can make a strong yet delicate and lightweight fabric used for clothing. Its high strength makes it very suitable for making cordage, rope, floor mats, wicker furniture and table linen, etc.

Sisal

Sisal is obtained from the leaves of a plant belonging to the Agave family. It is mostly grown in Africa, Central America and the West Indies. It is harvested from the leaves when the crop is four years old. The fibre is separated from the fleshy part of the leaf and further processed to remove chlorophyll, pectins and other non-cellulosic substances. The fibres are smooth, straight and yellow in colour. They can be dyed in bright colours using both direct and acid dyes. They have poor resistance to salt water and therefore not used in making maritime ropes. Their main usage includes ropes, twine and brush bristles. Sisal fibre is also used for upholstery, carpets, customs rugs, wall-coverings, background material for furnishings and heavy-duty commercial applications. It is a durable fibre but it has a tendency to shed and fade, with a susceptibility to waterborne stains.

Kapok

Kapok is obtained from the seed hairs of the Java Kapok tree. The tree grows to a height of 50 feet and produces a seed pod similar to cotton plant. The fibres can be dried and the seeds are easily removed by shaking, therefore ginning is not necessary. Kapok fibres are extremely soft, light and buoyant due to their hollow structure. They are very difficult to spin, so most of them are used as filling, padding and stuffing material. Kapok is non-allergenic and very suitable for filling pillows. It is also used as filling in upholstered furniture, mattresses and life-preservers.

Banana

Banana fibres are obtained from the leaves, shoots and pseudostem (trunk) of the banana tree belonging to Musa genus grown in over 120 countries. The banana tree is actually a large herb which reaches a height of 20–25 feet at maturity. The trunk or pseudostem is a cylinder of leaf-petiole sheaths. The fibres are extracted from the plant after the fruit is harvested. It can be extracted by either the Japanese method or the Nepalese method. Japan has been using banana fibres since the thirteenth century AD for making clothing and household items. The Japanese method uses carefully pruned leaves and shoots to get soft fibres. The Nepalese method utilises the trunk to obtain banana fibres. The outermost, decaying layers of the trunk are used for this purpose and are soaked in water to dissolve away the chlorophyll, pectins and other impurities leaving behind cellulose fibres.

Banana fibres are also called banana silk yarns as they are similar to silk yarns being somewhat shiny and lightweight. The fibres are flat, ribbon-like possessing high strength, high moisture absorption, average fineness and little elongation. The fibres can also be blended easily with cotton or man-made fibres. Banana fibres are used mainly to make clothes, rugs, saris, kimonos, home furnishings, high quality currency/security paper, packing cloth for agriculture, ropes, cables, etc.

NATURAL PROTEIN FIBRES

All protein fibres are composed of amino acids which are joined together in polypeptide chains of large molecular weights. They can be categorised into three basic groups:

1. *Animal hair fibres*: They include fibres obtained from animal's hair, mainly sheep's wool. Other animals like alpaca, camel, cashmere goat, llama, vicuna and angora goat are also used for their hair. Fur from animals like rabbit, mink, beaver and horsehair may also be used.
2. *Fibres formed from extruded filament*: The only important fibre in this category is silk which is produced by silkworm caterpillars. Other unusual materials like spider silk and byssus from mussels are also being experimented to obtain fibres.

3. *Regenerated fibres from vegetable or animal protein*: Some materials like milk, corn and soybean are being used to produce regenerated protein fibres.

For textile usage, wool and silk are the most commonly used natural protein fibres. All protein fibres contain carbon, hydrogen, oxygen and nitrogen. Additionally, wool also contains sulphur.

Wool

History

Wool was probably the first fibre to be used as a textile material. Early hunting people undoubtedly utilised the fleece from the animals they hunted. It was said to be first worn as sheepskin with the hair (pelt), and later fibres were used to make matted or felted fabrics. Thus, it was in use long before man discovered how to spin yarn from fibre and form fabric. Mesopotamia is believed to be the first place where domesticated herding of sheep was done and therefore the birthplace of wool. Clay tablets or *ur* dating to 2000 BC mention the production of wool fleece and the weaving of wool fabrics. The Old Testament and Jewish law also describe the use of wool in the ancient holy lands. In 2000 BC, the use of wool was also recorded in Egypt where many fragments of wool fabrics dating from 4000–3500 BC have been found preserved because of its climate. Some fragments of wool fabrics have been found from the graves of Scythian nomads of Eurasia, which date back to 500–300 BC.

In ancient India, wool was an important material for the vedic people. References to woollen cloths embroidered with gold can be found in the *Mahabharata*. The Indus valley people are believed to have used wool around the third millennium BC. The statues found at the archaeological sites give sufficient proof that hand-spun and hand-woven woollen shawls were being used back then. However, no actual textile fragments have been found because of the saline nature of the soil in the region which affects wool. Later, with the patronage of the Mughal kings, the Kashmir Shawl industry gained great popularity in the regions of Persia and modern-day India, Afghanistan, Russia and Europe. Nowadays many other items like Pashmina shawls, Shahtush, Chadder, Lohis, Gabba (embroidered carpet rug) and Namdas (embroidered felt rugs) are popular in India as well as in export markets.

Out of the many varieties of animal hair being used today, the fleece of sheep is the most widely used. Australia, which produces 27.4 per cent of the world's wool, is the largest producer. Other major producers include China, United States and New Zealand. Australia also produces the largest amount of the most valuable wool obtained from Merino sheep, about 43 per cent of the global production.

The woollen industry in India is still small and scattered with the top seven producers being Rajasthan, Jammu and Kashmir, Karnataka, Andhra Pradesh, Gujarat, Uttar Pradesh and Himachal Pradesh. India has the world's third largest sheep population of about 6.4 crore sheep. However, the average annual yield of wool per sheep in India is just 0.9 kg as compared to the global average of 2.4 kg. In 2012, India produced 44.4 million kg wool, of which 85 per cent was carpet grade, 5 per cent apparel grade and 10 per cent for making rugs and blankets. Thus, India is inadequate in the production of wool for apparel use and is dependent on imported raw material. India also produces a small quantity of specialty fibres from the Cashmere goat, Pashmina Goat and Angora rabbits for producing superior and luxurious fabrics. The industry employs about 12 lakh people in the organised industry sector, 112 lakh in sheep rearing and 3.2 lakh weavers in the carpet sector. The organisations dealing with the wool industry in India are Central Wool Development Board (CWDB), Central Sheep and Wool Research Institute (CSWRI), Wool and Woollen Export Promotion Council (WWEPC), Indian Woollen Mills Federation and Wool Research Association (WRA). Other international bodies related to the industry are International Wool Secretariat (IWS) and International Wool Textile Organization. In addition to these bodies, the Woolmark Company has registered certification marks which are carefully controlled and regulated to be used by authorised licensees. They indicate the wool content and product type of the articles bearing these marks.

Production

Approximately 200 different breeds and crossbreeds of sheep are used for producing wool fibre. As mentioned above, the merino variety of sheep produces the most valuable wool. It is very fine, strong and elastic though the fibres are relatively short, ranging from one to five inches. This wool accounts for about 30 per cent of the total global wool production. In order to provide the best quality wool for consumers, manufacturers control wool production carefully and scientifically. Raising sheep with care can produce better quality wool. This involves inoculating them against disease, chemical application to prevent infestation by insects and feeding a balanced diet. Sheep are generally kept in small herds, and in clean, fenced-in shelters. Climatic conditions and grazing area (pasture) also affect the quality of wool produced.

Processsing

Shearing

Wool can be obtained from sheep by shearing the wool fibre from the animal when it is alive (Image 2.5) or pulled from the hide after the animal has been slaughtered for meat. Shearing can be done mechanically (using clippers) or chemically. Shearing is carried once a year in most places, in early spring. The length of the fibre ranges from 1.5 to 15 inches in this method.

Image 2.5: Wool shearing

Source: 'Schafscheren in Orvelte (Prov. Drenthe)' by (Wikimedia Commons; CC by 3.0 SA).

Chemical or biological shearing may be done using chemicals similar to those used in cancer chemotherapy. This causes hair falling out after about two weeks of treatment, producing wool that is slightly longer than sheared wool with less physical damage. This method is said to have no lasting side effects on the animal.

Grading

Each fleece is removed as a single piece and rolled into a bundle before being sent for grading. In grading the fleece is judged for overall fibre fineness, length, fibre type and quality. After grading, the wool is sorted according to the location of the fibre on the animal's body. The wool is sent for further processing after being packed in bales, each containing only one grade of wool.

Scouring

Raw wool or grease wool contains impurities in the form of sand, dirt, grease and dried sweat (suint) which could account for anywhere between 30 and 70 per cent of the weight. It is imperative to remove these impurities by a scouring method to get clean wool suitable for further processing. Scouring is a treatment with warm alkaline soap solution. The removed grease is recovered as lanolin and is widely used in the cosmetic industry in the manufacture of creams, soaps, etc.

Types of Wool

The sheared wool is called 'fleece' or 'clipped' wool. Fleece that is sheared from sheep at 8 months of age or younger is called 'lamb's wool', which is much softer and finer in quality. Wool taken from the hide of a slaughtered animal is called **pulled wool**. Pulled wool is usually of inferior quality; it is less lustrous and elastic, and absorbs dyes less evenly. Pulled wool can be separated from the hide by one of two methods: treatment with a depilatory that loosens the fibres and allows it to be pulled away from the skin without damaging the hide, or loosening with the help of bacteria on the root of the fibre. Pulled wool is usually mixed with fleece wool by manufacturers before processing. **Virgin wool**

Figure 2.5: Chemical composition and molecular arrangement in wool

is the wool not used before or wool which has not been processed or woven before. **Recycled wool** refers to wool reclaimed from new fabric or fabric used by consumers. In order to make existing wool items into new fabric, each item must be shredded back into fibres. Recycled wool is also known as **shoddy wool.**

Figure 2.6: α keratin structure of wool

Wool and other speciality hair fibres are protein fibres made up of peptide chains or amino acid residues joined by amide linkages. Wool fibres consist of complex, cross-linked keratin protein made up of 18 different amino acids. Cystine (sulphur containing amino acid), arginine, serine, glutamic acid and leucine are the major amino acids present. Molecular chains are held together by strong cysteine (disulphide) and salt linkages (Figure 2.5). The protein chains are coiled in a helical shape like a spring (α keratin structure) (Figure 2.6). The chains have intramolecular hydrogen, ionic and disulfide bonds, linking the coils of the helix. When wool is stretched or pulled, its cross links help to recover its original shape, and recovery becomes difficult if the cross links are broken or damaged. The polypeptide chain arranged in a helix-like shape gives wool its high elasticity. On stretching, the helix structure converts into the ladder structure which reverts back to the helical structure on relaxation.

Properties

Microscopic Properties

The longitudinal section of wool fibre shows characteristic overlapping scales pointing towards the tip of the fibre (Image 2.6). The **scales** are actually the epidermis or cuticle cells of wool. These scales may be small or broad in size depending on the fineness of the fibre. The number of scales varies from 300 per inch in the case of coarse wool to 700 per inch for fine wool. The cross-section shows that wool is a cylindrical fibre consisting of the cuticle,

Image 2.6: Longitudinal view of wool fibre

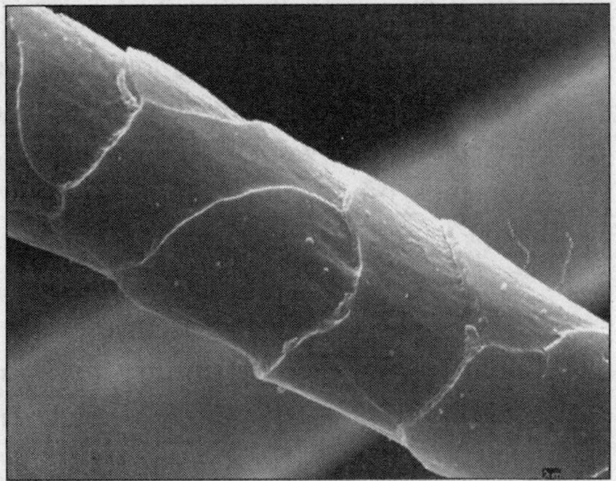

Source: Authors.

cortex and the central canal or medulla (Figure 2.7). The cortex makes up to 90 per cent of the fibre mass. It is divided into two parts which contain different cell types. The reaction to moisture and temperature is markedly different on the cortical cells of the two sides of wool. The crimp in wool is because of this difference in cell types. Melanin pigment is found in natural coloured wool.

Figure 2.7: Cross-sectional view of wool fibre

Source: Drawn by the authors.

Physical Properties

- *Shape and appearance*: Wool fibres vary greatly in length, from 1.5 inches (short fibres) to 15 inches (long fibres). The length of the fibre depends on the breed of sheep, growing conditions and the time gap between shearing. Typically, fine wool for apparel purposes is about 1.5–5 inches long while carpet-grade wool is 5–15 inches long. The width of wool fibres ranges from 15–17 microns for fine merino wool to 24–34 microns for coarse wool. Wool fibres are elliptical in shape and have a natural crimp in them.
- *Colour and lustre*: The colour and lustre of wool depends on the breed of the animal and the fineness of the fibre. After scouring, most wool appears yellowish-white or ivory in colour but wool is also available in grey, brown and black colours. The lustre also depends on the fineness of wool, with finer fibres being more lustrous and silky in appearance.
- *Tenacity*: Wool is weaker than many other textile fibres. The dry strength of wool ranges from 1–1.7 grams/denier. It loses strength when wet, the wet strength is 0.7–1.5 grams/denier. However, optimum fibre weight and type will result in satisfactory end-use products.
- *Elongation and elastic recovery*: Wool has excellent elongation and elastic recovery. Wool can elongate to 20–40 per cent under standard conditions. When wet, elongation may

extend up to 70 per cent. Elastic recovery is 99 per cent at 2 per cent extension; it reduces to 50 per cent at 10 per cent elongation, making it one of the most elastic fibres.

- *Resiliency*: Wool has excellent resiliency and it springs back to shape after creasing, crushing or folding. However, durable creases or pleats can be set into wool under specific conditions which cause new cross linkages in the polymer.
- *Density*: Wool is a heavy fibre with a density of 1.3 g/cm^3.
- *Moisture regain and absorbency*: Wool is hygroscopic in nature and it has a standard moisture regain of 13.6–16 per cent. At saturation, wool can absorb 29 per cent of its weight in moisture. While absorbing moisture it also liberates heat making it comfortable to wear in cold, damp climate. Due to the presence of scales, wool also exhibits some hydrophobic characteristics such as not absorbing liquid spills and rain quickly, thus allowing the fabric to remain dry.
- *Dimensional stability*: Wool fibres do not have good dimensional stability. The scaly structure of the fibre is such that it contributes to shrinking and felting during processing, use and care. Thus under conditions of heat, moisture and agitation, the scales interlock with each other causing the fibres to become entangled and matted. This causes the fabric to shrink in length and increase slightly in thickness. The resulting shrinkage is called **felting shrinkage** (Image 2.7). However, this unique property of wool can be carefully controlled to produce felt fabrics directly from wool fibres. Woven wool fabrics may also show relaxation shrinkage in addition to felting shrinkage.

Image 2.7: Felting of wool fibre

Source: 'Felt cloth' by Danielle Keller (Wikimedia Commons; CC0 Public Domain).

Chemical Properties

Keratin in wool is very sensitive to alkalis. Even a weak alkali like 5% sodium hydroxide solution will dissolve wool. Therefore most commercial detergents cannot be used for laundering wool, but neutral or mild detergents may be used. Undiluted sodium hypochlorite bleach will also damage the fibre. However, most organic solvents used in dry cleaning and stain removal do not damage wool fibres.

Wool is resistant to the action of dilute acids, but concentrated mineral acids like sulphuric acid and nitric acid will damage the fibre.

Chlorine bleach is harmful to wool and therefore hydrogen peroxide is normally used to bleach wool. In the early days, wool was bleached by exposing moist wool to fumes of sulphur dioxide produced by burning sulphur. This bleaching process of wool using sulphur is called 'stoving'.

Thermal Properties

Wool burns slowly with a slight sputtering sound, giving off the odour of burning hair. It self-extinguishes on removal of flame leaving a black, crushable bead-like residue. At temperatures greater than 130°C, wool will slowly decompose and turn yellow. Dry heat above 300°C will cause wool to char and disintegrate.

Biological Properties

Wool protein attracts insects like carpet beetles and larvae of some moths. Special finishes like mothproofing can be applied to prevent insect attacks. Wool is somewhat resistant to bacteria, but wool stored in damp conditions will be attacked by mildew. Food stains on wool attract pests, therefore wool should be stored in a clean, dry, air-tight state.

Other Properties

- *Resistance to sunlight*: Prolonged exposure to sunlight will cause breakdown of the disulphide bonds of cystine in the wool molecule, thus oxidising and degrading the fibre.
- *Static charge*: Wool has low heat and electrical conductivity, making it insulating and warm. However, no static charge problem is faced even in very dry conditions.

Use and Care

Wool finds usage in apparel, home furnishings and industrial uses. The warmth, resiliency, absorbency and elasticity of wool make it ideal for winter apparel, rugs, carpets, blankets, wall hangings, tapestry and upholstery fabrics. Specialty wool fibres and furs are used for interlinings, upholstery, luxury coatings, suits, dresses, sweaters, shawls and furnishing fabrics. In the industry, felts are commonly used for filtration and medical textiles. Wool is also used in geotextile applications such as wool mulch mats for landscaping and weed control. Tiny wool balls are used to clean oil-spills, as they can absorb up to 40 times their weight of oil.

Wool can be dry cleaned and pressed easily (when steam ironed at not more than 200°C) but demands a lot of care during washing. Wool can be laundered at home if we use lukewarm water and mild, neutral soaps. During washing, minimum agitation must be done to prevent felting. Moreover, only gentle pressing—and not wringing—should be done to squeeze out excess water. The garment should be dried flat. Chlorine bleaches should be avoided and hydrogen peroxide must be used instead. However, all care labels and instructions should be read and followed accordingly. Wool can be stored after brushing and cleaning to prevent insect and fungi attacks. Woven wool garments can be stored on hangers while knitted wool garments like sweaters should be folded and stored flat in drawers to prevent sagging.

SPECIALTY HAIR FIBRES AND FURS

Mohair

Mohair fibre is obtained from the Angora goat which is native to Tibet. The natural colour of Mohair ranges from yellow to greyish white, which turns white and silky upon cleaning. It is fine, light, fluffy, highly lustrous, smooth and slippery. It has similar physical and chemical properties as wool. It is highly abrasion-resistant due to the lower number of scales present on the fibre, thus making it more resistant to felting or shrinking and also easier to launder.

Cashmere

Cashmere fibre is obtained from the Cashmere (Kashmir) goat, which is native to the Himalayan mountain region of India, China and Tibet. Only the fine, downy undercoat is used for Cashmere fabrics, which is obtained by combing the fleece hairs during shedding season. The natural colour of the fibre varies from grey, brown to white in rare cases. These fibres are finer than wool but have similar chemical and physical properties. It is very soft and lustrous, but very scarce, making it a very expensive fibre. It is often used in blends with wool, silk and cotton.

Alpaca

This fibre is obtained from the fleece of Alpaca which is native to Peru, Ecuador, Bolivia and Argentina. The colour of this fibre varies from white, grey, brown and black. The fibre is long, fine, soft, silky and strong with a glossy lustre. It is used mainly for apparel, handicrafts, rugs and plush upholstery.

Camel Hair

This fibre can be obtained from the two-humped Bactrian camel of Central Asia or one-humped Dromedary camel of the Arabic desert. It can be gathered when the animal sheds its hair or through shearing. The soft, fine, downy undercoat called 'noils' is best suited for making apparel. However, noils can be used in blends with the outer coarse, bristly hairs. Camel hair is usually tan or light brown in colour, which is difficult to bleach and is therefore dyed in darker colours only. It is lightweight and highly insulating but weaker in strength.

Llama

This fibre is obtained from the Llama which is native to the South American Andes region. It produces a fleece similar to the Alpaca, which is fine, soft and lustrous. The colours range from mostly blacks and browns to light colours at times. It is weaker than alpaca or camel hair but can be used in blends.

Vicuna

Vicuna, a close relative of the alpaca and llama, is a wild animal found in the very high altitudes of the Andes mountains. The fibre is naturally brown or tan in colour which is hard to dye. The fibre is mostly gathered by hunting and killing the vicuna, and is thus scarce. This combined with its extraordinary softness and fineness makes it one of the most expensive specialty hairs in the world.

Quiviut or Musk Ox

This fibre is obtained from the downy undercoat of the Quiviut/Musk ox. The fibre is similar to cashmere in texture and softness. It is very warm and expensive and thus may be used in blends with cashmere or silk.

Fur

Fur is the animal pelt, which consists of the skin with the hair and needs tanning and processing to convert it into fur. It was probably the earliest known clothing of mankind. Fur can be obtained from beaver, rabbit, chinchilla, fox, mink, muskrat, nutria and raccoon. Fur articles are expensive and are held in high regard. However, in recent years, due to the endangerment of many species, many strict regulations have come up for fur trade. Imitation synthetic fur is also being used to combat these issues.

Silk

History

The history of silk is based on both fact and myth and it is difficult to isolate truth from fiction and legend. The story told most frequently is that of the discovery of silk by the Chinese princess Hsi Ling Shi around 2700 and 2600 BC. It is said that while investigating what was damaging the imperial mulberry grove, she came across tiny silk worms that were devouring the leaves and then crawling from the leaf to spin shining, pale, almost white cocoons. The princess then gathered a handful of cocoons and accidentally dropped one into her cup of hot tea. The princess noticed that the cocoons softened and she could draw a slender, tiny filament into the air from this cocoon.

For around 3000 years China successfully guarded the secret of silk and its manufacture, holding a monopoly on the silk industry. However, other countries managed to obtain silk worms by devious means. In India, silk has been referred to in the great epics. Valmiki in his *Ramayana* says that Sita was clad in silk when she accompanied Ram to the forest. The country-made silks were called by names like *ketaja*, *patta* or *kauseya*. The Chinese silks were called *china-shuka* or *chinapatta*. *Pitamber* is another very superior class of *dhoti* (piece of fabric meant for wrapping around the body by both men and women) made of pure silk—generally of pink or yellow or the colour of sunflower—worn on festival days or at meal time by many Hindus. Tasar (*kosa* or *kaushikvastra*, *pattavastra*) is also indigenous to India. Patronised by royal courts, Indian sericulture has had a long and chequered past. A traveller from Damascus, Abul Abbas Ahmad, has recorded the manufacture of silk in India. He writes that about 400 silk weavers were employed in a manufactory set up by Sultan Muhammad Tughlak (AD 1325–50). He also mentions distribution of 1,00,000 silk dresses manufactured in Delhi every year by the Sultan (Dhantyagi 1996).

The main countries producing silk are China, India, Thailand, Brazil and Vietnam. India is the second largest producer of silk with the distinction of being the only country to produce all four commercially known varieties of silk—Mulberry, Tasar, Eri and Muga. India produced approximately 23,060 tons of raw silk in 2012. It provides employment to about 7.65 million people. The main states producing mulberry silk are Karnataka, Andhra Pradesh, West Bengal and Tamil Nadu. Tasar silk is mainly produced by Jharkhand, Chhattisgarh, Manipur, Himachal Pradesh, Uttar Pradesh, Assam and Meghalaya. Eri silk is produced in Assam and the northeastern states while Muga silk is primarily produced in Assam. The prominent organisations working for the silk industry are the Central Silk Board (CSB), Silk Mark Organization of India (SMOI) and Indian Silk Export Promotion Council (ISEPC).

Production

The cultivation of silk under controlled conditions to produce filament silk fibre is called **sericulture.** The silk fibre is produced by the larvae of a wide variety of silkworms or moths. Mulberry silk is produced by the moth of *Bombyx mori* species. The four basic stages of the development of *Bombyx mori* for cultivation of silk are:

1. *Laying of eggs*: A silk moth lays about 700 eggs, and each egg is about the size of a pinhead and has a soft place on one end permitting easy hatching of the larvae. The eggs may be stored in a controlled environment, at low temperatures for long periods of time unless required otherwise.
2. *Hatching of eggs*: The eggs are brought out of storage and warmed slightly when a supply of mulberry leaves is available. In three to seven days, the eggs hatch into caterpillars or larvae which are about one-fourth of an inch long. The larvae emerge and begin to feed on the tender mulberry leaves. Between hatching and spinning a cocoon, the larva sheds its outer skin four times and after the fourth mottling it eats for 10 more days and increases in size. In total it eats for about 35 days and grows as much as 3 inches long and increases to about 10,000 times the weight of the newly-hatched larva.
3. *Spinning of a cocoon*: The adult caterpillar begins to spin a cocoon by extruding silk through two tiny orifices or spinnerets in its head. The cocoon is formed by movement of the head similar to figure '8'. Two strands of fibres (fibroin) are spun together. These fibroins are held together by a gummy substance which is known as sericin. The fibroin and sericin are produced by two glands within the silkworm. A single strand of silk is called 'brin', and the dual strand is called 'bave'. After the cocoon is completed, the silkworm enters the dormant chrysalis or pupa stage. The pupa must be killed before an adult silk moth emerges from the cocoon and breaks the continuous silk filament.
4. *Emerging of the silk moth*: Only those moths selected as breeding stock are permitted to complete the growth cycle. The remaining cocoons are subjected to dry heat (stifling) which kills the pupa inside the cocoon. These cocoons can be then stored for long periods of

time. In addition to the sericulture of *Bombyx mori*, the larvae of other moths spin cocoons of tasar silk or wild silk. This silk from wild species is coarser, stronger and of short fibre. The fibre is often tan or light brown in colour and cannot be bleached without causing some damage. The most important moths producing tasar silk are *Antheraea mylitta* and *Antheraea pernyi* which feed on oak leaves.

Processing

Reeling

The whole, unbroken cocoons are sorted according to colour, texture, size and shape. A manufacturing plant called Filature is used to unwind silk filaments from the cocoons. Several cocoons are placed in a container of water heated to about 60°C. This warm water softens sericin, the gum that holds the filaments of silk together. Very little gum is actually removed in reeling. The surfaces of the cocoon are then brushed lightly by hand to find the ends of the filaments. These ends are then collected and threaded through a guide, and wound onto a wheel called a reel, hence the name reeling. As the reeling continues, a skilled operator adds or lets off fibres as needed to make a smooth, uniform-sized yarn. These are then packed into bales for shipping.

Throwing

While the fibres are combined and pulled onto the reel, twist can be inserted to hold the filaments together. This process is called throwing and the resulting yarn is called thrown yarn. Several types of silk yarn can be produced depending on the number of filaments combined together and the amount of twist inserted during the throwing process. The four major types are:

- *Singles*: It is a strand of three to 10 filaments collected together and held by low, medium or high twist. They can be used as warp or weft in knitted or woven fabrics.
- *Tram silk*: It is a ply yarn formed by combining two or three singles having a low twist. They are of average strength and are used as weft yarns.
- *Organzine*: It is a ply yarn with a medium twist. It is used as warp yarns in weaving. Crepe organzine is used in making crepe and chiffon fabrics.
- *Grenadine*: It is made up of two to three singles with a tight twist. The ply twist is inserted in the direction opposite to the single yarn twist.

Degumming

Sericin remains on the fibres during reeling and throwing. Frequently, it is left on through the fabric construction process, if so desired. The sericin gum makes the silk filament stiff and also provides some protection during processing. Before final finishing, the sericin gum can be removed by boiling the fabric in a solution of soap and water. Raw silk fabrics do not have sericin removed.

Weighting of Silk

During the late nineteenth century, silk fabric were treated with metallic salts like tin to add body, weight, better covering power, drape and dyeability to the final fabric. This process is called weighting of silk. However, silk can absorb metallic salts even more than its own weight resulting in excessive weighting. This eventually leads to breaking or 'shattering' of the silk fabric.

Chemical Composition and Molecular Arrangement

Silk is essentially made up of fibroin (70–75%), the fibre-forming protein, and sericin (20–25%), the protein gum around it. Fibroin is less complex than keratin and is composed of at least 17 amino acids, 90 per cent of which is glycine, alanine, serine and tyrosine (Figure 2.8). Very insignificant amounts of cystine or sulphur linkages are present in silk. There is very little cross-linking between fibroin protein chains but hydrogen bonds are formed, producing a highly crytalline structure.

Figure 2.8: Chemical structure of silk fibroin

Gly Ser Gly Ala Gly Ala

Note: Gly: Glycine, Ser: Serine, Ala: Alanine.

Properties

Microscopic Properties

In the longitudinal view, degummed cultivated silk looks like smooth, transparent rods, but silk that has not been degummed has a rough, irregular surface. In the cross-sectional view, silk appears triangular with two fibres (brins) lying with their flat sides together (bave) (Figure 2.9). Raw/wild silk will appear darker in colour and may have some longitudinal striations.

Figure 2.9: Silk filament

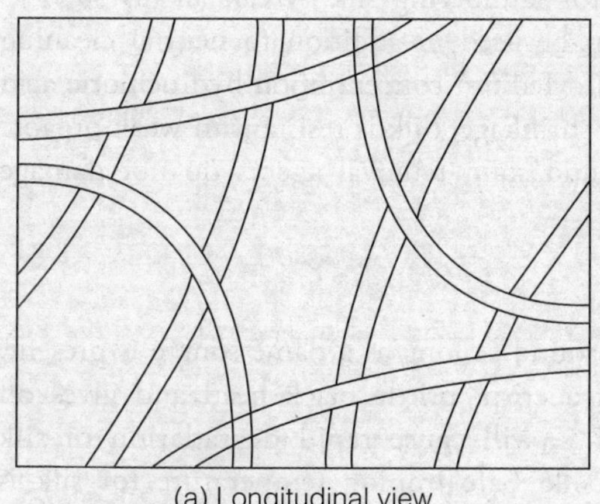

(a) Longitudinal view

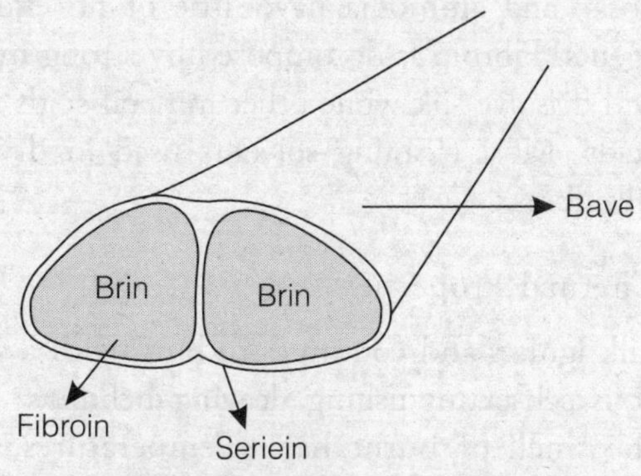

(b) Cross-sectional view

Source: Drawn by the authors.

Physical Properties

- *Shape and appearance*: Silk is the only natural filament fibre, which can reach 900 to 1200 meters in length. A single silk brin is very fine with a diameter of 9–11 micron.
- *Colour and luster*: Cultivated mulberry silk is usually off-white to cream in colour but can also be yellow, light brown or almost pink. Tasar silk is darker in colour ranging from tan to light brown due to tannin in oak leaves. Silk has a soft lustre but wild silk may have a duller lustre due to the irregular surface.
- *Tenacity*: Silk is one of the strongest natural fibres with a dry tenacity of 2.4–5.1 grams/denier. It becomes weaker when wet with a wet strength of 2–4.3 grams/denier.
- *Elongation and elastic recovery*: Silk has an average elongation of 10–25 per cent at breaking point. Wet elongation may increase up to 35 per cent. It has good elastic recovery of 92 per cent at 2 per cent elongation.
- *Resiliency*: Silk has medium resiliency, which is less than that of wool due to the absence of cystine linkages. Therefore, creases and wrinkles do not hang out as quickly as in wool. However, silk has better resiliency than cotton.
- *Density*: The density of silk ranges from 1.25 to 3.24 g/cm^3.
- *Moisture regain and absorbency*: Silk is quite absorbant. It is less absorbent than wool but more absorbent than most other fibres, making it comfortable to wear and easy to dye. The standard moisture regain of silk is 11 per cent which increases to 25–35, per cent at saturation.
- *Dimensional stability*: Silk is dimensionally stable with a good resistance to stretching or shrinking when laundered and dry-cleaned.

Chemical Properties

Alkalis can damage fibroin and heated sodium hydroxide will dissolve silk. Therefore, strong alkaline soaps and detergents should not be used for laundering silk. Weak alkalis such as borax and ammonia have little or no effect and can be used in addition to neutral cleaning agents. Fibroin is decomposed by strong mineral acid. Medium concentration hydrochloric acid will dissolve silk while other mineral acids may cause shrinkage. Silk is resistant to weak organic acids. Most cleaning solvents used in dry-cleaning and stain removal agents do not damage silk fibres.

Thermal Properties

Silk ignites and continues to burn with a sputtering sound as long as a flame source is present. It is self-extinguishing, leaving behind a residue of a crisp, brittle black bead and gives off the smell of burnt hair. Temperatures above 177°C will cause rapid degradation of silk but prolonged exposure at 135°C does not affect silk. Safe ironing temperature for silk is 120–135°C.

Biological Properties

Silk is fairly resistant to mildew, many bacteria, fungi and clothes moth. However, it can be attacked by rot-producing bacteria. Also, silk can be damaged by carpet beetles, therefore they should be stored carefully.

Other Properties

The UV rays in sunlight can hasten the decomposition of silk more easily than that of wool. Weighted silk is more prone to damage from natural ageing.

Use and Care

Silk is mainly used for luxury apparel, home textiles and some industrial textiles. It is said to be the queen of textile fibres due to its luxurious and expensive appearance. Silk is mostly used in saris, suits, formal dresses, lingerie and accessories like ties, scarves, stoles and handbags. Silk is also used in making fine drapery, upholstery, bed spreads, cushion covers, rugs and carpets. Silk has been used for long to make luxurious and exclusive saris all over India. The famous silk saris of India, mostly named after their places of production, include Brocades (Banaras), Ikats (Orissa), Patolas (Gujrat), Tanchoi (Gujarat), Balucharis, Kanchipuram, Dharmavaram, Pochampalli, Chanderi, Bhagalpur and Paithanis. Silk is also used for medical sutures, embroidery threads, typewriter's ribbon and even racing bicycle tyres.

Silk fabric requires special care as it is expensive. Dry cleaning is usually the preferred method. However, silk can also be hand washed at home as long as proper care is taken in handling the fabric. Warm or cold water with mild detergent must be used and any stains present should be cleaned prior to washing. Non-chlorine bleach should be used if required. Brushing and wringing the garment should be avoided and silk should not be dried in the sun or tumble-dried. Rolling in a towel can squeeze excess moisture out. If stiffening is required, gum arabic solution can be applied. Ironing can be done satisfactorily using a wet muslin cloth on top of silk.

Varieties of Silk

Mulberry Silk

In India, the major mulberry silk producing states are Karnataka, Andhra Pradesh, West Bengal, Tamil Nadu and Jammu & Kashmir which together account for 92 per cent of the country's total mulberry raw silk production. Mulberry silk is used mainly for Banarasi saris, zari saris, chinnon and chiffon scarves, organza and embroidered garments.

Tasar (or Tussah) Silk

This kind of silk can be further categorised into two types—tropical tasar and oak tasar. Tropical tasar is a coarse silk, that is copperish, honey or tawny in colour. It is produced by the silkworm, *Antheraea mylitta* which mainly thrives on the plants Asan (*Terminalia elliptica*) and Arjun (*Terminalia arjuna*). The rearing is mostly done in nature on trees. It is less lustrous than mulberry silk but has its own feel and appeal. Tasar culture is the mainstay for many a tribal community in India. In India, tasar silk is mainly produced in the states of Jharkhand, Chhattisgarh and Orissa, besides Maharashtra, West Bengal and Andhra Pradesh.

Oak tasar is a finer variety of silk produced by the silkworm, *Antheraea proyeli J.* which feeds on leaves of the oak tree. In India, this variety is produced in the sub-Himalayan belt, in the states of Manipur, Himachal Pradesh, Uttar Pradesh, Assam, Meghalaya and Jammu & Kashmir. China is the major producer of oak tasar in the world and the Chinese variety is produced from another species of silkworm, *Antheraea pernyi.*

Tasar is used in saris which can have zari-work, embroidery or prints. The Kostha weavers of Chhattisgarh make tasar saris known as mailooga and gamchha. Tasar blended with wool is used for making shawls, stoles, scarves, mufflers, jackets, dresses and painting canvas.

Eri Silk

It is also known as endi or errand silk. It is a multivoltine silk spun from open-ended cocoons, unlike other varieties of silk. Eri silk is produced by the *Philosamia ricini* variety that feeds mainly on castor leaves. It produces white-coloured silk similar to mulberry. It is a domesticated silkworm and Eri culture is a household activity as the protein-rich pupae are also considered a delicacy among the tribals. In India, this culture is practised mainly in the northeastern states. It is also found in Bihar, West Bengal and Orissa. The silk is used indigenously for the preparation of chaddars (or wraps) for use by the tribal people. Good thermal properties make it suitable for making shawls, jackets and blankets. It can be easily blended with cotton, wool, other silks, jute, ramie and even synthetics, and thus a large variety of blends can be produced. It is also used in home furnishings as it is durable, comfortable and strong.

> **Univoltine silk** is obtained from silkworms such as *Bombyx mori* that give only one crop of silk in a year. **Multivoltine silk** is obtained from silkworms, such as *Bombyx bengalensis*, which give as many as eight crops as they pass through successive generations in a year.

Muga Silk

It is a golden-yellow silk produced by the semi-domesticated multivotine variety of silkworm, *Antheraea assamensis.* It has the highest tensile strength among all natural fibres. These silkworms feed on the aromatic leaves of Som (*Machilus bombycina*) and Soalu (*Litsea polyantha*) plants and are reared on trees. Muga culture is specific to the state of Assam and is an integral part of the tradition and culture of that state. The Muga silk, a high-value product, is used in products like saris, mekhalas and chaddars. It is also used in articles like stole, shawls, tie, scarf and even as a zari substitute. As it is durable, moisture-absorbent and stain-free, it also finds use in home furnishings.

OTHER NATURAL FIBRES

In addition to the natural fibres obtained directly from plant and animal sources there are other fibres obtained from nature such as rubber (from rubber plant sap) and asbestos (mineral fibre).

Rubber

Natural rubber is obtained from the *Hevea* species trees. The thick, gummy sap of the tree is collected by making an incision in the trunk. This liquid is further processed to convert it into usable form. Charles Goodyear discovered in 1839 that the properties of rubber can be enhanced by heating it with sulphur, a process known as vulcanising. Vulcanised rubber has increased strength and elasticity while maintaining flexibility at cold temperatures without turning brittle and hard. Rubber was converted to fibre form in the 1920s. It was discovered that liquid rubber (latex) could be extruded to fibre form which had high elongation and elastic recovery. Often, rubber fibres are used as a central core for other fibres like cotton to impart elasticity and flexibility. Rubber yarns are used in under garments, swimwear, foundation garments, elastic tape, support hose, surgical elastic bandages, tops of socks and also in decorative stitching. In the foam form it is used as filling in mattresses, pillows, etc.

Rubber has high flexibility, elasticity, pliability, strength and also imparts impermeability to air and water. It has a density of 1 g/cm³. It is resistant to cutting and tearing while also being resistant to many chemicals. It is damaged by high temperature (greater than 93°C) causing it to turn brittle and lose pliability. It can be damaged by body oils and petroleum solvents. Other factors like exposure to sunlight and ageing can also cause deterioration.

Care must be taken in laundering rubber by keeping washing and drying temperatures below 60°C and using neutral soaps and detergents. Proper cleaning should be done to remove body oils and dirt. Dry-cleaning must be avoided in rubber fabrics.

Asbestos

Asbestos is the only mineral used as a textile fibre in the form that it is mined. Asbestos mineral is found in the fibrous veins of serpentine or amphibole rock. The use of asbestos has been recorded since the early days of Greek and Roman civilisations. It has been written about by Pliny the Elder and Marco Polo describing its surprising ability to not burn even in direct flame. This highlights the most important property of asbestos fibre, that it is completely fireproof. Asbestos was commercially developed in the nineteenth century after the discovery of large deposits of asbestos in Canada and South Africa. There are many varieties of asbestos but only the chrysotile variety is used as a textile fibre.

Asbestos fibres have good strength, flexibility, toughness and average length. It has a silky texture with a soft to harsh handle. It usually occurs in white, amber, grey or green colour. It has low conductivity and can tolerate extremely high temperatures. The fibre remains undamaged even at temperatures of 3315°C if exposed for a brief period. However, the water bond within the molecule dries out at 593°C and asbestos fibre will fuse at 1520°C if held in that temperature for a period of time.

After the 1970s, many serious health hazards came to the fore regarding asbestos. It has been classified as a potential carcinogen (cancer-causing). Asbestos can cause many lung diseases as it can break away into really small particles that can be easily inhaled or ingested. Due to these reasons, alternatives to asbestos fibres are being developed and asbestos usage is actively being discouraged.

SUMMARY

- Natural cellulosic fibres are those which are obtained from plant sources. Cellulose is the main building block of these fibres.
- The cellulose molecule is made up of many cellobiose units. A cellobiose unit is made up of two glucose units.
- Cotton and linen are the most commonly used cellulosic fibres.
- All protein fibres are composed of amino acids which are joined together in polypeptide chains.
- Wool and silk are the most commonly used natural protein fibres.
- Animal hair fibres include fibres obtained from animal's hair, mainly sheep's wool.
- Silk fibre is formed from extruded filament which is produced by silkworm larvae.
- The various types of silk produced in India are mulberry, tasar, muga and eri.
- Other fibres obtained from nature are rubber (from rubber plant sap) and asbestos (mineral fibre).

KEY WORDS

Bast fibres: The fibres obtained from the stem or bast portion of a plant.

Retting: The process of removal of outer woody covering to obtain flax with the help of bacteria or chemicals.

Felting shrinkage: Under conditions of heat, moisture and agitation, the scales on the surface of wool interlock with each other causing the fibres to become entangled and matted resulting in shrinkage.

Degumming: The removal of sericin by boiling silk in soap and water solution.

Weighting: Treatment of degummed silk fabric with metallic salts of tin to add body and weight.

EXERCISES

1. List the major producers of cotton in the world.
2. What are the properties of cotton that make it suitable for use in hot, humid climates?
3. What is retting? Explain the different types of retting.
4. What are the similarities between flax and jute?
5. Name the four varieties of silk found in India.
6. What are the differences between wool and silk?
7. Describe the production of mulberry silk.
8. What are the care instructions for wool and silk?
9. Explain the phenomenon of felting in wool.

REFERENCES

Collier, Billie J., Martin J. Bide and Phyllis G. Tortora. 2009. *Understanding Textiles*. New Jersey: Pearson Education Inc.

Dhantyagi, S. 1996. *Fundamentals of Textiles and their Care*. Fifth edition. New Delhi: Orient Longman Limited.

Hudson, P. B., A. C. Clapps and D. Kness. 1993. *Joseph's Introductory Textile Science*. Sixth edition. New York: Harcourt Publishers.

Kadolph, J. Sara. 2009. *Textiles*. Tenth edition. New York: Pearson Education.

Joseph, M. L. 1986. *Introductory Textile Science*. Fifth edition. New York: CBS College Publishing.

Kozłowski, Ryszard M. 20212. *Handbook of Natural Fibres: Processing and Applications*, vol 2. Philadelphia: Woodhead Publishing.

Mather, R. R., R. H. Wardman. 2011. *The Chemistry of Textile Fibres*. London: RSC Publishing.

Sekhri, S. 2011. *Textbook of Fabric Science: Fundamentals to Finishing*. First edition. New Delhi: PHI Private Learning Private Limited.

ONLINE SOURCES

www.indiansilk.kar.nic.in/silk.html (accessed 13 April 2014)

http://cotcorp.gov.in/current-cotton.aspx?pageid=4 (accessed 13 April 2014)

http://www.woolmark.com/about-woolmark/understanding-our-brands (accessed 13 April 2014)

http://www.ibef.org/exports/indian-silk-industry.aspx (accessed 13 April 2014)

http://www.csb.gov.in/silk-sericulture/silk/ (accessed 13 April 2014)

http://www.ahimsasilks.com/aboutus1.htm (accessed 13 April 2014)

http://articles.economictimes.indiatimes.com/2002-04-16/news/27351520_1_bt-cotton-transgenic-cotton-american-bollworm (accessed 13 April 2014)

http://edugreen.teri.res.in/explore/bio/Btcot.htm (accessed 13 April 2014)

http://msme.gov.in/WriteReadData/DocumentFile/ANNUALREPORT-MSME-2012-13P.pdf (accessed 13 April 2014)

http://msme.gov.in/Chapter%206-Eng_200708.pdf (accessed 13 April 2014)

http://coirboard.nic.in/about_coirfibre.htm (accessed 13 April 2014)

http://www.teonline.com/knowledge-centre/banana-fiber.html (accessed 27 February 2017)

www.smallb.sidbi.in/sites/default/.../banana_fibre_extraction_and_weaving.pdf (accessed 27 February 2017)

www.gujagro.org/agro-food-processing/banana-fibre-processing-13.pdf (accessed 27 February 2017)

http://www.swicofil.com/products/010banana.html (accessed 27 February 2017)

http://ojs.cnr.ncsu.edu/index.php/BioRes/article/view/BioRes_03_1_0155_PectinolyticEnzymes_Retting/174 (accessed 27 February 2017)

http://www.hindawi.com/journals/isrn/2013/186534/ (accessed 27 February 2017)

3

MAN-MADE FIBRES

HIGHLIGHTS

- Production and classification of man-made fibres
- Fibres from natural polymers
  - Viscose rayon
  - Cuprammonium rayon
  - High wet modulus rayon
  - High tenacity rayon
  - Polynosic rayon
  - Cellulose acetate
  - Lyocell
  - Modal
  - Rubber
- Fibres from synthetic polymers
  - Acrylic
  - Polyamide fibres
    - Nylon
    - Aramids
  - Polyester
  - Polyolefins
  - Elastane (spandex fibre)
- Fibres from inorganic materials
  - Mineral fibre
  - Glass fibre
  - Carbon fibre
  - Ceramic

The era of man-made or manufactured fibres began in the early 1900s with the commercial production of rayon fibre. The development of X-ray technology in the 1920s and 1930s helped scientists characterise synthesised fibres in terms of building blocks and molecules, and with this the manufactured fibre industry grew rapidly. In 1931, nylon was developed by DuPont de Nemours and Company and subsequently, the share of man-made fibres (MMFs) increased tremendously.

GROWTH AND DEVELOPMENT OF MAN-MADE FIBRES

History

The very idea of creating an artificial fibre was first documented in 1664. The English naturalist Robert Hooke predicted that it would be possible to duplicate silk by artificial means. Unfortunately, no fibre could be produced for more than two centuries. The first commercial-scale production of a manufactured fibre was achieved by the French chemist Count Hilaire de Chardonnet in 1884. In 1889, his fabrics of 'artificial silk' caused a sensation at the Paris Exhibition. Two years later, he built the first commercial rayon plant at Besancon, France, and became known as the 'Father of the rayon industry'. In 1893, Arthur D. Little of Boston invented another cellulosic product, acetate, and developed it as a film. The first commercial textile uses for acetate in fibre form were developed by the Celanese Company in 1924. In September 1931, the American chemist Wallace Carothers focused his work on a fibre referred to simply as '66', a number derived from its molecular structure—and now known as nylon, the 'miracle fibre'. By 1938, Paul Schlack of the I.G. Farben Company in Germany, polymerised caprolactam and created a different form of the polymer, identified simply as nylon '6'. The arrival of nylon revolutionised the fibre industry. Rayon and acetate had been derived from plant cellulose, but nylon was synthesised completely from petrochemicals. In the 1940s, three new manufactured fibres were being produced: Dow Badische Company (today, BASF Corporation) introduced metalised fibres; Union Carbide Corporation developed modacrylic fibre; and Hercules, Inc. developed olefin fibre. By the 1950s, a new fibre, 'acrylic', was added to the list of generic names, as DuPont began production of this wool-like product. Meanwhile, polyester, first examined as part of Wallace Carothers' early research, was attracting new interest at the Calico Printers Association in Great Britain. There, J. T. Dickson and J. R. Whinfield produced a polyester fibre by condensation polymerisation. In the summer of 1952, the term 'wash and wear' was coined to describe a new blend of cotton and acrylic. The term eventually was applied to a wide variety of manufactured fibre blends. Commercial production of polyester fibre transformed the 'wash and wear' novelty into a revolution in textile product performance. The production of manufactured fibre reached new heights in the 1960s. These fibres were further modified to offer greater comfort, provide flame resistance, reduce clinging, release soil more easily, easier dyeability and better blending qualities. New fibres were introduced to meet special needs. Spandex, a stretchable fibre; aramid, a high-temperature-resistant polyamide; and para-aramid, a fibre with outstanding strength-to-weight properties.

In India, the man-made fibre spinning units were started in the 1920s and 1930s when pure silk was the most expensive fibre. A major breakthrough occurred when, in 1950, a factory was established by Travancore Rayons Ltd. (Rayapuram, Kerala) to manufacture synthetic fibre. All the powerloom and handloom units were importing synthetic fibre to make fabric until then. Soon National Rayon Co. at Mumbai and the Sirsilk Ltd. were established at Hyderabad. By 1960, Century Rayon Ltd. at Kalyan (Maharashtra) and the Gwalior Rayon Silk Manufacturing Company at Nagda had also been commissioned.

Today, the man-made fibres produced in India include polyester (staple and filament), viscose (staple and filament), acrylic (staple), nylon 6 (filament) and polypropylene (staple and filament). India is the second largest producer of polyester (staple and filament fibre) and viscose filament fibre. It is the third largest manufacturer of viscose staple fibre and eighth largest manufacturer of acrylic staple fibre. However, the specialised man-made fibres like acetate, tri-acetate, cuprammonium rayon, nylon 66, nylon 11, spandex, poly vinyl alcohol (PVA) and modacrylic are not being manufactured in India. In India, viscose staple fibre is produced only by Grasim Industries. Reliance Industries produces about 64 per cent of polyester staple fibre and 48 per cent of polyester filament yarn. Century Enka and JCT Ltd. produce about 85 per cent of nylon 6 filament yarn. Century Rayon and Indian Rayon Corporation produce about 80 per cent of viscose filament yarn (GoI n.d.).

PRODUCTION OF MAN-MADE FIBRES

Man-made fibres/filaments are manufactured using chemical spinning processes. All spinning processes to produce man-made fibres are based on the following three steps:

1. Preparation of spinning solution or dope
2. Extrusion of the dope through the spinneret to form a fibre
3. Solidifying the fibre by coagulation, evaporation or cooling

The **dope**, a flowing solution, is prepared by dissolving the raw material (natural or synthesised polymer) in a suitable solvent or by melting the polymer. The dope is then forced or pumped through the holes in a spinneret, which is similar to a shower head. This process is known as **extrusion**. As filaments emerge from the holes of the spinneret, the liquid polymer is converted first to a rubbery state and then solidified. An untwisted rope of a number of these filament fibres is called a tow. This process of extrusion and solidification of endless filaments is called **chemical spinning**. There are three methods of spinning filaments of manufactured fibres: wet, dry and melt spinning.

Wet Spinning

Wet spinning is the oldest chemical spinning process. It is applied to polymers which do not melt or which cannot be dissolved in a volatile solvent. Therefore, a non-volatile solvent is used to convert the raw material into a solution and later the solvent must be removed by chemical means. The spinneret is submerged in a chemical/spinning bath and as the filaments emerge they precipitate in the bath and solidify. The spinneret is then washed to remove the chemicals deposited on their surface and then dried (Figure 3.1). Viscose rayon is produced

Figure 3.1: Wet spinning

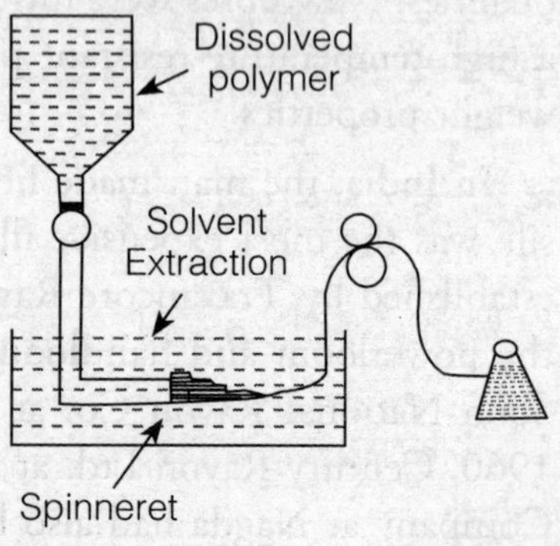

Source: Drawn by the authors.

by this process. This method of wet spinning is more complex than the other two and is used when spinning by the other two methods is not possible.

Dry Spinning

Dry spinning is used for fibre forming substances that dissolve in a volatile solution. The dope is extruded into an evaporating cabinet filled with hot air. The fibre is solidified by evaporating the solvent (Figure 3.2). The solvent vapours can be removed and condensed for reuse. Since the filaments do not come in contact with a precipitating liquid, there is no need for drying. This process is used in the production of acetate, triacetate, acrylic, modacrylic and spandex.

Figure 3.2: Dry spinning

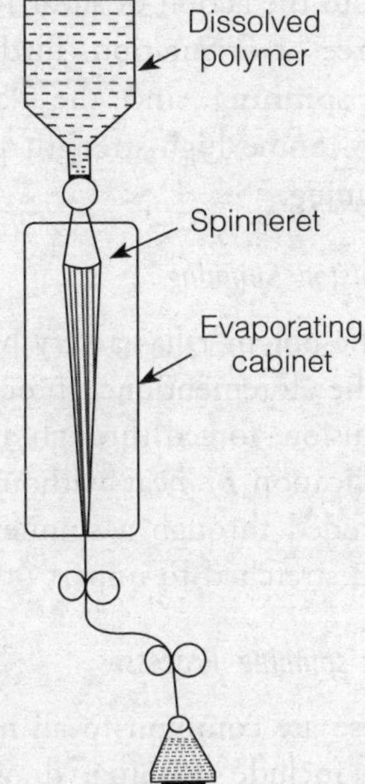

Source: Drawn by the authors.

Melt Spinning

Melt spinning is used for those fibre forming substances that can be melted without getting decomposed. In this method, the polymer is first melted in an autoclave, then extruded through the spinneret and finally solidified by cooling. Solid polymer chips are dropped from the hopper into the autoclave where heat melts the solid polymer into a viscous liquid. The liquid is then pumped through filters to remove any impurity and then delivered to the spinneret at a controlled rate of flow. When the liquid polymer emerges from the spinneret hole, a cool stream of air is passed over the fibre causing it to harden (Figure 3.3). Nylon, olefin and polyester are produced by this method. An advantage of melt spun fibres is that they can be extruded in different cross-sectional shapes—viz., round, trilobal, pentagonal, octagonal, etc.

Figure 3.3: Melt spinning

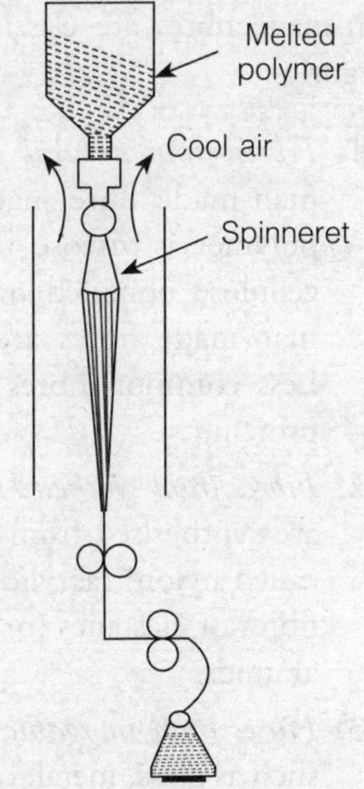

Source: Drawn by the authors.

Special Spinning Methods

Gel Spinning

Gel spinning is a special process used to obtain fibres with high strength or other special properties. The polymer of an extremely high molecular weight is dissolved and a very viscous (thick) solution is obtained. This highly viscous solution is the **gel**. The polymer chains in the gel are bound together at various points. This produces strong inter-chain forces in the resulting filaments that can significantly increase the tensile strength of the fibres. In addition, the liquid crystals are aligned along the fibre axis due to the action of shear forces during extrusion. The filaments emerge with an unusually high degree of orientation, further enhancing strength. The process can also be described as **dry-wet spinning**, since the filaments first pass through air and are then further cooled in a liquid bath. Some high-strength polyethylene and aramid fibres are produced by the method of gel spinning.

Emulsion Spinning

If the polymer has a very high melting point or is insoluble it cannot be extruded through any of the aforementioned processes, it is spun by emulsion spinning. The polymer is made into an emulsion, forced through a narrow tube and then fused by application of heat without melting. The polymer is then extruded through a spinneret into a coagulation bath and then stretched to impart orientation.

An **emulsion** is a fine dispersion where solid particles are not completely dissolved in the solvent.

Post-spinning Processes

These are common to all methods of chemical spinning and include washing, drawing (stretching) to improve orientation (Figure 3.4) and heat setting for thermoplastic fibres to impart dimensional stability.

Figure 3.4: Effect of drawing on polymer chain orientation in a fibre

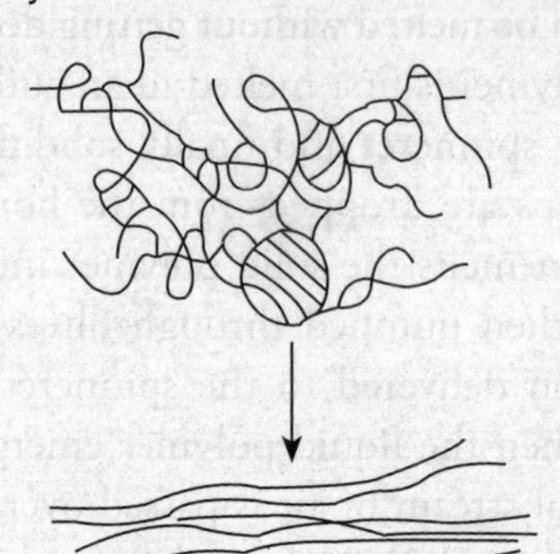

Source: Drawn by the authors.

CLASSIFICATION OF MAN-MADE FIBRES

Man-made fibres are classified into the following three categories:

1. *Fibres from natural polymers*: The most common man-made fibre made from naturally occurring polymer is viscose rayon. Viscose is made from cellulose obtained mostly from the wood pulp of various trees. Other cellulose based man-made fibres are cupramonium rayon, acetate and triacetate, lyocell and modal. Less common fibres of this type are made from rubber, alginic acid and regenerated protein.
2. *Fibres from synthetic polymers*: These are organic fibres that are made from polymers that are synthesised from petrochemicals. The most common are polyester, polyamide (often called nylon), acrylic and modacrylic, polypropylene, segmented polyurethanes or elastic fibres or elastanes (or spandex in the USA), and speciality fibres such as high-performance aramids.
3. *Fibres from inorganic materials*: The inorganic man-made fibres are made from materials such as glass, metal, carbon or ceramic. These fibres are very often used to reinforce plastics to form composites.

FIBRES FROM NATURAL POLYMERS

There are two categories of man-made cellulosic fibres: regenerated cellulosic fibres and regenerated modified cellulosic fibres. Both the fibres are made from cotton linters and wood pulp. A **regenerated fibre** is produced by dissolving and extruding a natural polymer, or its chemical derivative, as a continuous filament. After this fibre formation process the chemical nature of the natural polymer is either retained or regenerated. Examples of regenerated cellulosic fibres are the various types of rayon—viscose rayon, cuprammonium rayon, high wet modulus rayon, high tenacity rayon and polynosic rayon. Modified fibres are made by chemically modifying, dissolving and extruding a natural polymer as a continuous filament. Examples of this kind are secondary acetate and triacetate. Since the fibre-forming substance for these fibres is not cellulose but chemically-modified cellulose (i.e. cellulose acetate, an ester), these are called **regenerated modified fibres**.

Viscose Rayon

Figure 3.5: Chemical structure of rayon

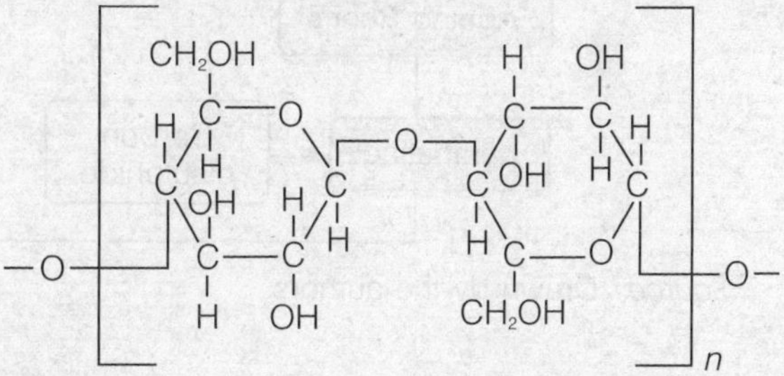

Rayon fibres are made from cellulose that has been reformed or regenerated. Figure 3.5 shows the chemical structure of rayon. Rayon is 100 per cent cellulose and has the same chemical composition as natural cellulose. However, it consists of cellulose of lower degree of polymerisation (DP) than cotton cellulose. The DP of viscose polymer is (300–450). The various varieties of rayon fibres are: viscose rayon, cuprammonium rayon, high wet modulus rayon, polynosic rayon and high tenacity rayon.

Production

Viscose rayon is produced from cellulose which is majorly obtained from wood pulp. Cotton linters (short cotton fibres) can also be used. The wood pulp, taken from the trees of eucalyptus, beech and pine, is treated to remove lignin and resin. The pulp is then pressed and cut into sheets. More than 95 per cent of these sheets is made up of cellulose. The various steps involved in the process of manufacturing viscose are as follows (Figure 3.6):

1. *Steeping*: Sheets of cellulose pulp are immersed in 17–20% aqueous NaOH (caustic soda) solution at the temperature range of 18–25°C in order to swell the cellulose fibres and convert cellulose to alkali cellulose (Eqn 3.1).

$$(C_6H_{10}O_5)_n + nNaOH \longrightarrow (C_6H_9O_4ONa)_n + nH_2O \qquad \ldots (3.1)$$

2. *Pressing*: The swollen mass of alkali cellulose is pressed to a wet weight equivalent of 2½–3 times the original pulp weight to obtain an accurate alkali to cellulose ratio.
3. *Shredding*: The pressed alkali cellulose is shredded mechanically to yield finely divided, fluffy particles called 'crumbs'. This step increases the overall surface area of the alkali cellulose mass, thereby increasing its ability to react in the steps that follow.

Figure 3.6: Steps in the production of viscose rayon

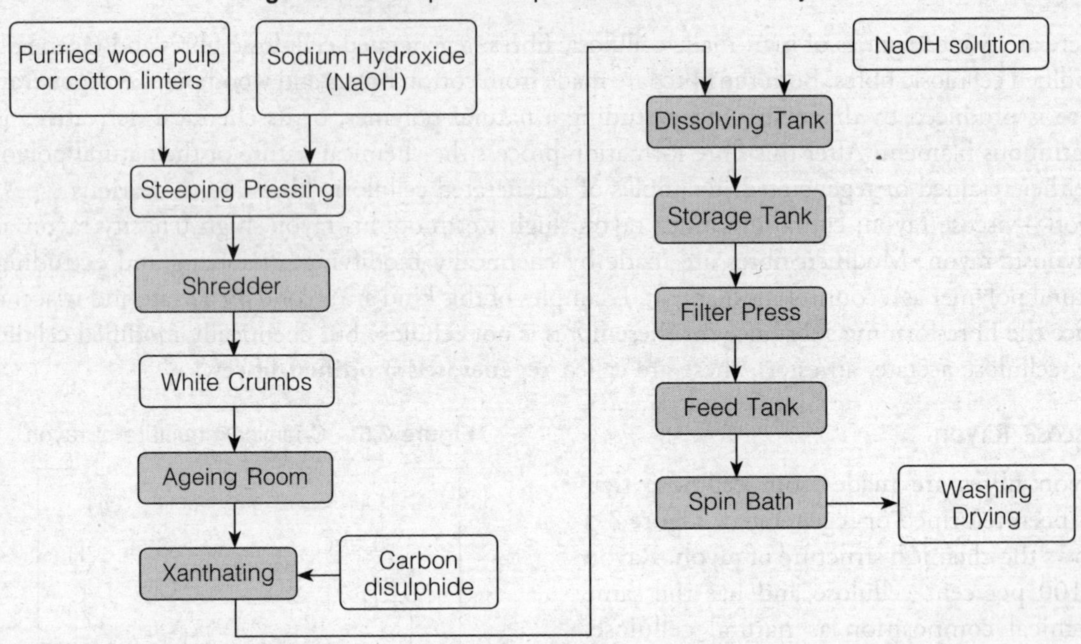

Source: Drawn by the authors.

4. *Ageing*: The alkali cellulose is aged under controlled conditions of time (for 50 hours) and temperature (between 18°C and 30°C) in order to depolymerise the cellulose to the desired degree of polymerisation to get a solution of the right viscosity and cellulose concentration. In this step the average molecular weight of the original pulp is reduced.
5. *Xanthation*: The aged alkali cellulose crumbs are placed in vats and allowed to react with carbon disulphide under controlled temperature (20–30°C) to form an orange-coloured compound called cellulose xanthate (Eqn 3.2).

$$(C_6H_9O_4ONa)_n + nCS_2 \longrightarrow (C_6H_9O_4O\text{–}SC\text{–}SNa)_n \quad \ldots (3.2)$$

6. *Dissolving*: Xanthate crumbs are dissolved in dilute NaOH solution at 15–20°C under high-shear mixing conditions to obtain a viscous orange-coloured solution. The large xanthate substituents on the cellulose force the chains apart, reducing the inter-chain hydrogen bonds and allowing water molecules to solvate and separate the chains, leading to the dissolution of the otherwise insoluble cellulose. Since the cellulose xanthate solution has very high viscosity, it is termed **viscose** in this stage.
7. *Ripening*: Viscose is allowed to stand for a period of time to 'ripen'. Two important processes occur during **ripening**—redistribution and loss of xanthate groups. The reversible xanthation reaction allows some of the xanthate groups to revert to cellulosic hydroxyls and free CS_2. This free CS_2 can then escape or react with other hydroxyl groups (–OH) in the other parts of the cellulose chain. In this way, crystalline regions are gradually broken down and

a more complete solution is achieved. The colour of the ripened viscose solution is gold with the consistency similar to that of honey.

8. *Filtering*: The viscose solution is filtered to remove undissolved materials that might disrupt the spinning process or cause defects in the rayon filament. At this point, delustering agents or pigments for colouring the fibres can be added.
9. *Degassing*: Bubbles of air entrapped in the viscose are removed through the process of degassing prior to extrusion, so that voids or weak spots in the fine rayon filaments are avoided.
10. *Spinning* (*wet spinning*): The viscose solution is metered through a spinneret into a spin bath containing 10% sulphuric acid (for acidification and regeneration of sodium cellulose xanthate to cellulose), 18% sodium sulphate (for rapid coagulation of viscose), and 1% zinc sulphate (which controls the rate of regeneration). Once the cellulose xanthate is neutralised and acidified, rapid coagulation of the rayon filaments occurs (Eqn 3.3).

$$(C_6H_9O_4O\text{–}SC\text{–}SNa)_n + (n/2)H_2SO_4 \longrightarrow (C_6H_{10}O_5)_n + nCS_2 + (n/2)Na_2SO_4 \qquad \ldots (3.3)$$

11. *Drawing*: The rayon filaments are stretched while the cellulose chains are still relatively mobile. This causes the chains to stretch out and orient along the fibre axis. As the chains become more parallel, inter-chain hydrogen bonds form and give the filaments the properties necessary for use as textile fibres. Stretching is vital to get the desired tenacity and other properties of rayon.
12. *Washing*: The freshly regenerated rayon contains many salts and other water soluble impurities which are removed through one of several washing techniques.
13. *Cutting*: If the rayon is to be used as staple fibre, the group of filaments (i.e. tow) is passed through a rotary cutter to provide a fibre which can be processed in much the same way as cotton.

Properties

Rayon fibres are normally white in colour. The lustre of rayon can be modified by the addition of the delustering agent titanium dioxide to the solution before the fibres are extruded. The cross-section of rayon fibre is an irregular circle with serrated edges. Lengthwise lines called striations are seen in the longitudinal section. Figure 3.9 gives the cross-section and longitudinal sections of all man-made fibres. Viscose rayon has silk-like aesthetics with superb drape and feel. Its cellulosic base contributes many properties similar to those of cotton or other natural cellulosic fibres. The strength of viscose rayon is low because of its lower polymer chain length as compared to cotton and also because the physical structure is different.

There is a considerable decrease in strength when the fibre is wet. Viscose has a high percentage of –OH groups present in amorphous regions that contribute to its hygroscopic behaviour. Thus, due to high moisture regain, rayon is breathable, comfortable to

> A **pill** (or bobble) is a small ball of fibres that forms on a piece of cloth because of washing and wearing of fabrics that causes loose fibres to push out from the surface.

wear and can be easily dyed in vivid colours. Due to low elastic recovery and resiliency, rayon wrinkles badly. It does not build up static electricity, and there is no pilling unless the fabric is made from short, low-twist yarns. As a cellulosic fibre, rayon is flammable and burns like paper. Properties of viscose rayon are given in Table 3.1.

Uses and Care

Rayon is comfortable, soft to the skin, and has moderate dry strength and abrasion resistance. One of rayon's strengths is its versatility and ability to blend easily with fibres such as polyester, acrylic, nylon, acetate, cotton, wool to reduce cost, or for lustre, softness or absorbency and the resulting comfort.

The strength of rayon fibre is low, thus the products should not be twisted or wrung following laundering. Rather they should be padded by using a towel or a gentle cycle in the washing machine can be used. Since it is not affected by organic solvents, dry cleaning does not damage rayon fabrics.

Major Domestic and Industrial Uses

- *Apparel*: Blouses, coats, dresses, jackets, lingerie, linings, millinery, rainwear, slacks, sports shirts, sportswear, suits, ties, work clothes.
- *Home furnishings*: Bedspreads, blankets, carpets, curtains, draperies, sheets, slipcovers, tablecloths, upholstery.
- *Other*: Industrial products, medical and surgical products, non-woven products, tyre cord.

Cuprammonium Rayon

Production

Cupprammonium rayon is produced by a solution of cellulosic material in aqueous ammonia (NH_3) and copper sulphate ($CuSO_4$) at low temperature in a nitrogen atmosphere. It is put in mixer along with caustic soda (NaOH). The step of ripening or ageing is not required for cuprammonium production. The solution is filtered and then wet spun. Filaments are neutralised in sulphuric acid, followed by washing, lubrication, drying, twisting into yarn and winding. This is a more expensive process than that of viscose rayon. The cross-section of a fibre of cuprammonium rayon is almost round. The main disadvantage of the cuprammonium process is the toxicity of copper sulphate. It must be fully recovered from the process, and as a result large-scale production is limited.

Properties and Uses

Its properties are quite similar to those of viscose rayon. The fibre can be made into very lightweight fabrics. A good conductor of heat and fairly absorbent, it is especially suitable for use in warm-weather clothing.

High Wet Modulus Rayon

One of the drawbacks of viscose rayon is its very low wet strength. This limits its usage as a substitute for cotton. This limitation led to the development of High Wet Modulus (HWM) rayon. HWM rayon does not exhibit drastic reduction in strength when wet. HWM rayon is produced using the same process as viscose but with some modifications in the spinning conditions and bath composition—i.e. the addition of certain chemicals in the spinning bath that slow down the rate of regeneration of cellulose in the extrusion bath. Apart from these, higher draw ratios are also employed to improve the degree of orientation and achieve much higher tenacities in dry as well as wet conditions. A comparison of properties of viscose rayon and HWM rayon is given in Table 3.1.

> **Draw ratio** is the extensibility that can be subjected to the fibre during a process.

High Tenacity Rayon

High Tenacity (HT) rayon was developed for high performance and industrial applications rather than for apparel. For many years HT rayon was used for the manufacture of tyre cords before it was substituted by nylon and polyester. HT rayon possesses superior tensile strength as compared to regular viscose and has a greater percentage of crystalline regions which translate to greater strength. For the production of HT rayon certain modifications are made in the composition of the spinning bath, like increasing the quantity of zinc sulphate (up to about 4 per cent). This helps retard the coagulation of viscose. A greater stretch is also imparted to improve the orientation. Slow regeneration of cellulose and stretching of rayon will lead to greater areas of crystallinity within the fibre and therefore increased strength.

Polynosic Rayon

These fibres have a very high degree of orientation, achieved as a result of very high stretching (up to 300 per cent) during processing along with DP values as high as 800. They have a unique fibrillar structure, high dry and wet strength, low elongation (8–11 per cent), relatively low water retention and very high wet modulus.

Lyocell

Lyocell is a new generation of cellulosic fibre that was first commercially produced in 1992, by Acordis Cellulosic Fibres in the US. The US Federal Trade Commission (FTC) defines lyocell as a fibre 'composed of cellulose precipitated from an organic solution in which no substitution of the hydroxyl groups takes place and no chemical intermediates are formed.' The FTC classifies lyocell as a subcategory under rayon. Lyocell fibres are manufactured by wet spinning process. The cellulose is directly dissolved in the solvent N-methylmorpholine n-oxide (NMMO) and water. The solution is then filtered and spun through spinnerets to make the filaments, which are spun into water. The NMMO solvent is recovered from this aqueous solution and reused. Lyocell fibres, like other cellulosics, are soft, strong, moisture-absorbent and biodegradable. They have a dry strength higher than other cellulosics and approaching that of polyester. They also retain 85 per cent of their strength when wet. They are wrinkle resistant and can be laundered easily.

Table 3.1: Properties of viscose and HWM rayon

	Viscose Rayon	*HWM Rayon*
Physical properties		
Shape and appearance	Man-made fibres can be manufactured in any length and diameter. The lustre can be modified by the addition of the delustering agent titanium dioxide.	
Tenacity	Dry: 1.0–5.0 g/d Wet: 0.5–1.5 g/d	Dry: 4.0–5.0 g/d Wet: 2.0–3.0 g/d
Moisture absorbency (65% RH)	13%	11.5%
% Elongation at break (65% RH)	24–27%	15–23%
Elasticity	Poor, wrinkles badly	Poor
Resiliency	Low	Better than viscose rayon
Density	1.5 g/cm^3	Not available
Electrical conductivity	No static charge	No static charge
Chemical properties		
Effect of acids	Carbonised by hot acids. Cold concentrated acids cause gradual fibre disintegration.	
Effect of alkalis	No damage	No damage
Effect of organic solvents	No damage	No damage
Effect of bleaches	No damage	No damage
Effect of sunlight	Not greatly harmed by sunlight	Not greatly harmed by sunlight
Biological properties	Attacked by silverfish and termites, but generally resists insect damage	
Thermal properties		
Safe ironing temperature	110–150°C	Not available
Melting point	Not applicable (fibres decompose before melting	Not applicable (fibres decompose before melting)
Burning behaviour	It ignites quickly, burns freely and has an afterglow. Has an odour similar to burning paper.	

Note: RH: relative humidity
Source: Compiled by the authors.

Lyocell fibres are mostly used for apparel fabrics, especially outerwear. End uses of lyocell include dresses, coats, slacks etc. Tencel® is a trade name of Lyocell produced by the Lenzing Fibers Pvt. Ltd.

Modal

Modal, a type of rayon, is made by spinning reconstituted cellulose, in this case often from beech trees. They are made by a modified viscose process with a higher degree of polymerisation and modified precipitating baths. The process produces fibres with improved properties such as better wear, higher dry and wet strengths and better dimensional stability. The modal fibres are smooth and soft, making them very suitable for clothing such as intimate apparel. They can be used alone or with other fibres (often cotton or spandex). They are highly absorbent and quickly soak up water and thus the fabric remains dry.

Acetate

Acetate was the first thermoplastic (heat sensitive) fibre to be introduced to consumers. Like rayon, acetate is often used as a substitute for silk, but it is much weaker than either rayon or silk and is generally used for apparel that will not be worn often. The fibre-forming substance in this case is cellulose acetate. Cellulose acetate and cellulose triacetate fibres are classified as regenerated modified cellulosic fibres. The difference between acetate and triacetate fibres is in the number of cellulose hydroxyl groups that are acetylated. When a substance is formed from cellulose in which almost all three hydroxyl groups per glucose ring have been acetylated (i.e. more than 92 per cent of the groups), the material is called cellulose triacetate (or just triacetate). If approximately 2.4 out of three hydroxyl groups per glucose ring are acetylated (between 75 and 92 per cent), the substance is called cellulose acetate (or simply acetate) (Figure 3.7).

Figure 3.7: Chemical structures of cellulose acetate and cellulose triacetate

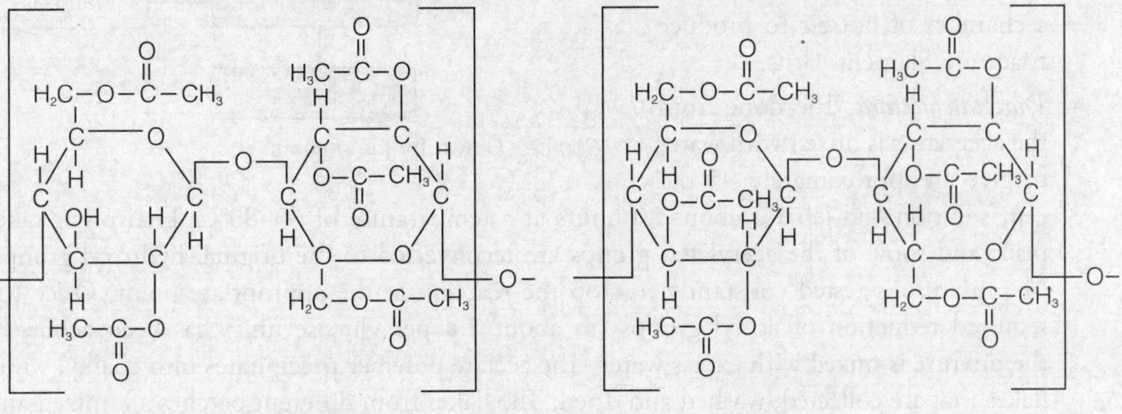

(a) Cellulose acetate

(b) Cellulose triacetate

Production of Acetate and Triacetate

In the production of acetate fibres, triacetate is formed first and therefore it is also known as primary acetate. The triacetate is then partially hydrolysed to obtain diacetate which is also known as secondary acetate or acetate. The process involves the following steps (Figure 3.8):

1. Pure cellulose is pretreated with a small amount of glacial acetic acid at about 35°C for one hour. The acid pretreatment causes swelling of cellulose material and facilitates the acetylation process.
2. The pretreated material is added to the acetylation mixture of acetic acid, acetic anhydride and sulfuric acid (which acts as the catalyst for the reaction). Proper stirring is required to

ensure uniform acetylation. The reaction is allowed to continue for about 8 hours till a clear viscous solution is formed. During this period the temperature has to be maintained by cooling because the reaction is exothermic and high temperature can cause degradation of cellulose.

Figure 3.8: Steps in the production of acetate

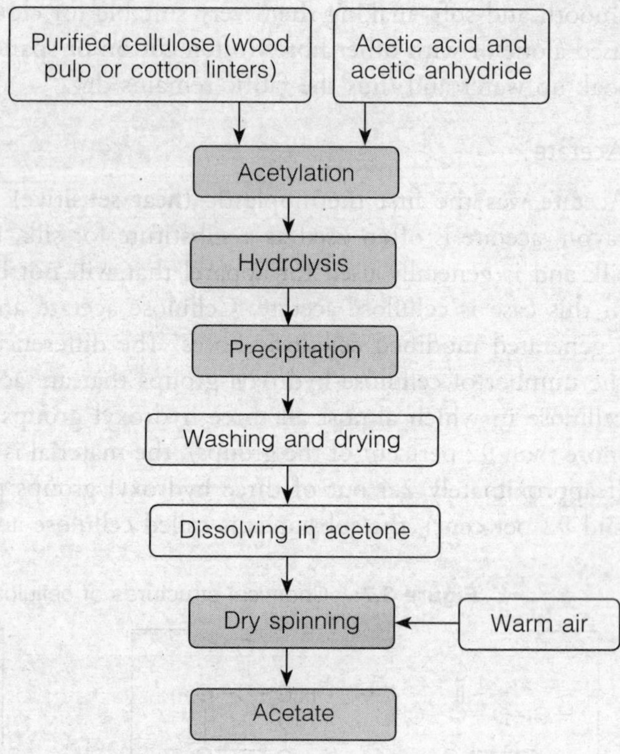

Source: Drawn by the authors.

3. *Triacetate*: For production of triacetate, the viscous solution is mixed with water. Triacetate precipitates out in the form of white flakes. These flakes are washed, dried and then dissolved in methylene chloride containing about 10% methanol. The polymer dope thus formed is dry spun in a chamber of hot air to produce triacetate filament yarns.
4. *Diacetate/acetate*: The dope from the acetylator is mixed with water to give an approximately 95 per cent solution and left for about 20 hours at a temperature of 60–80°C. Hydrolysis takes place and some of the acetylated groups are reconverted to the original hydroxyl groups. The mixture is tested constantly to stop the reaction at the appropriate point. Once the required reduction of acetyl groups (to about 2.4 per glucose unit) has been achieved, the mixture is mixed with excess water. The acetate polymer precipitates into chalky, white flakes that are collected, washed and dried. The flakes from different batches are mixed and dissolved in acetone to form the spinning solution. The solution is filtered and extruded through the spinneret in a chamber of hot air that evaporates acetone.
5. The filaments are given slight stretch to improve orientation and the resultant strength.

Properties of Acetate and Triacetate

If acetate and triacetate have not been treated to decrease lustre, both fibres have a bright appearance and good lustre. Fibres are white unless they have been solution dyed. In microscopic appearance both are very similar. They have a clear and irregular multilobed shape in cross-section. The longitudinal appearance shows broad striations. Both acetate and triacetate have very low strength. Both are weaker when wet. Acetate has poor elastic and wrinkle recovery whereas triacetate has relatively higher elastic recovery and is more resilient. Acetate fabrics are fast drying, resistant to wrinkling and shrinkage. They give excellent drape and moderate moisture absorbency. Both the fibres tend to build up static electricity Properties of acetate and triacetate fibres are given in Table 3.2.

Table 3.2: Properties of acetate and triacetate fibres

	Acetate	*Triacetate*
Physical properties		
Shape and appearance	Man-made fibres can be manufactured in any length and diameter. The lustre can be modified by the addition of the delustering agent titanium dioxide.	
Tenacity	Dry: 1.2–1.4 g/d Wet: 1.0–1.3 g/d	Dry: 1.2–1.4 g/d Wet: 1.0–1.3 g/d
Moisture absorbency (65% RH)	6.5%	3.2%
% Elongation at break (65% RH)	25–35%	25–35%
Elasticity	Poor	Better than acetate
Resiliency	Low	Better than acetate
Density	1.3 g/cm^3, lighter in weight than rayon	
Electrical conductivity	Poor for both fibres, they tend to build up static electricity.	
Chemical properties		
Effect of acids	Damaged	
Effect of alkalis	Damaged	
Effect of organic solvents	Dissolves in acetone, phenol and chloroform	
Effect of bleaches	Bleached with hydrogen peroxide	
Effect of sunlight	Causes a loss of strength and deterioration	Moderate resistance to sunlight
Biological properties	May be attacked by mildew if stored incorrectly	More resistant to mildew attack
Thermal properties		
Safe ironing temp	130°C	150°C
Melting point	175°C	235°C
Burning behaviour	Both fibres are thermoplastic, they will soften and melt with the application of heat.	

Source: Compiled by the authors.

Uses and Care

Fabrics from either type of fibre require careful handling in laundering or dry cleaning. Acetates tend to wrinkle badly, and therefore any kind of twisting should be avoided, gentle machine wash cycle should be used. Triacetate performs a little better than acetate in this respect. Any type of detergent may be used. Bleaches must be used with care. The fibres are also affected by some organic solvents. Acetate will dissolve in acetone, so nail polish removers should be used with caution. Both fibres are thermoplastic, and they will soften and melt with the application of heat. Medium to low ironing temperatures should be used. Acetate may be attacked by mildew if incorrectly stored, triacetate is more resistant. Extended exposure to sunlight should be avoided.

Both acetate and triacetate have a luxurious feel and appearance. Therefore, they are popularly used in a variety of applications, but mostly for apparel and home furnishings.

Major Domestic and Industrial Uses

- *Apparel*: Blouses, dresses and foundation garments, lingerie, linings, shirts, slacks, sportswear
- *Home furnishings*: Draperies, upholstery
- *Other*: Cigarette filters, fibre fill for pillows, quilted products

FIBRES FROM SYNTHETIC POLYMERS

Acrylic

Acrylic fibres are the third largest class of synthetic fibre after polyester and nylon. Commercial acrylic fibre was first developed by DuPont in USA as Orlon while modified acrylic (modacrylic) fibre was first developed by Union Carbide as Dynel. Other trade names are Acrilan, Zefran, Creslan and Courtelle for acrylic and Teklan and Verel for modacrylics. BISFA (Bureau International pour la Standardisation des Fibres Artificielles; an international association of man-made fibre producers) defines acrylic fibres as 'fibres composed of linear macromolecules having in the chain at least 85 per cent (by mass) of acrylonitrile repeating units' (CIRFS n.d.). Typical comonomers in acrylic are vinyl acetate or methyl acrylate. The modacrylic fibre chains contain at least 50–85 per cent by mass of acrylonitrile. The comonomers vinyl chloride, vinylidene chloride or vinyl bromide used in modacrylic give the fibre flame-retardant properties.

Production of Acrylic

The starting materials for acrylonitrile ($CH_2{=}CH{-}CN$) are propylene and ammonia, which are reacted with oxygen in the presence of catalysts (Eqn 3.4). The acrylonitrile is then polymerised to produce polyacrylonitrile (PAN). Pure acrylonitrile is very inert with no dye sites. Therefore, comonomers, usually with anionic groups as mentioned above, are added during polymerisation reaction to introduce reactive and dye sites. PAN is then spun into fibres from a solution in a solvent. Two process routes are used: (*a*) wet spinning, in which the fibres

Figure 3.9: Longitudinal and cross-sectional views of various fibres

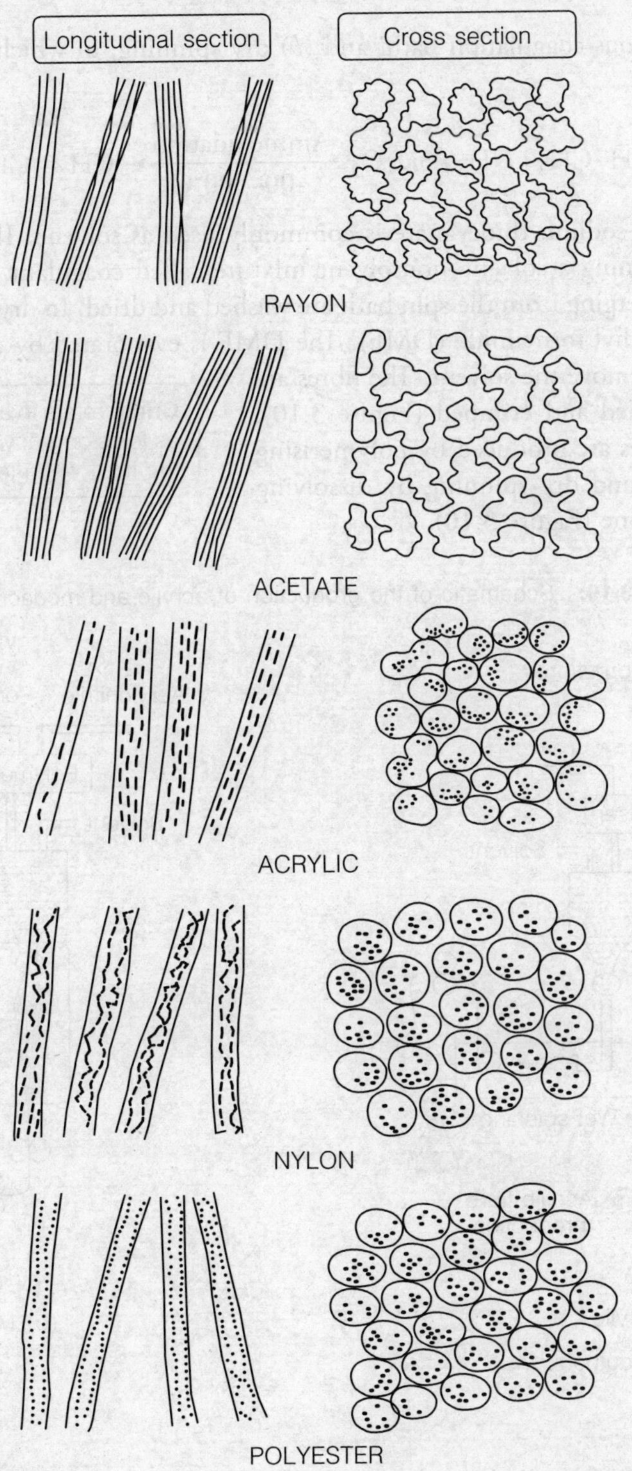

Source: Drawn by the authors.

are spun in an aqueous coagulation bath, and (*b*) dry spinning, in which the fibres are spun in hot air.

$$CH_2{=}CH{-}CH_3 + NH_3 + 3/2\ O_2 \xrightarrow[400\text{–}500^\circ C]{\text{Ammoxidation}} CH_2{=}CHCN \qquad \ldots (3.4)$$

In wet spinning, sodium thiocyanate is commonly used as solvent. The fibres are spun into a liquid bath containing a solvent–nonsolvent mixture called coagulant. Nonsolvent is usually water. The fibres emerging from the spin bath are washed and dried. In dry spinning, the polymer is dissolved in dimethyl formamide (DMF). The DMF is evaporated by circulating an inert gas at about 300°C to remove the solvent. The fibres are then stretched, washed and crimped (Figure 3.10). The modacrylic fibres are produced by polymerising various copolymers and dry spinning by dissolving the polymer in acetone (Figure 3.10).

Crimp refers to any kind of texture, such as folds or grooves, which is imparted to the fibre when texturisation is done.

Figure 3.10: Schematic of the production of acrylic and modacrylic fibres

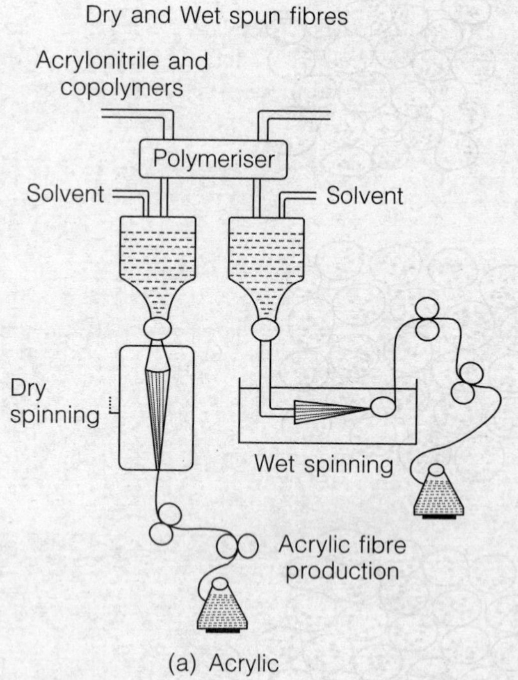

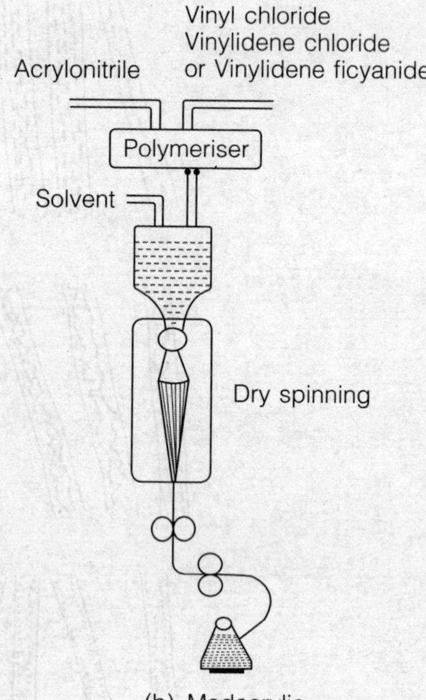

Source: Drawn by the authors.

Properties

The longitudinal view of acrylic fibre shows a rod-like structure with a uniform diameter, while the cross-section is either dumb-bell, corn mushroom or round shaped. They are mostly of staple length and are dull or semi-dull lustre. Acrylic is a hydrophobic fibre, has low moisture regain of 1.5 per cent and thus, dries quickly. It has fair dry strength of 2.0–2.7g/denier and weakens on wetting. The elongation of acrylic ranges between 20 and 55 per cent. Its elastic recovery is medium to high. Acrylic fibres have good resiliency. Bulky fabrics are especially resilient and lofty. If heat set properly, they do not shrink nor stretch. Acrylic, in general, has good heat resistance and can be effectively heat set. The fibre degrades and decomposes before melting. Once ignited, they burn with a yellow flame and drip in molten drops, which are hot enough to ignite combustible substances. On the other hand, modacrylic fibres have excellent flame resistance. They are difficult to ignite and self-extinguishing when the flame source is removed. Acrylics have good resistance to most mineral and organic acids. Weak or dilute acids have no destructive effect. Strong or concentrated acids cause a loss of strength and cold concentrated nitric acid will dissolve the fibre. They have good resistance to weak alkali, but concentrated alkali will cause damage. Acrylics have remarkable resistance to sunlight and other climatic conditions. Detergent and soaps have no deleterious effect on them. Acrylics show static build up that increases at low humidity. Micro-organisms like mildew or bacteria have no effect on acrylics. Properties in brief are also given in Table 3.3.

Uses and Care

Acrylic fibres do not require special care and maintenance. Most acrylic fibres can be washed and dried in home laundering equipment. However, for machine drying, the setting has to be either low or medium. Any type of detergent or soap and bleach can be used. Dry cleaning is not recommended for acrylic fibres.

Knitted products of acrylic fibres are quite popular such as sweaters and other apparel items. These are also used for home furnishings. As acrylic fibres are warm, light weight and soft, they are also popular in sportswear. The major end use of modacrylic fibres has been as part of a blend for children's sleepwear and other items of apparel that must meet flame resistant regulations.

Major Domestic and Industrial Uses

- *Apparel*: Dresses, infant wear, knitted garments, ski wear, socks, sportswear, sweaters
- *Fabrics*: Fleece and pile fabrics, face fabrics in bonded fabrics, simulated furs, jerseys
- *Home furnishings*: Blankets, carpets, draperies, upholstery
- *Other*: Auto tops, awnings, hand-knitting and craft yarns, industrial and geotextile fabrics

Polyamide Fibres

Nylon

Nylon fibres (i.e. nylon 6, nylon 66, Nylon 11 or others) are made up of linear macromolecules whose structural units are linked by the amide (–NH–CO–) group (Figure 3.11) and are therefore also known as polyamides. BISFA defines polyamide fibre as 'a fibre composed of linear macromolecules having in the chain recurring amide linkages, at least 85 per cent of which are joined to aliphatic or cycloaliphatic units' (CIRFS n.d.). In nylons, the structural units are essentially aliphatic.

Figure 3.11: Chemical structures of nylon 6 and nylon 66

$[NH(CH_2)_5CO]_n$	Nylon 6
$[NH(CH_2)_6NHCO(CH_2)_4CO]_n$	Nylon 66

The word nylon was coined by DuPont as a name for its aliphatic polyamide fibre which was commercially launched in 1938. Since the name was deliberately not registered as a trademark, nylon fibre became an internationally accepted generic name for fibres based on linear polyamides. The generic name polyamide fibre has the same meaning as nylon fibre, but the term 'nylon fibre' is used principally in countries that derive their fibre technology directly or indirectly from the USA, and 'polyamide fibre' in countries that derive their fibre technology from Germany.

The numbers 66 and 6 after the word nylon indicate the number of carbon atoms in the raw materials used for producing these fibres. Nylon 7 and nylon 11—which are variants of the structure of nylon 6—are also produced but as films and not as fibres.

Production of Nylon

Nylon 66 is produced from the polycondensation of 1, 6-diaminohexane, the traditional name of which is hexamethylenediamine, and hexanedioic acid, which is often called adipic acid. In the case of nylon 6, the monomer is caprolactam.

Various steps in the production of nylon 66 (Figure 3.12) and 6 (Figure 3.13) are as follows:

1. *Polymerisation*:

   (a) *Nylon 66*: Hexamethylenediamine and adipic acid are combined to prepare nylon salt. The salt is then transferred to an autoclave. Condensation polymerisation takes place in the autoclave in air-free atmosphere. The water which is also produced during polymerisation is allowed to escape from the autoclave. If a dull fibre is required, then the delusterant titanium dioxide is added during this step. The molten polymer that forms is extruded from the tank as a ribbon, several inches in width. The material is quenched in cold water which reduces the size of the crystals formed.

   Hexamethylene diamine: $NH_2(CH_2)_6NH_2$; Adipic acid: $COOH(CH_2)_4COOH$

   Nylon 66 salt: $H_2N\ (CH_2)_6\ NHOC(CH_2)_4\ COOH$;

   Nylon 66 polymer: $(–HN\ (CH_2)_6\ NHOC(CH_2)_4\ CO–)n$

   (b) *Nylon 6*: Caprolactum is melted, filtered and then heated under high pressure in an autoclave, leading to condensation polymerisation. Alternately, caprolactum and water (10 per cent on weight of caprolactum) are heated to a high temperature. The polymer is formed and steam that is produced is allowed to escape. The polymer is made into nylon chips, washed and dried.

Caprolactum: $(CH_2)_5$ CONH; Nylon 6 Polymer: $(–HN(CH_2)_5CO–)_n$

2. *Spinning*: Both nylon 66 and nylon 6 are melt spun. The melting point of nylon 66 is 260°C and that of nylon 6 is 221°C. The molten nylon forms a pool of liquid that is filtered to remove any impurities. This melted nylon passes through orifices in a spinneret and filaments are formed. Cold air is blown across the filaments causing them to harden. The filaments are then stretched to orient the molecules and coated with spin finish which is a mixture of lubricants, wetting agents and antistatic agents. This finish helps reduce the friction and build-up of static charge on the fibres during subsequent processing.
3. *Drawing*: In this state nylon is not very strong and lustrous. In order to orient the molecules in the nylon fibre and thereby increase both strength

Figure 3.12: Steps in the production of nylon 66

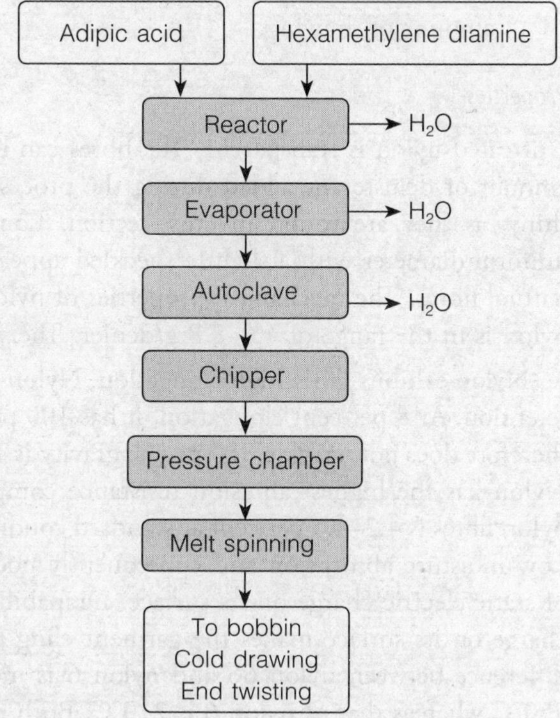

Source: Drawn by the authors.

Figure 3.13: Steps in the production of nylon 6

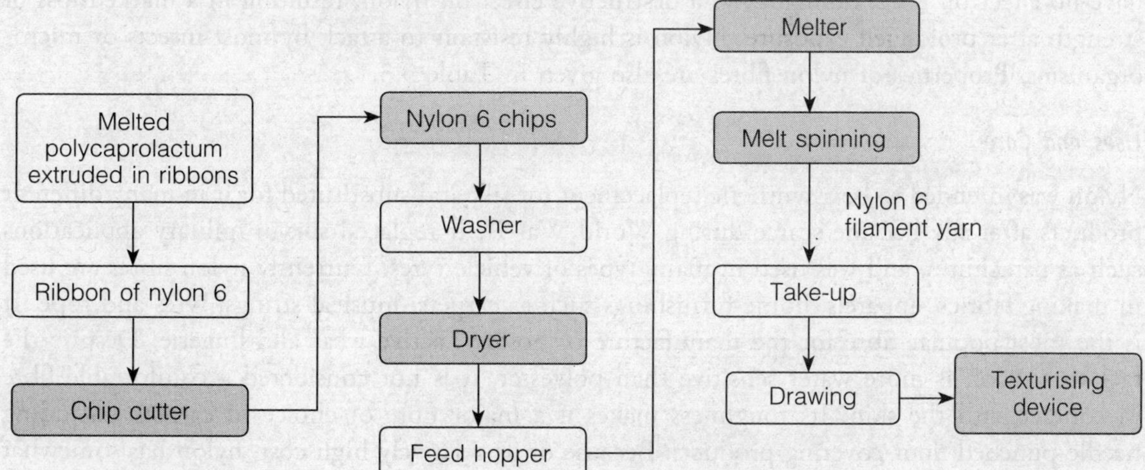

Source: Drawn by the authors.

and lustre, the cooled fibre is drawn or stretched in its length. The fibres are drawn four to six times their original length. This is also important to improve the crystallinity of the fibres. The degree of crystallinity may vary between 60–80 per cent depending on the amount of drawing.

Properties

Untreated nylon is transparent. The fibres can be bright, semi-dull or dull depending upon the amount of delusterant added during the process. Most of the nylon filaments are smooth and shiny, as they are round in cross-section. Longitudinal structure shows transparent fibres of uniform diameter with a slightly speckled appearance (Table 3.3). Nylon is stronger than many natural fibres. The mechanical properties of nylon 66 and nylon 6 are quite similar. Tenacity of nylon is in the range of 4.6–8.8 g/denier. The wet strength reduces to 3.5–8.0 g/denier.

Nylon exhibits fairly high elongation. Nylon is highly elastic and therefore has excellent shape retention. At 8 per cent elongation, it has 100 per cent elastic recovery. Its resilience is good and therefore does not wrinkle. Its specific gravity is 1.14 and therefore is lighter than cellulosic fibres. Nylon has the highest abrasion resistance compared with other fibres. Moisture absorption of nylon fibres is 4.2–4.5 per cent at standard conditions, fairly low. It dries quickly after laundering. Low moisture absorption and consequently poor electrical conductivity result in accumulation of static electric charge on its surface. Drapability of nylon is excellent. However, static electric charge on its surface makes the garment cling to body. It is in the thermal properties that the difference between nylon 66 and nylon 6 is most apparent. The melting point of nylon 66 is 250°C whereas that of nylon 6 is 210°C. Both melt in flame and form a grey gummy substance that hardens as it cools. Because nylon is heat sensitive or thermoplastic, it can be heat set during processing, so that it will retain its shape. Nylon is substantially resistant to alkalis, but acids, whether mineral or organic, will destroy the fibre. Dry-cleaning solvents, soaps and detergents have no effect on fibre. Sunlight has a destructive effect on nylon, resulting in a marked loss of strength after prolonged exposure. Nylon is highly resistant to attack by most insects or micro-organisms. Properties of nylon fibres are also given in Table 3.3.

Uses and Care

Nylon was intended to be a synthetic replacement for silk and substituted for it in many different products after silk became scarce during World War II. It replaced silk in military applications such as parachutes, and was used in many types of vehicle tyres. Currently nylon fibres are used in making fabrics, apparels, home furnishings such as carpets, musical strings, tyres and rope. It is the most popular fibre for the manufacture of hosiery, active wear and lingerie. Despite the fact that nylon is more water sensitive than polyester, it is not considered a comfortable fibre in contact with the skin. Its toughness makes it a major fibre of choice in carpets, including needle-punched floor-covering products. Because of its relatively high cost, nylon has somewhat limited use in non-woven products. It is used as a blending fibre in some cases, because it conveys excellent tear strength.

Most items made from nylon can be machine-washed and tumble-dried at low temperatures. Bleaches can be used when needed. Since nylon 6 has a much lower melting point than nylon 66 garments made from it must be ironed with considerable care.

Aramids

The second group of polyamides are aramid fibres. In the aramids 85 per cent or more of the amide linkages are attached to two aromatic rings (Figure 3.14). This difference in percentage causes significant differences between aramid and nylon. Nomex and kevlar are examples of aramids.

Figure 3.14: Molecular structure of kevlar

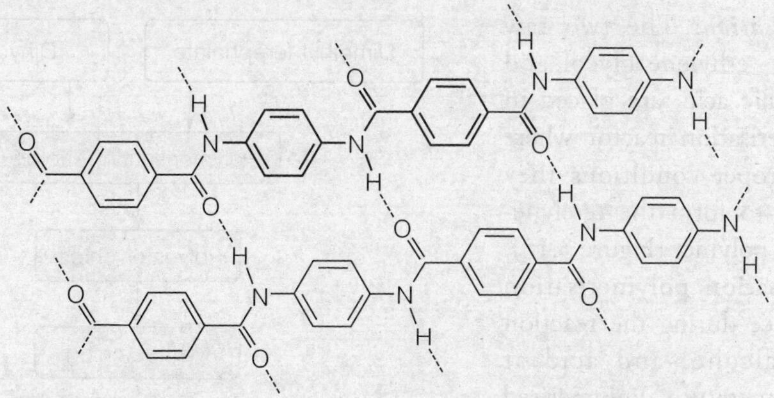

Properties of Aramids

- Highly flame-resistant, they have no melting point and have extremely low combustibility
- High strength
- High resistance to stretch
- Good flex and abrasion resistance
- Good dimensional stability
- Maintain shape and form at high temperatures
- Excellent resistance to insects and micro-organisms
- Not a good conductor of heat

Major Domestic and Industrial Uses

Aramid fibres are used in industrial and military clothing, industrial hot-air filtration fabrics, military helmets, protective vests, structural composites for aircraft and boats, sailcloth, tyres, ropes and cables, mechanical rubber goods, marine and sporting goods. Highly temperature resistant papers are made from Nomex. A variety of consumer goods like carpets, drapery fabrics, upholstery fabrics and ironing board covers are also made from Nomex. Kevlar is used in bullet-resistant garments and coated fabrics.

Polyester

According to BISFA polyester is 'a fibre composed of linear macromolecules having at least 85 per cent by mass of a diol and terephthalic acid' (CIRFS n.d.). The first polyester, polyethylene terephthalate (PET), was made in the UK in 1941 and has become by far the world's major man-made fibre. Other polyesters such as polybutylene terephthalate (PBT) and polytrimethylene terephthalate (PTT) are also made but in much smaller quantities.

Production of Polyester

Polyester fibres are made in a way very similar to polyamide. The steps in the production of polyester are as follows (Figure 3.15):

1. *Polymerisation*: The two raw materials, ethylene glycol and terephthalic acid, are placed in a polymerisation reactor where under proper conditions they combine to form the terelyne-polyester polymer (Figure 3.15). Condensation polymerisation takes place during the reaction of the alcohol and acid at high temperature. Polymerised material is extruded in the form of a ribbon. The ribbon hardens and small polyester chips are cut from the ribbon.
2. *Spinning*: The chips are dried to remove any residual moisture and are then melt spun.
3. *Drawing*: Fibres are hot stretched to about five times their original length which decreases their width. The drawn fibres is either wound onto cones as filaments or crimped and cut into staple length.

Figure 3.15: Steps in the production of polyester

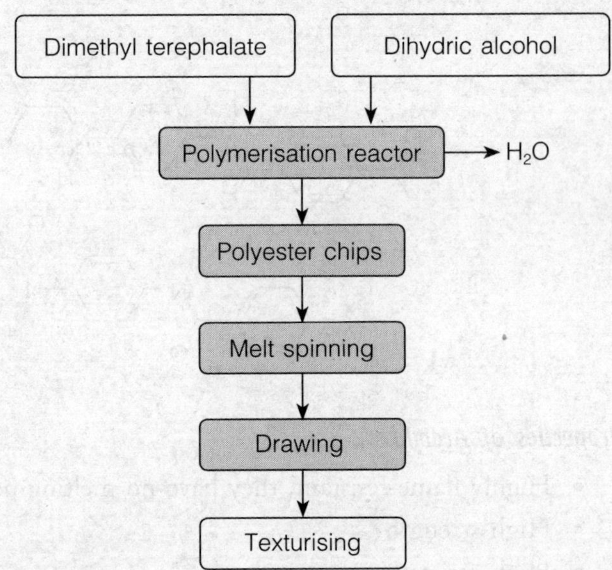

Source: Drawn by the authors.

Figure 3.16: Polymerisation reaction to form polyester

Terephthalic acid
Ethyleneglycol
$- H_2O$
Polyethylene Terephthalate (polymer)

Properties of Polyester

Polyester fibres are dull or semi-dull in appearance due to the addition of delustering agents in the spinning solution. The fibre is generally round in cross-section and uniform and rod-like with a smooth surface in the longitudinal view.

The strength of polyester varies from 2.5 g/denier to 9.5 g/denier. There is no loss of strength when wet. Elastic recovery is superior to cellulosic fibre but is slightly inferior to nylon. The resilience of polyester is excellent. Further, heat-setting of fabric can stabilise fibres and yarns so that they need little or no pressing to retain a smooth appearance. Polyester has low moisture regain of 0.4 per cent under standard conditions. Therefore, moisture has negligible effect on its strength. For the same reason static electric charge accumulates on the fabric. Owing to low moisture regain, it requires special dyeing and finishing processes. It is hydrophobic and therefore is also oleophilic and consequently retains greasy soil to some extent but water-borne stains and soil can be easily removed. Polyester fabrics have satisfactory draping quality and very good abrasion resistance. Polyester melts between 237°C and 272°C depending on the type. It burns while in flame but is self-extinguishing when removed from the flame. It produces dark smoke with an aromatic odour. It forms a fawn-coloured bead that is uncrushable. Fabrics made of polyester filaments can be ironed between 135°C and 149°C. Polyesters are better conductors of heat compared to acrylics, as smooth polyesters have fewer air pockets and consequently less insulation. However, hollow polyester fabrics, bulked types and crimped fibre fills have better insulating property.

A fibre is **oleophilic** if it has a strong affinity for oils rather than water.

Polyester has good resistance to weak alkalis even when hot, they have moderate resistance to strong alkalis even at room temperature and are degraded only at higher temperatures. Boiling weak acids have no effect on polyesters. Polyesters have good resistance even to concentrated acids at room temperature. Prolonged exposure to boiling hydrochloric acid causes destruction of the fibre and 96 per cent sulphuric acid disintegrates the fibre. Polyesters have good resistance to household bleaches and oxidising agents. Polyester has good resistance to sunlight when placed behind glass windows, but direct sunlight weakens it on prolonged exposure. Micro-organisms and bacteria do not affect polyesters as also insects. But insects such as beetles when trapped between folds of fabric cut their way through fabric only as a means of escape.

Uses and Care

Polyester fibres and fabrics have many uses. They have good moisture transport through wicking and dry quickly, and are thus popularly used for apparel including sportswear for men, women and children. Polyester is often used in outerwear because of its high tenacity and durability. Blends of wool, cotton, rayon or linen with polyester fibres are commonly available. In blended fabrics polyester contributes easy maintenance, strength, durability, abrasion resistance and wrinkle resistance. Polyester has industrial uses as well, such as in making carpets, filters, synthetic artery replacements, ropes and films. It is widely used in home furnishing fabrics for upholstery, draperies and floor coverings.

Care requirement of polyester fabrics are minimum. Most materials made from polyester can be machine-washed and machine-dried. Ironing should be done at moderate temperature, because

Table 3.3: Properties of nylon, polyester, acrylic and modacrylic

	Nylon 6	*Nylon 66*	*Polyester*	*Acrylic*	*Modacrylic*
Physical properties					
Shape and appearance	The fibres can be bright, semi-dull or dull depending on the amount of delusterant added to the fibre		Polyester fibres are dull or semi-dull in appearance due to delustering agents added to spinning solution	They are mostly staple, with dull or semi-dull lustre due to addition of delusterant.	
Tenacity	Dry: 4.9–8.5 g/d Wet: 4.2–8.0 g/d	Dry: 4.6–8.8 g/d Wet: 4.0–7.6 g/d	2.5–9.5 g/d There is no loss of strength when wet.	Dry: 2.0–2.7 g/d Wet: 1.6–2.2 g/d	Dry: 2.0–3.1 g/d Wet: 2.0–3.1 g/d
Moisture absorbency (65% RH)	2.5–5%	4–4.5%	0.4%	1.5%	2.5–4%
% Elongation at break (65% RH)	16–50%	19–40%	9.5–75%	20–60%	35–45%
Elasticity	Very Good	Very Good	Excellent	Moderate	Moderate
Resiliency	Very Good	Very Good	Excellent	Good	Very Good
Density	1.13 g/cm^3	1.14 g/cm^3	1.38 g/cm^3	1.16–1.18 g/cm^3	1.30–1.37 g/cm^3
Electrical conductivity	Poor	Poor	Poor	Static build-up that increases at low humidity	Poor
Chemical properties					
Effect of acids	Poor resistance to acids	Poor resistance to acids	Can withstand weak acids, but destroyed by strong acids	Strong or concentrated acids will cause loss of strength	Good to excellent resistance
Effect of alkalis	Good resistance	Good resistance	Good resistance	Good resistance to weak alkalis, concentrated alkalis cause damage.	Good resistance
Effect of organic solvents	Resistant	Resistant	Resistant	No damage	No damage

(*Contd.*)

Table 3.3: (*Contd.*)

	Nylon 6	*Nylon 66*	*Polyester*	*Acrylic*	*Modacrylic*
Effect of bleaches	Resistant	Resistant	Resistant	No damage	No damage
Effect of sunlight	Good resistance	Good resistance	good resistance to sunlight behind glass, but direct sunlight weakens it after prolonged exposure	Resistant	Resistant
Biological properties	Excellent resistance	Excellent resistance	Resistant	Resistant	Resistant
Thermal properties					
Safe ironing temp	150°C	151–175°C	121°C	<160°C	<150°C
Melting point	210°C	250°C	240–290°C	190°C	150–190°C
Burning behaviour	Melts, thermoplastic and forms a bead that is non-crushable	Melts, thermoplastic and forms a bead that is non-crushable	Melts, thermoplastic and forms a bead that is non-crushable	Once ignited, burns with yellow flame and drips in molten drops, which are hot enough to ignite combustible substances on which they fall	Does not support combustion, self-extinguishing

Source: Compiled by the authors.

high temperature can increase the possibility of shrinkage. Most items made from polyester can be dry-cleaned. Soaps and detergents can easily be used for washing polyester fabrics.

Polyolefins

A polyolefin belongs to a class of polymers produced from a simple olefin (or alkene; C_nH_{2n}) as a monomer. The two common polyolefins are polyethylene (Figure 3.17a) which is produced by polymerising ethylene, and polypropylene (Figure 3.17b) which is made from propylene. The BISFA defines polyethylene fibre as 'a fibre composed of linear macromolecules of unsubstituted saturated aliphatic hydrocarbons' and polypropylene fibres as 'a fibre composed of linear macromolecules made up of saturated aliphatic carbon units in which one carbon atom in two carries a methyl side group' (CIRFS n.d.), Polyolefin fibres were developed in the 1950s and 1960s. Polyethylene in fibre form was first developed by Imperial Chemical Industries of England while polypropylene was developed by Montecatini of Italy. Both these polyolefins are very important in plastic moulding and for making plastic sheets, but both are also spun into synthetic fibres on a large scale.

Figure 3.17: Chemical structures of polyethylene and polypropylene

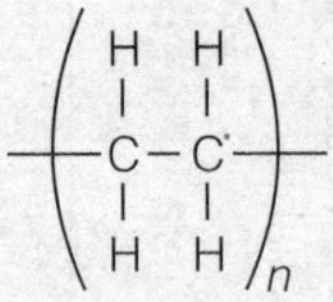

$$\left(\begin{array}{cc} H & H \\ | & | \\ -C- & C- \\ | & | \\ H & H \end{array}\right)_n$$

(a) Polyethylene

$$\left[\begin{array}{c} CH_3 \\ | \\ -CH - CH_2- \end{array}\right]_n$$

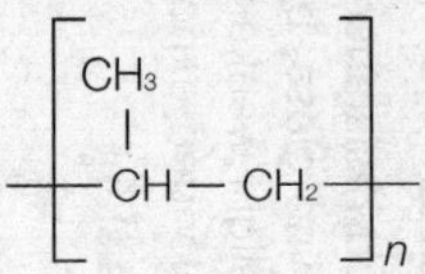

(b) Polypropylene

Production of Olefins

Polyolefins are produced by polymerising olefin raw materials under pressure with a catalyst. While polyethylene fibre is made from polymerising ethylene, polypropylene is from propylene. They are both melt spun. Usually, polymer granules are fed to an extruder which melts the polymer which is then pumped through a spinneret. The filaments are cooled in an air stream and are then drawn or stretched to six times the spun length. Since the fibres are difficult to dye, coloured pigments are often added to the polymer stream before extrusion. An alternative process is to produce a film, cut the film into strips and then fibrillate the individual strips before winding onto a package.

Properties

Both polyolefin fibres resemble glass rods in both longitudinal and cross-sectional views. They both have a density less than 1.0. They have good strength but it varies with the degree of polymerisation and molecular orientation. They do not absorb moisture, which is an advantage in many end uses, but without modification they cannot be dyed. Their melting points are around 130°C for polyethylene and 160°C for polypropylene, which is low in comparison to nylon or polyester but high enough for textile applications. They have a high resistance to chemical attack. They lose strength on prolonged exposure to sunlight. They have good resistance to crushing. They burn slowly and give off a sooty, waxy smoke. They are highly resistant to alkaline and acidic substances, except for oxidising acids. They exhibit static build-up. They are seldom attacked by biological agents. Properties of olefin fibres are given Table 3.4.

Uses and Care

Polypropylene fibre consumption has grown rapidly during the past decade. This is largely due to its acceptance as a carpet fibre (as it has good resistance to crushing) and the growth in the non-woven end-uses, especially disposables and geotextiles where polypropylene is now the dominant fibre. These floor coverings can easily be cleaned. The end uses of polyolefins include ropes, tapes, twines and fishing nets. Apparel (blended with other fibres) such as sportswear, knitted sweaters, suits and coats are also being made.

Polyolefin fibres launder well, dry quickly and require little ironing. Dry-cleaning is not recommended.

Box 3.1: Elastomeric fibres

An elastomer is a polymer with the physical property of elasticity. Fibres that possess extremely high elongations at break and that recover fully and rapidly from high elongations are known as elastomeric fibres. These fibres are all used in specialised applications where high elasticity is necessary within the textile structure. Elastomeric fibres include the cross-linked natural and synthetic rubbers, spandex fibres (segmented polyurethanes), anidex fibres (cross-linked polyacrylates) and the side-by-side biconstituent fibre of nylon and spandex (Monvelle).

Rubber

Rubber is a manufactured fibre in which the fibre-forming substance is comprised of natural or synthetic rubber. Rubber fibres from natural sources have been known for over 100 years. Natural rubber is derived from coagulation of the gum from *Hevea brasiliensis* and is primarily-polyisoprene, a diene polymer. (Details of natural rubber have been discussed in the previous chapter.)

Rubber fibres exhibit excellent elastic properties but are sensitive to chemical attack, thereby limiting their usefulness. As a fibre rubber dates back to the 1920s when the US Rubber Company found that fine rubber filaments could be used as a central core for other fibres, i.e. any fibre such as cotton can be wrapped around the rubber core, and thus, an elastic yarn can be obtained. Synthetic rubber is more common in these elastic fabrics than natural rubber. Neoprene, a type of synthetic rubber made from polychloroprene, is used as an elastomeric fibre or a supported elastic film. It is resistant to acids, alkalis, alcohols, oils, caustic soda and solvents. It is found in protective gloves and clothing, industrial hoses and belts and coatings for wiring. Rubber has been replaced in many uses by spandex but it continues to be used in narrow elastic fabrics, such as narrow elastic tapes in the waist bands of many garments.

Elastane (Spandex Fibre)

BISFA defines elastane as 'a fibre composed of at least 85 per cent by mass of segmented polyurethane $(–NH–COO–)_n$ which, if stretched to three times its unstretched length, rapidly reverts substantially to the unstretched length when the tension is removed' (CIRFS n.d.). Therefore, elastane yarns are primarily characterised by a property not found in nature—the ability to recover from extraordinary stretch. Although elastane was first synthesised in 1937 in Germany, it was not commercialised as a fibre until 1958. Elastane fibres are better known today under their trade names—Lycra (manufactured by DuPont) and Dorlastan (by AsahiKASEI of Japan). While the generic term preferred in the US is elastane, the fibre is known as spandex in many other parts of the world. In India, the Thapar group was the first to launch spandex in 1991 under the trade name Elyxa. Elastane polymer chains have two kinds of segments. The rigid segments which impart strength to the fabric and the flexible, amorphous segments which lie coiled up like a spring when the filament is in a relaxed state. On being subjected to stretch, these coiled up structures straighten up.

Properties

The appearance of elastomeric fibres under the microscope varies considerably. Lycra has a dark, spotted longitudinal appearance with dog-bone-shaped cross-section. Other spandex fibres may be rounder than Lycra.

Natural rubber has an excellent elongation (500–600%) and recovery, but its low tenacity of 0.34 g/d and low dye-acceptance limits its use in lightweight garments. Spandex is superior to natural rubber in strength and durability. It is highly stretchable, light weight and durable. It can be repeatedly stretched over 500 per cent without breaking and recovers instantly. Elastanes resist perspiration and cosmetic oils, are easily washable and are dyeable (Table 3.4).

Uses and Care

The main end uses for the yarns are garments and other products where stretch, comfort and/or fit are important. Typical examples are sports and leisure wear, swimwear, elastic corset fabrics and stockings.

INORGANIC FIBRES

There are many inorganic fibres, including glass, carbon, metal and ceramic. They are used particularly in the industrial fibre sector.

Table 3.4: Properties of polyolefins and spandex

	Polyethylene	*Polypropylene*	*Spandex*
Physical properties			
Shape and appearance	Can be controlled	Can be controlled	Length and diameter can be controlled
Tenacity	1–1.5 g/d	1.5–3.5 g/d	0.7–1 g/d
Moisture absorbency	0%, but can transport moisture by wicking	0–0.5%	Less than 1%
Elongation	20–80%	15–50%	500–800%
Elasticity	Excellent	Excellent	600–800%, Excellent
Resiliency	Good	Excellent	Excellent
Density	0.9 g/cm^3 lighter than water		1–1.2 g/cm^3
Electrical conductivity	No static build-up	No static build-up	Static build-up
Chemical properties			
Effect of acids	Resistant	Resistant	Good resistance
Effect of alkalis	Resistant	Resistant	Conc. alkalis at high temperature cause degradation
Effect of organic solvents	Good resistance	Resistant	Good resistance
Effect of bleaches	Good resistance	Resistant	Not resistant to chlorine bleaches
Effect of sunlight	Gradual decline in strength		Good resistance
Biological properties	Good resistance	Resistant	Good resistance
Thermal properties			
Safe ironing temp	<75°C	<75°C	<150°C
Melting point	130°C	160°C	150°C
Burning behaviour	Burns slowly	Flammable and melts on burning	Burns and forms a gummy residue

Source: Compiled by the authors.

Glass

Glass, a mineral fibre, is the most important inorganic fibre. It is produced by melting glass pellets in an electric furnace at around 1500°C. The molten glass passes through small holes in a plate at the base of the furnace. After cooling in air it is wound up on a package. Alternatively, it can be spun centrifugally to form a web.

Several types of glass fibres are produced. They have the following properties in common: high moduli, high rot resistance, low moisture uptake, are brittle and have low breaking extensions.

Glass is used extensively for insulation in the form of a felt and also for reinforcing plastics to make boats, caravans, automobile parts, etc. Other lesser uses are flame-resistant curtains and decor fabrics.

Carbon

BISFA defines carbon fibre as 'a fibre containing at least 90% by mass of carbon obtained by thermal carbonisation of organic fibre precursors' (CIRFS n.d.). The common raw materials used to make carbon fibres are polyacrylonitrile (PAN) or pitch. Pitch is the name for any of a number of viscoelastic polymers such as tar, resin, etc. If pitch is used the process consists of extrusion, oxidation and graphitisation, while if PAN is used the tow is oxidised, carbonised and finally graphitised. Carbon fibres are characterised by having high moduli and high strength. They are also brittle and have a low density. The main end uses are as reinforcement fibres in composites for the aircraft and aerospace industry and sports goods.

Advancements in fibre science and technology are being made at a fast rate in the twenty-first century, new fibres and textile materials are being developed in association with other fields such as Information Technology (IT) and biotechnology. Optical fibres, integrated circuit boards for mobile phones and wearable computers are examples where fibre textile technology has interfaced with the IT revolution. Bacteria are being used to produce cellulose and polyesters, and soon there could be more industrial bioplants producing fibres. Many chemical fibres and textiles in the twentieth century are being developed by the approach called **biomimetics**, the mimicking of biochemical processes in nature in fibre-creating synthetic processes. These fibres and textiles could well themselves possess bio-functions in the twenty-first century.

SUMMARY

- Man-made fibres can be categorised as regenerated and synthetic fibres.
- Man-made fibre/filaments are produced by chemical spinning which includes the process of extruding the polymer extrusion through the spinneret. The different types of chemical spinning methods are melt spinning, dry spinning and wet spinning.
- Rayon is a manufactured regenerated cellulose fibre. Rayon is known by the names 'viscose rayon' and 'artificial silk' in the textile industry.
- Cellulose acetate and cellulose triacetate are classified as regenerated modified cellulosic fibres.
- Acrylic fibres are composed of linear macromolecules having in the chain at least 85 per cent (by mass) of acrylonitrile repeating units.
- A polyamide fibre is composed of linear macromolecules having in the chain recurring amide linkages, at least 85 per cent of which are joined to aliphatic or cycloaliphatic units.
- Polyester fibre is composed of linear macromolecules having a chain at least 85 per cent by mass of a diol and terephthalic acid.

KEY WORDS

Regenerated fibre: A regenerated fibre is formed when a natural polymer, or its chemical derivative is dissolved and extruded as a continuous filament, and the chemical nature of the natural polymer is either retained or regenerated after the fibre-formation process.

Synthetic fibre: Synthetic fibres are those where the polymers are synthesised from petrochemicals.

Dope: The spinning solution or dope is prepared by dissolving the raw material (natural or synthesised polymer) in a suitable solvent or by melting the polymer.

Tow: Tow is an untwisted rope of thousands of filament fibres.

BISFA: Bureau International pour la Standardisation des Fibres Artificielles (the international bureau for the standardisation of man-made fibres).

HWM: High Wet Modulus Rayon

EXERCISES

- What are synthetic and regenerated fibres. Give examples.
- What are the disadvantages of rayon?
- What is high tenacity rayon?
- Identify the differences between nylon 66 and nylon 6.
- Explain the production process of acrylic fibre.
- Why are aramid fibres popular for army clothing?
- Enumerate the properties of polyester and relate them to end use.
- How do acrylic and modacrylic fibres compare?
- Why is spandex used for foundation garments?

REFERENCES

Corbman, B. P. 1983. *Textiles: Fiber to Fabric*. Sixth edition. New York: Mcgraw-Hill Book Company.

CIRFS. n.d. 'European Man-Made Fibres Association is the representative body for the European man-made fibres industry.' Available at http://www.cirfs.org/Home.aspx (accessed 2 February 2017).

Government of India (GoI). n.d. 'Section II: Man Made Fibers' in *National Fibre Policy for Textiles and Garments*. New Delhi: Ministry of Textiles, GoI. Available at https://www.westbengalhandloom.org/national_fibre_policy/Fibre_Policy_Sub_%20Groups_Report_dir_mg_d_20100608_2.pdf (accessed 2 February 2017).

Hollen, N. and J. Saddler. *Textiles*. New York: Macmillan.

Joseph, M. J. 1988. *Essentials of Textiles*. Philadelphia: Saunders College Publishing.

Kadolph, S. J. 2009. *Textiles*. New York: Pearson Publications.

Mather, R. R. and R. H. Wardman. 2011. *The Chemistry of Textile Fibres*. London: RSC Publishing.

Tortora, P. G. 2009. *Understanding Textiles*. New York: Macmillan.

ONLINE SOURCES (all accessed February 2017)

en.wikipedia.org
www.fibresource.com
www.textilelearner.blogspot.in
www.nptel.ac.in
www.osf1.gmu.edu

4

YARNS

HIGHLIGHTS

- Methods of yarn construction
- Yarn number or count
- Yarn twist
- Types of yarn
- Sewing thread
- Yarn defects

In the previous chapters we have learned about various kinds of natural and man-made fibres, their processing methods, properties, uses and care. Once processed, the fibres are then made into yarns by using various kinds of spinning methods. There are various mechanical and chemical methods of spinning, and they are chosen on the basis of the properties of the fibre. These yarn construction methods, yarn properties such as yarn number and twist, and the various types of yarns have been dealt with in detail in this chapter.

PROCESS OF YARN CONSTRUCTION

A **yarn** is a linear assembly of fibres or filaments formed into a continuous strand with or without twist, with high flexibility, reasonable amount of tensile strength and good tactile characteristics. Based on the process of yarn-making and the kind of fibre used, yarns can be of the following types:

- Spun yarns made of short-length fibres
- Filament yarns made from long continuous fibres. These could be:
  - *Multifilament*: Composed of a number of filaments laid together with or without twist
  - *Monofilament*: Composed of a single filament
- Tape or network yarns which are not made from conventional fibres but slit/split polymer films

Spun Yarns

Spun yarns are composed of a number of staple fibres held together with or without twist, though the former is most commonly used. Most natural fibres such as cotton, wool, linen, jute, except silk, can be made into spun yarns. Man-made filament fibres can also be processed into spun yarns by cutting the long continuous lengths into staple lengths. Twisting short-length fibres makes the ends of fibres protrude from the surface of the yarn, and gives it a dull, fuzzy appearance. Long staple lengths produce less protruding ends as compared to shorter lengths. Fabrics made of spun yarns are warmer and more comfortable on the skin. They can shed soil/dirt easily but are prone to pilling.

Spun yarns can be manufactured by any one of the following mechanical spinning methods:

1. Conventional spinning methods
   i. *Cotton system*: Uses short staple fibres in the range of 0.5–2 inches
   ii. *Woollen system*: Uses fibres of length more than cotton but < 2.5 inches
   iii. *Worsted system (or long staple system)*: Uses fibres of length > 2.5 inches
   iv. *Flax system*
2. Unconventional spinning methods
   i. Open-end spinning
   ii. Friction spinning
   iii. Twist-less spinning
   iv. Self-twist spinning
   v. Vortex spinning
3. Filament fibres cut into staple lengths can be processed by the following methods
   i. Tow-to-top system
   ii. Tow-to-yarn spinning

Conventional Yarn Spinning

The process of yarn-making which consists of converting a stock of fibres into a yarn by imparting an optimum amount of twist is called conventional yarn spinning. The twist acts as the cohesive factor which binds the fibre of short lengths. Cotton system, wool system and worsted system are the most commonly used conventional yarn spinning methods.

Cotton System

The following is the sequence of steps of the cotton spinning system (Figure 4.1):

1. *Opening, cleaning and blending (blow room operations)*: The raw material for a spinning mill is a bale which is a highly compressed form of the fibre. The first operation in spinning demands opening of the bale. This includes two operations—opening out and breaking apart. In **opening out**, fibre to fibre individualisation is achieved. This also helps in the

process of cleaning the bale of any trapped impurities such as dry leaves, twigs or metal pieces. The quality of the product greatly depends on the efficiency of opening and cleaning. Following this, blending is done. **Blending** involves mixing together fibres originating from different bales to maintain uniformity in the final product. Blending improves the functional characteristics such as strength, drape, lustre, texture, etc. and helps reduce cost by mixing cheaper raw materials with expensive ones without affecting the final product.

Figure 4.1: Cotton yarn spinning

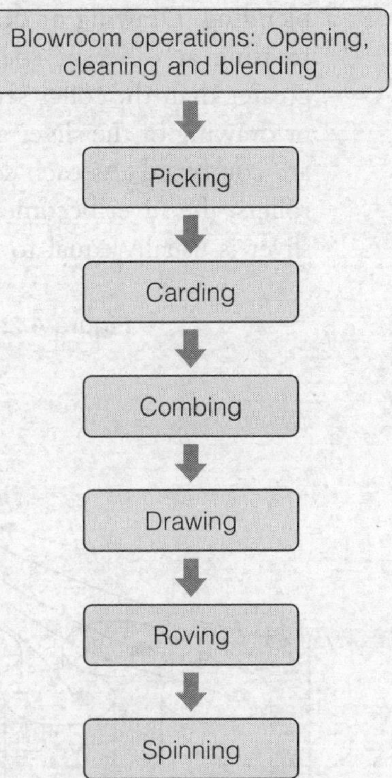

Source: Drawn by the authors.

2. *Picking*: The main purpose of picking is to receive fibres which are opened, cleaned and blended and convert them into a sheet or lap. This is called a picker lap. Therefore, a picker collects the fibres after the blow room operations and assembles them in the form of a sheet which is suitable to be sent for the next step of carding. The aim here is to produce a lap as uniform as possible so as to produce a uniform yarn. The picker lap is rolled into big rolls and transported to the carding section. Picking also brings about further cleaning and parallelisation of fibres.
3. *Carding*: The main objectives of carding are to further clean and disentangle the fibres, remove short fibres, straighten the long fibres and convert the picker lap into thick soft ropes called slivers (a sliver is an assembly of linear fibres). The carding machine consists of a main cylinder/drum covered with wires or small hooks. As the lap moves on the main cylinder it comes in contact with flats which are a series of rectangular metal toothed strips which are chained together to form a loop. The lap gets compressed between the flats and main cylinder. The flat and the cylinder have opposite wire direction as well as speed direction. This results in carding action. Maximum individualisation of fibres is achieved in this region and short fibres, dirt and dust are further removed by this action. At the end is the sliver-forming apparatus. The guide collects the wide web into a single mass which after passing through the condensing tube gets converted into slivers and is piled up inside the sliver can.
4. *Combing*: Combing is not a fundamental process in making a yarn and is used only when yarns of superior quality are desired. To improve the yarn properties, the yarn may be combed to further remove short fibre lengths, increase parallelisation and bring about further cleaning. Combing results in an even yarn with improved strength, less hairiness, increased lustre and better dye uptake.
5. *Drawing or drafting*: The objectives of drawing are thinning of sliver, improvement in the fibre arrangement through straightening and parallelisation and reducing the possibility

of unevenness by coupling of slivers from different origins. This is also a good stage for blending. Drawing or drafting is carried out on a draw frame which has a series of rollers rotating at different speeds (Figure 4.2). The speed of every consecutive roller pair is greater than the roller set preceding it. This difference in the speed brings about thinning or drawing of the sliver at the first draw frame. Several card slivers (normally six to eight) are combined. As each set of successive rollers is moving faster than the preceding set of rollers, the sliver becomes thinner as it is drawn forward. The final thickness of the drawn sliver is usually equal to a single card sliver.

Figure 4.2: Diagrammatic representation of a draw frame

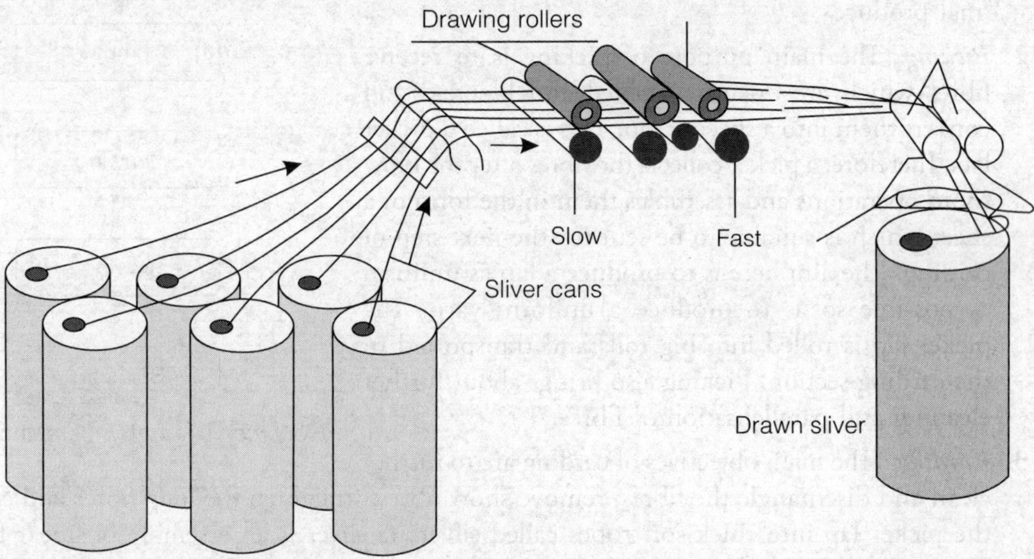

Source: Drawn by the authors.

6. *Roving*: The drawn sliver moves into the roving frame which thins down the sliver to one-eighth of its original diameter and also gives it a slight twist. This makes the strand longer, finer and firmer. It is now referred to as a 'roving'. The slivers from the drawing frame are fed into a 3x3 drafting zone. As in previous process, every consecutive roller pair is moving faster than the roller set preceding it. The first set of rollers moves at a relatively low speed, the middle one at intermediate speed and the final one at a speed ten times that of the first set of rollers. This drafts out the sliver further reducing its diameter and also brings about some parallelisation. The strand then reaches the flyer which is rotating at a constant speed with the help of a spindle. Due to this, the length of the strand between the last set of rollers and the flyer gets a small amount of twist (Figure 4.3). The roving is then wound onto the bobbin/package which is driven by an external motor. This roving is then taken for ring spinning to convert it into the final yarn.
7. *Ring spinning*: Ring spinning is the process of conversion of roving into final yarn. The roving travels through three zones in this process—first through the drafting zone, then

Figure 4.3: Diagrammatic Representation of Roving

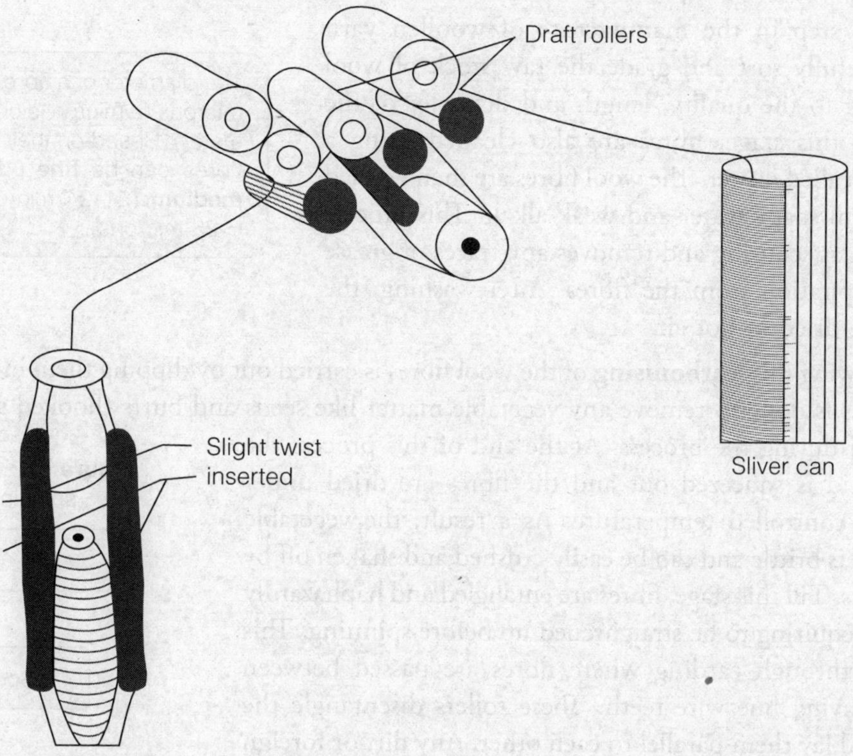

Source: Drawn by the authors.

through the twisting zone and finally through the winding zone. Drafting is carried out with the help of three sets of rollers. These lead to further drawing and thinning of the roving. Thinned out roving then passes through a lappet guide to a spinning frame and is then guided towards a U-shaped guide known as the traveller. The traveller moves freely around the bobbin in a circular way. The zone between the lappet guide and the traveller is called the twisting zone and that between the traveller and the bobbin is called the winding zone. Finally, the yarn is wound on the bobbin (Figure 4.4).

Figure 4.4: Ring spinning

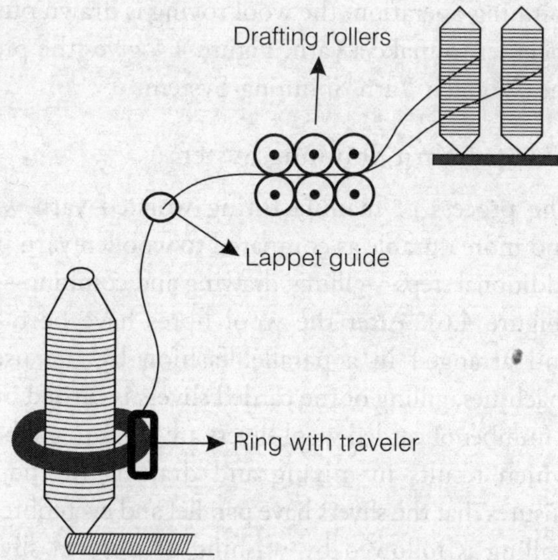

Source: Drawn by the authors.

Woollen Yarn Spinning System

The first step in the manufacture of woollen yarn is to carefully sort and grade the raw stock of wool according to the quality, length and diameter of the fibre. At this stage, fibres are also cleaned using a machine called duster. The wool fibres are then washed with warm soapy water and weak alkali. This process is known as scouring and removes any traces of grease and perspiration from the fibres. After washing, the fibres are dried in hot air.

> The diameter of a fibre is measured in microns (a micron is one-millionth of a meter). Based on their diameter, wool fibres can be **fine** (18–20 microns), **medium** (21–24 microns) or **coarse** (>25 microns).

Following this, **carbonising** of the wool fibres is carried out by dipping them in dilute sulphuric acid. This is done to remove any vegetable matter like seeds and burrs (hooked seeds) which are burnt off during the process. At the end of this process the excess acid is squeezed out and the fibres are dried under carefully-controlled temperature. As a result, the vegetable matter gets brittle and can be easily crushed and shaken off by the rollers. Till this stage, fibres are entangled and haphazardly placed, requiring to be straightened up before spinning. This is done through carding where fibres are passed between rollers having fine wire teeth. These rollers disentangle the fibres and lay them parallel to each other. Any dirt or foreign matter is also removed due to this action. After carding, a light twisting operation called roving is carried out just before spinning. This action helps hold the slivers in place. In the spinning operation, the wool roving is drawn out and twisted together to make a yarn. Figure 4.5 gives the process flow of the Woollen Yarn Spinning System.

Figure 4.5: Woollen yarn spinning system

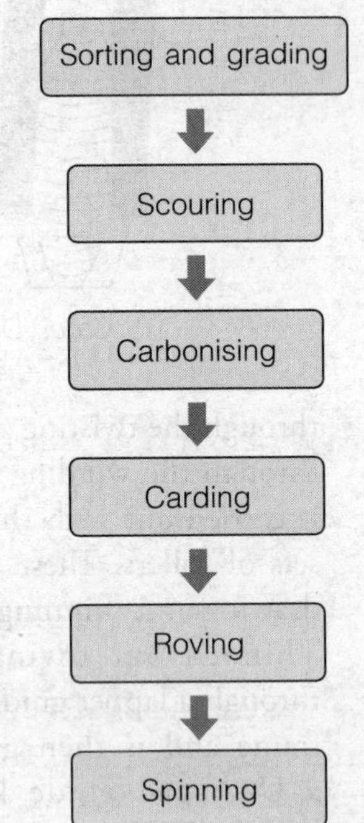

Source: Drawn by the authors.

Worsted Yarn Spinning System

The process of manufacturing worsted yarn which is finer and more durable as compared to woollen yarn involves three additional steps—gilling, drawing and combing—after carding (Figure 4.6). After the wool fibres have been disentangled and arranged in a parallel fashion by the use of carding machines, gilling of the carded slivers is carried out. In gilling, a number of carded wool slivers are fed into gill boxes together which results in mixing and drafting of the slivers. This ensures that the slivers have parallel and even fibre distribution. Gilling is followed by washing of fibre in sliver form and then again gilling it for the second time.

> A **gill box** (or **gill**) is the machine that further aligns the fibres in the carded slivers and blends together slivers from different cards.

At this stage fibres are ready for the combing process which further straightens the fibre, removes short length fibres and also cleans the fibre by further removing any loose impurities. The short length fibres which get removed during combing are known as **noils**. These fibres can be used as raw material for woollen yarns. Presently, two systems are being used for preparing wool slivers for combing process: (*a*) oil combed system, which is used for longer fibres (4–7 inches) where oil is added during the combing process and as a result, cohesion between fibres increases, facilitating leaner, smoother and stronger yarn; and (*b*) dry combed system which was initially used only for short length fibres but is now used for both short and long fibres, and where the resultant yarn is fuzzy in appearance and suited for soft fuller worsted yarns.

The wool tops—i.e. long length fibres arranged in a highly parallel manner—are now subjected to drawing or drafting. It is an advanced operation which draws out the sliver, doubles and redoubles it together in progressive stages. A slight twist is also imparted at this stage so that the wool slivers are converted into roving which is then spun into worsted yarn.

Figure 4.6: Worsted yarn spinning system

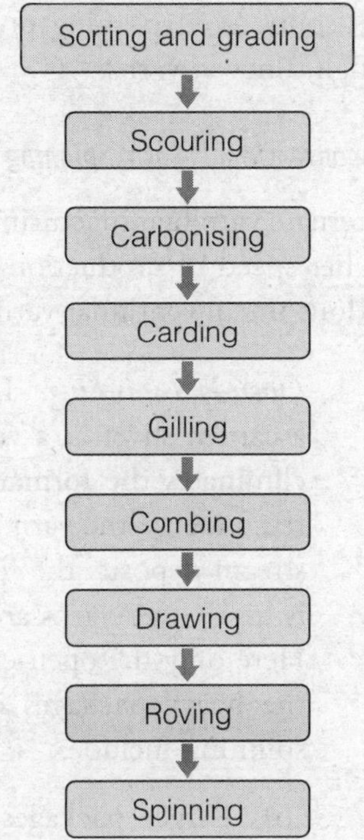

Source: Drawn by the authors.

Table 4.1: Differences between woollen and worsted yarn

Woollen yarn	*Worsted yarn*
Spun from short staple wool fibres (1–3 inches long) and have medium to coarse diameter.	Spun from high-quality long staple wool fibres (>3 inches) having fine diameter.
The partially parallel arrangement of fibres increases air spaces between fibres, and this provides warmth.	Fibres are arranged in a more parallel manner with minimum air spaces.
Short fibres in woollen yarns are held together with low to medium twist which gives a soft, fuzzy and matte look to the yarn.	The longer lengths of fibre in worsted yarns are tightly twisted to produce a smooth lustrous effect.
Fabrics (suede, tweed flannel, broadcloth, etc.) are bulky, have low tensile strength, are less durable and do not crease well.	Fabrics (gabardine, serge, etc.) are wrinkle- and dirt-resistant, give a crisp tailored look to the garment, are more durable with high tensile strength, and drape and crease well.

Source: Compiled by the authors.

Flax System

The flax system of yarn processing comprises hackling, carding, drawing and spinning. Hackling is a simple combing process which removes the fibres from the non-fibrous material, straightens the fibres and separates tows from lines. **Tows** are yarns made from shorter and less-parallel fibres, while **lines** are yarns made from longer and parallel fibres. The fibres can be dry-spun or wet-spun. Wet-spinning produces finer and more uniform yarns while dry spinning produces rough, uneven yarns.

Unconventional Yarn Spinning

Alternate yarn manufacturing processes offer advantages over the conventional method by offering higher speed of production and reduction in labour costs, energy and space requirements. The various unconventional yarn spinning systems are:

1. *Open-end spinning*: The basic process of open-end spinning involves removing fibres from a carded sliver at a very high speed and re-condensing it with a twist to form a yarn. It eliminates the formation of the roving. Air, water or mechanical rotor may be used to re-condense the yarn. Rotor air-jet spinning process is more commonly used where an air stream deposits the fibres on the inner surface of the mechanical rotor. When the speed is high, finer yarns are produced and when the speed is slow, coarser yarns are produced. Here only the open-end rotor is rotated to impart the twist. Therefore, the need to rotate the heavy packages is eliminated. Advantages of this method over conventional ring spinning include:

   (i) Larger packages of yarn produced and knots are eliminated, thus yarn is more uniform.

   (ii) Very high rates of production (may reach up to four times that of ring spinning) and savings on space and power.

   (iii) Improved abrasion resistance, absorption and better dye uptake.

   (iv) Better distribution of fibres in the blends.

   Disadvantages include weaker yarns with a coarser feel. It is also not possible to produce very fine yarns. This spinning method is therefore used for preparation of denim, pile yarn fabrics and for preparing the base fabric for laminated fabrics.

2. *Friction spinning*: In this method, as the sliver enters the system, the fibres are separated and spread onto a roll. Carded fibres are transported by air current to a friction zone consisting of two perforated rolls. The perforated rolls have a vacuum core which makes sure that the fibres stick to the cylinder surface. As the rollers rotate, the fibres enter the nip of two rolls. The rollers move in the same direction and the friction between them twists the fibre. The fibre alignment is controlled by the feed angle. A seed yarn is fed in which initiates the process of yarn formation. The speed of rotation of rollers is very high and therefore the speed of production is very high. These yarns are more even, have much less lint, are softer and lofty but are weaker than the conventional yarns. Thus, this technique is commonly used for woollen/acrylic yarns used for knits.

3. *Self-twist spinning*: The principle behind this method of spinning is to impart twist in alternate directions along the length of the yarn. If such yarns are held close to each other side by side and then released, they untwist and form a 2-ply yarn. Eight rovings are fed side by side through a pair of rollers which are rotating and oscillating. The rovings are placed about three-fourth of an inch apart as they emerge from the roller nip. The rollers perform the function of imparting twist in opposite directions by their sideways movement and the roving moves forward because of their rotation. All the eight strands get twisted in the S and Z directions cyclically. As they emerge from roller nip, they are made to unite in pairs. To unite them, one strand keeps going straight and the other takes a right angle path to meet at a point. This ensures that opposite direction twist areas face each other and when they are released they untwist and result in four 2-ply yarns.
4. *Twist-less spinning*: In this method, cotton roving is boiled in an alkaline solution to wash away all fatty substances and to bring it to a thoroughly wet stage. In this wet state, it is drafted between a pair of rollers. One of the rollers is supplied with a suspension of inactive starch grains. The suspension of starch grains is picked up by the wet cotton strand. The strand is then wound onto a package in a twist-less state and steamed at 110°C for one hour. During this time, starch grains swell and spread over the fibre as an adhesive film. Subsequently, it is dried in an oven at 100°C. The yarns produced are flat, ribbon-like, lustrous and slightly bulky.
5. *Vortex spinning*: Vortex spinning is based on the air jet spinning technology. The vortex-spun yarn has a two-part structure: core and sheath. The yarn has a centre of parallel fibres around which an outer layer of fibres are wrapped. In the vortex system, drafted fibres are introduced into a spindle orifice by an air vortex. As they enter and pass through the orifice, fibres are twisted by the swirling air. High-speed or vortex air currents wrap the fibres and yarn twist develops as the fibres swirl around the spindle.

Filament Yarns

Filament yarn consists of filament fibres (very long continuous fibres) which are either twisted together or only grouped together. Filament yarns could be either monofilament (single fibre extruded from spinneret) or multi filament. Thicker monofilaments are typically used for industrial purposes. All man-made fibres are produced largely as filament yarns while silk is a natural filament. Sometimes, extruded filament fibres could be cut into short staple lengths and then processed by systems used for making natural fibres into yarns. Since filament yarns have hardly any protruding ends and their surface is smooth, they resist pilling. Round cross-sections cause compact packing of the yarns in fabric formation, resulting in reduced bulk and cover.

Spun Yarns from Filament Fibres

Tow is composed of bundles of filaments that have been extruded simultaneously through spinnerets having large number of holes. Many filaments pulled together form a large sliver of fibres. Manufacturers process tow into spun yarns by the following methods:

Tow-to-Top System

The opening cleaning and carding steps are eliminated in this process and the filament tow is cut into staple lengths and made into top (top is a sliver which is made up of staple fibres). It is then spun into a yarn by a conventional spinning process. The staple lengths can also be produced by stretching the tow such that it breaks at the weakest points.

Tow-to-Yarn Spinning

The extruded tow is directly passed through the drafting rollers. The rollers stretch the tow and break it into shorter lengths. It is directly taken for spinning, drawn out to yarn size and twist inserted.

Split Film or Tape Yarns

These yarns are produced from thin sheets or films of polymer that are cut into narrow strips or ribbons. The major steps in the production are extrusion, cooling, slitting, drawing and winding. Man-made fibres have the potential of converting them to tape yarns. The polymer is extruded through a slot to form a thin layer that resembles plastic wrap. They can be extruded through narrow slots which eliminate slitting or through a circular die to form a tube that is then slit. The film is cooled under controlled conditions. The sheet or tube is then slit into narrow tapes. The tapes are heated and drawn which extends it linearly and decreases its width.

YARN NUMBER OR YARN COUNT

Yarn numbering systems describe the relative size or fineness of a yarn as a mathematical relationship between the length and mass of a strand of yarn. In other words, the count of yarn is a numerical expression which measures the linear density or fineness of the yarn. Direct measurement of the diameter of a yarn poses a rather difficult problem. This is because yarns are only roughly circular in cross section and irregularity in thickness is unavoidable. Also, direct measurement of thickness will cause squashing or flattening of the yarn and hence make correct assessment impossible. **Yarn count** is defined as the number indicating the mass per unit length or the length per unit mass of yarn. Yarn count can be expressed in two different systems—direct and indirect systems. The definition above contains the phrases 'mass per unit length' and 'length per unit mass', indicating the two systems that are employed.

Direct System

In a direct yarn counting system the yarn number or count is the weight of a unit length of yarn. There is a direct relationship between the yarn number and the yarn size, i.e. as the yarn number increases, the yarn becomes thicker and coarser. Direct count can of three basic types:

(a) *Tex*: It is the weight in grams of 1000 m of yarn.
(b) *Denier*: It is the weight in grams of 9000 m of yarn.
(c) *Metric Count*: It is the weight in kilograms of 1 km of yarn.

Indirect System

It measures length per unit mass of the yarn where mass remains constant and the length varies. In the indirect system, the yarn number or count is the number of units of length per unit of weight. There is an inverse relationship between the yarn number and the fineness of the yarn. There are three basic types of indirect counts:

(a) *Cotton Count*: It is the number of hanks of 840 yards that weigh one pound.

(b) *Wool Count*: It is the number of hanks of 300 yards that weigh one pound.

(c) *Worsted Count*: It is the number of hanks of 560 yards that weigh one pound.

Hank is the term used to refer to a loose-coiled bundle of yarns of definite length.

Measurement of Yarn Count

Irrespective of the system of yarn numbering employed, two basic requirements for the determination of the yarn number are—an accurate value for the sample length and an accurate value for its weight. The methods used for determining the yarn number or count depends to a large extent on the form in which the yarn is available for testing. Two of the common balances for measuring yarn count are:

1. *Quadrant balance*: The quadrant balance is designed to directly indicate the yarn number when the lea of a specified length (120 yards) is placed on a hook. It measures the cotton count (Indirect Yarn Numbering System) or Tex (Direct Yarn Numbering system). There is a pointer which indicates the yarn count directly. To make the 120 yard lea for the quadrant balance, a Wrap Reel is required. This is a simple machine consisting of a reel, a yarn package creel, a yarn guide which has a small sideways traverse to spread the loops of yarn, a length indicator and a warning bell. The wrap reel may be hand- or motor-driven and some even may have a yarn tensioning device. The reel gives the optimum number of revolutions to produce a 120 yards lea. With the help of a wrap reel, fixed length leas can be drawn from a yarn package by winding the yarn around a drum and simply rotating a handle.

   A **lea** is a hank of yarn (loose coiled bundle) measuring 120 yards.

2. *Digital balance*: Modern laboratory balances indicate the weight by digital methods. The exact weight of a specific length of a yarn is determined by placing it in the digital balance. This reading can be extrapolated to the required length to determine the tex, denier or metric count (direct count).

YARN TWIST

Yarn twist is the measure of spiral turns given to a yarn in order to hold the constituent fibres together. Twist is calculated by counting the number of turns per unit length (generally an inch of yarn) and is expressed as 'twist per inch' or tpi or 'turns per metre' (tpm).

Direction of Twist

The direction of twist is indicated by the use of letters S or Z. A single yarn has **S twist** if, when it is held in the vertical position, the fibres inclined to the axis of the yarn conform to the direction of the central arm of the letter S. Similarly, **Z twist** yarns are those which when they are held in the vertical position, the fibres inclined to the axis of the yarn conform to the direction of the central arm of the letter Z (Figure 4.7).

Figure 4.7: Direction of S and Z twists

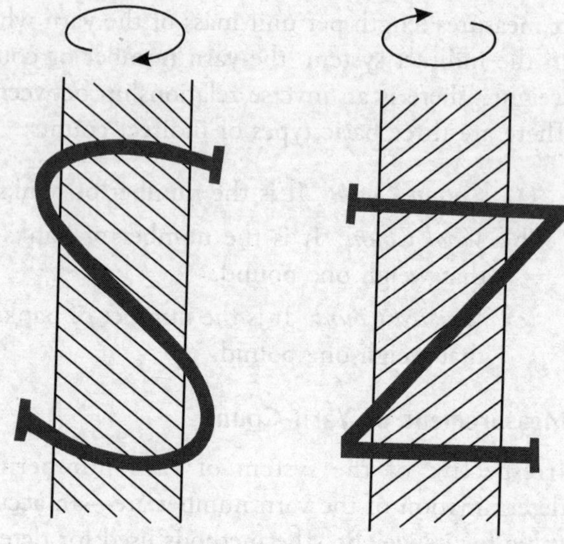

Source: Drawn by the authors.

Amount of Twist

The amount of twist in a given yarn at any stage of manufacture is denoted by number of turns or twist per unit length in a twisted condition at that stage. The amount of twist given to a yarn varies with the following characteristics:

1. *Length*: Short fibres require more twist than longer fibres.
2. *Fineness*: Finer yarns require more twist than coarser yarns.
3. *Stage of processing*: Twist also depends on the stages of manufacture the yarn undergoes, e.g. carded slivers require more twist as compared to the combed slivers.
4. *End use*: Yarns for knitting require lesser twist as compared to the yarns used for woven fabric. Similarly, embroidery threads require less twist as compared to sewing threads as the latter require more strength.

Increasing the twist increases the yarn strength until a point when there is perfect fibre-to-fibre cohesion. This is also termed as the **optimum twist**. Increasing the twist beyond this level makes the fibres perpendicular to the yarn axis thereby reducing the strength. Increase in twist also affects the yarn hairiness, comfort and pilling properties. Low-twist yarns are prone to more pilling but are more comfortable on the skin. Higher twist also increases the cost since the length of yarn is reduced and productivity is lower.

Balance of the Yarn

A yarn is said to be balanced if it has the right amount of twist to hold it in place and also does not allow it to twist upon itself. Moreover, a yarn is balanced if the individual fibres run parallel to the yarn length and in case there is any slope in the fibre, it is either over-twisted or under-twisted. Balance of the yarn is also an important consideration when two or more yarns are plied together. The component parts of the yarn, i.e. the single yarns used for plying, should be of the same thickness and the amount of twist inserted in ply yarn should be the same as singles but in opposite direction. The yarn thus produced would be well balanced and hang neatly into an elongated U-shaped loop when two ends of yarn are held close to each other.

Functions of Twist

1. It is the basic force which binds the fibres together
2. Adds strength
3. Alters appearance
4. Increases abrasion resistance
5. Imparts texture
6. Reduces bulk and adds elasticity

Twist affects the properties of a yarn in the following ways:

1. Ensures the close packing of all the components in a given cross-section, i.e. a high amount of twist ensures close packing as compared to a low amount of twist.
2. Twist reduces the cover or covering power of the yarn, so that as twist increases the covering power of the yarn decreases.
3. Twist affects the hairiness of the yarn, i.e. higher the twist, more the hairiness.
4. A high amount of twist in the yarn ensures higher abrasion resistance and higher tensile strength.
5. Low-twist yarns are more fluffy and warmer than high-twist yarns.
6. Low-twist yarns are high in lustre as compared to the highly-twisted yarns, which are dull in appearance.

CLASSIFICATION BASED ON YARN STRUCTURE

Based on the number of parts and their general structure, yarns are of the following three types:

(a) Simple yarns
  (i) Single yarn
  (ii) Ply yarn
  (iii) Cord yarn
(b) Complex yarns
(c) Textured yarns

Simple Yarns

Simple yarns are even in texture and size, and uniform in shape. They have an even twist per inch. Simple yarns are uniform in appearance and the fabrics made out of them are evenly textured. Such yarns are more durable and less prone to snagging and are easy to maintain. Crêpe yarn is one exception of simple yarns as it has a high amount of twist and is therefore unbalanced. Based on the number of strands that make up the yarn simple yarns can be of the following three types:

(i) **Single yarn** is a basic assemblage of fibres. When the yarn is untwisted, it disintegrates into fibres (Figure 4.8a).

(ii) **Ply yarn** is a combination of two or more singles, plied and twisted together and when the ply is untwisted, individual single yarns are obtained (Figure 4.8b). Each part of the yarn is a ply and a ply yarn is named on the basis of the number of single strands that make up the yarn. The ply yarn is twisted in the direction opposite to the twist in constituent single yarns.

(iii) **Cord yarn** is a combination of two or more ply yarns to obtain a composite structure (Figure 4.8c).

Figure 4.8: Simple yarns

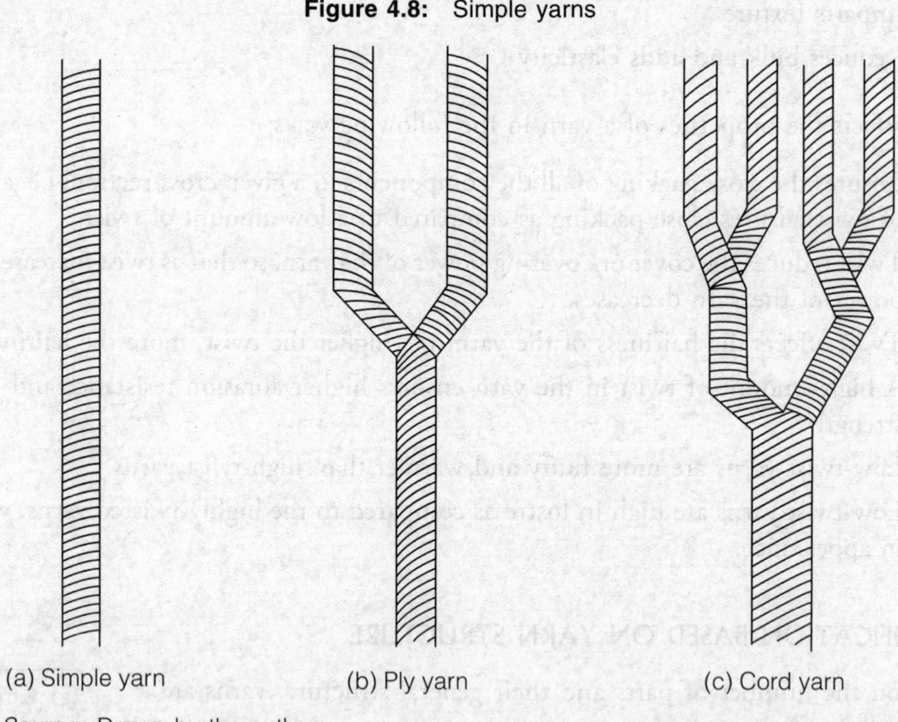

Source: Drawn by the authors.

Complex Yarns

Complex yarns are characterised by irregularities in size. They are uneven in texture and may have variable twist along their length. Most complex yarns are single or ply. Due to varied textures they provide for an interesting appearance to the fabric. Corded complex yarns consist of a base yarn, effect yarn and a binder yarn. The **base yarn** is responsible for giving length and stability in the yarn. The **effect yarn** is responsible for the design, texture and look of the yarn and fabrics made from it. The **binder yarn** holds the base and the effect yarns together (Figure 4.9).

Figure 4.9: Complex yarn

Source: Drawn by the authors.

Various types of complex yarns are:

1. *Slub yarn*: They are either single or 2-ply yarns. There are certain areas which are fluffy, soft and bulky. These areas have a low twist and are thus thicker than the adjacent areas (Figure 4.10).
2. *Thick and thin yarns*: These are similar to slub yarns. These are made from man-made filaments and are created by increasing the pressure in the spinneret intermittently to get thick or bulbous spots.
3. *Flock yarns*: These are single yarns in which small tufts of fibres are inserted at regular intervals to give a novelty, three-dimensional effect. The tufts are held in place by the twist in the single yarn.
4. *Loop/curve yarn*: These are 3-ply yarns. The base is a heavy and coarse yarn. The effect yarn is curled up in loops around the base and is held in place by the binder yarn (Figure 4.11).
5. *Boucle yarn*: This also like a loop yarn, but has tighter loops which project from the body of the yarn.
6. *Nub/spot yarn*: In the nub yarn the base is held stationary and the effect is wrapped around it several times to make it into a large segment or spot at regular intervals. It is a 2-ply yarn with the base and the effect yarn and rarely a binder yarn also (Figure 4.12).

Figure 4.10: Slub yarn

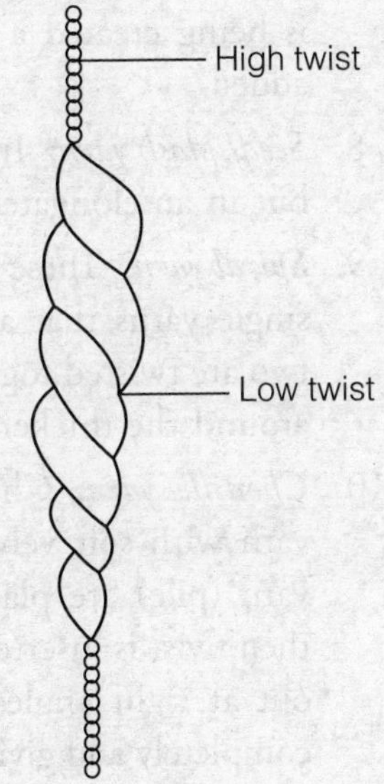

Source: Drawn by the authors

Figure 4.11: Loop/Curve yarn

Source: Drawn by the authors.

Figure 4.12: Nub/Spot yarn

Source: Drawn by the authors.

7. *Knot/knop yarn*: It is similar to the nub or spot yarn, the only difference being that while the spot is being created a tuft of a different colour yarn is added.
8. *Seed/splash yarn*: In this type of yarn, a nub is formed but in an elongated shape resembling a seed.
9. *Spiral yarn*: These yarns are made up of two or more single yarns that are of different sizes. When these two are twisted together, the fine yarn seems to spiral around the thicker yarn (Figure 4.13).
10. *Chenille yarn*: Chenille yarn is a short thick pile yarn with soft velvety appearance. Short lengths of yarn (pile) are placed between two core yarns and then twist is inserted. As a result, the pile yarns stand out at right angles from the core yarn covering it completely and giving a soft fuller look to the chenille yarn that is produced (Figure 4.14).
11. *Core-spun yarn*: In this type of yarn the base yarn is completely covered or wrapped by another yarn. The base or the core yarn adds to strength, whereas the effect yarn lends the desired properties.

Figure 4.13: Spiral yarn

Source: Drawn by the authors.

Figure 4.14: Chenille Yarn

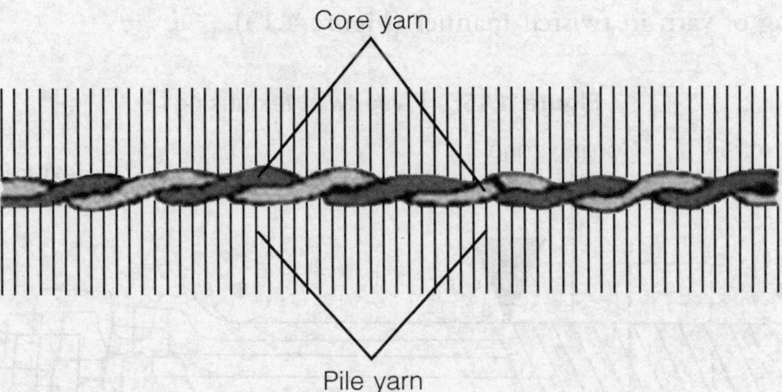

Source: Drawn by the authors.

Textured yarns

Textured yarns have permanent loops, coils, crimps or other distortions in an otherwise straight, smooth filament. They have improved properties like increased pill and crease resistance, better covering power, increased softness and warmth, etc. In textured yarns, the individual filaments are displaced from their natural closely-packed positions to various haphazard configurations. The process of **texturisation** modifies the geometry of the yarn by deshaping, deforming, twisting, bending and crimping them so as to increase their specific volume. It imparts stretch, bulk, improved handle and greater absorbency to the filaments. There are many methods of texturisation and all of them, except air-jet texturising, require thermoplasticity in the fibres.

Air-jet Texturising

Air-jet texturising process is the most versatile mechanical process of yarn texturising where a cold air stream is used instead of heat such that even non-thermoplastic fibres can be texturised. Yarn is over-fed into the texturising chamber into which compressed air is blown at a high speed. The air current separates the strands of multifilament yarns and entangles them by looping and curling with each other which hold the strands of yarn in place without any twist. The yarns produced are bulky, have low extensibility and their physical characteristics and appearance resemble spun yarns. Moreover, this method can also be adopted for producing blended yarn by passing different filaments fibres/yarns through the air jet.

False-twist Texturising

This is a continuous method of yarn texturising which involves twisting of thermoplastic yarn, heat setting and then finally untwisting the yarn to produce a textured effect. Yarn from the supply package is fed into the heating unit at controlled tension and simultaneously, a twisting device inserts a twist in the clockwise direction (Z-twist). The yarn is heat-set in the twisted form and when it cools down, it is twisted in the opposite direction, i.e. counter clockwise (S-twist). Due to this action of twisting and untwisting, the resultant yarn has no twist and hence this method is known as **false-twist texturising**. However, the yarn will retain the loose spiral shape formed due to heat setting of yarn in twisted manner (Figure 4.15).

Figure 4.15: False-twist texturising

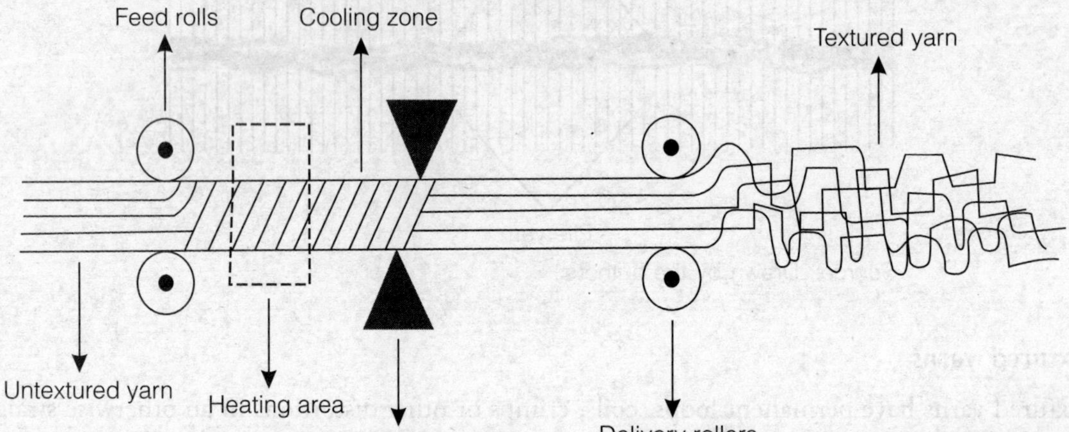

Source: Drawn by the authors.

Gear-crimping Process

Yarns are gripped between rotating intermeshing gears to give a saw-toothed appearance to the filament surface. The filaments to be textured are fed in a heated state between a set of gears. The withdrawn yarn has a similar effect embossed on its surface (Figure 4.16).

Figure 4.16: Gear-crimping method

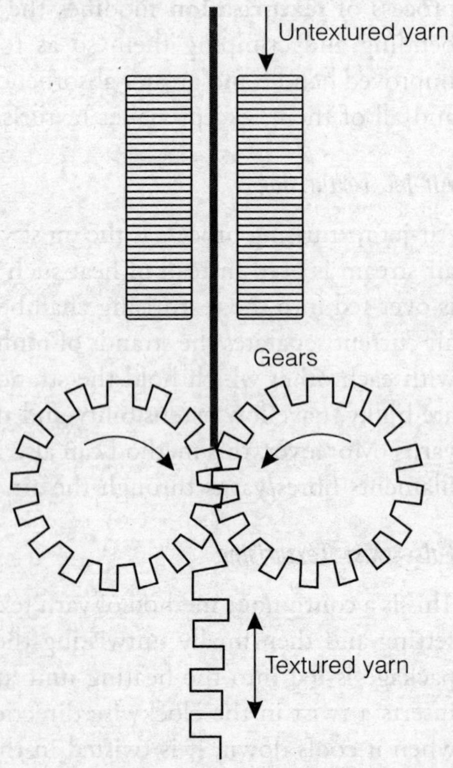

Source: Drawn by the authors.

Knife-edge Texturising or Edge Crimping

In this method, the thermoplastic filaments yarns are passed over a heated knife in stretched condition and then over a cold knife edge. The filament curls up to accommodate the longer length on one edge (Figure 4.17).

Stuffer-box Method

Stuffer-box texturising unit comprises two feed rollers and a tubular stuffer box with a heating device. In this method of texturising, the thermoplastic filament yarn is fed into a heated stuffer tube at a rate faster than the rate at what it is taken out. As a result, the

Figure 4.17: Knife-edge texturising/Edge crimping method

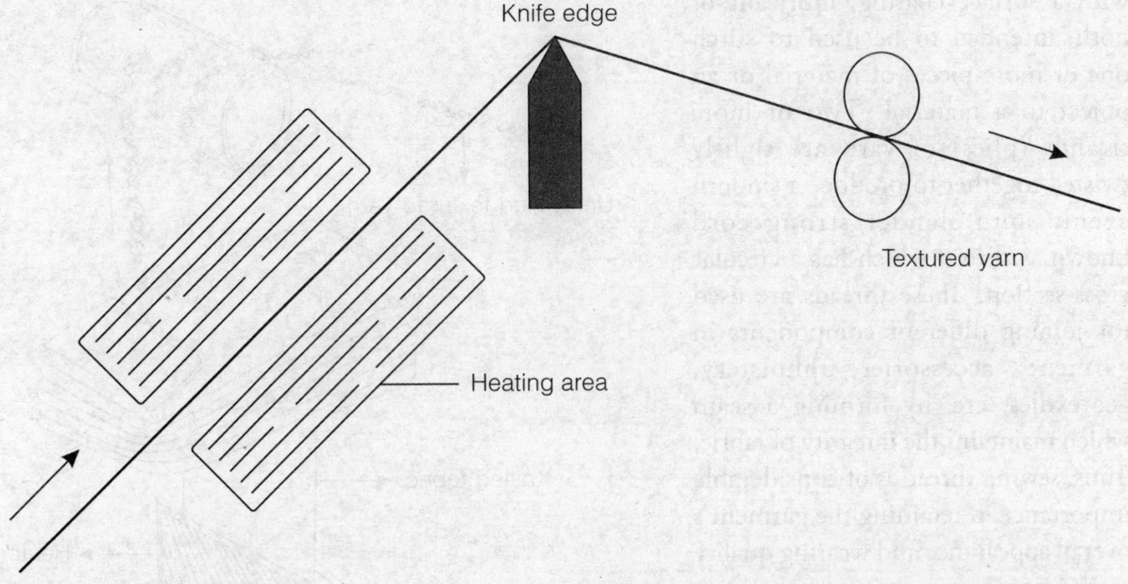

Source: Drawn by the authors.

compressed yarn inside the stuffer-box gets folded and bent at various angles which in turn is heat-set. When the yarn comes out of the tube, it has a crimp (Figure 4.18).

Figure 4.18: Stuffer-box texturising

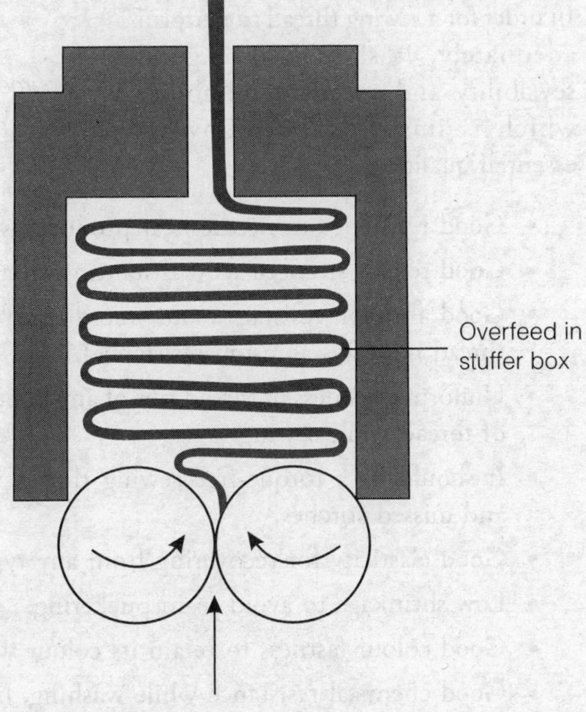

Source: Drawn by the authors.

Knit-deknit Technique

In this method of texurising, a thermoplastic filament yarn is knit into a tubular structure of narrow diameter. This knitted tube is then autoclaved to heat-set the yarn. After cooling, the yarn is unravelled and wound to produce a final package of textured yarn which has wavy configuration due to retention of the prominent looped structure (Figure 4.19).

SEWING THREAD

According to the definition given by American Society for Testing and Materials (ASTM), sewing thread is a flexible, small-diameter yarn or strand usually treated with a surface coating, lubricant or both, intended to be used to stitch one or more pieces of material or an object to a material. Two or more strands (plies) of yarn are tightly twisted together to produce a smooth evenly-spun slender strong cord known as thread which has a circular cross-section. These threads are used for joining different components in garments, accessories, upholstery, geotextiles, etc. by forming a seam which maintains the integrity of fabric. Thus, sewing thread is of considerable importance in retaining the garment's overall appearance and wearing quality in the long run.

Figure 4.19: Knit-deknit texturising process

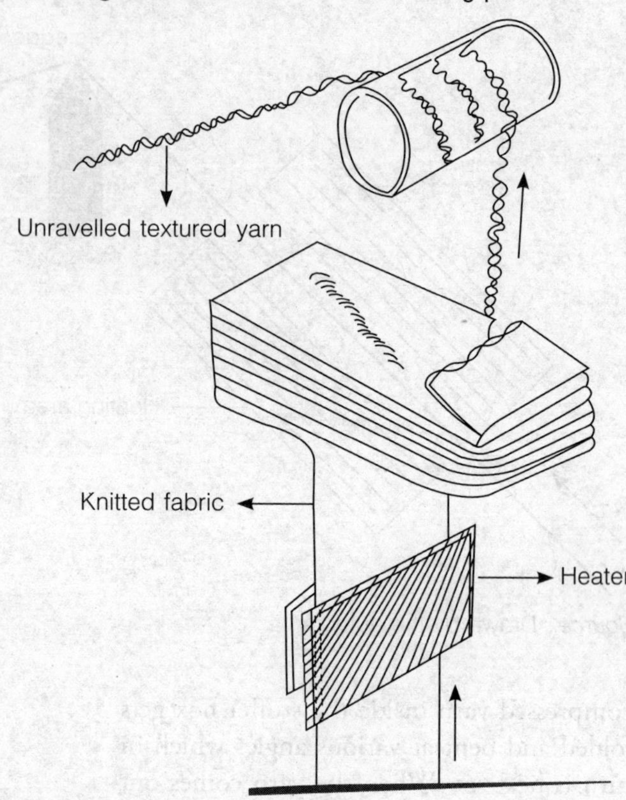

Source: Drawn by the authors.

Essential Qualities of Good Sewing Thread

In order for a sewing thread to perform adequately, it should have good sewability and excellent durability which is imparted by following essential qualities:

- Good resistance to needle heat produced while sewing.
- Good tensile strength to withstand wash and wear.
- Good abrasion resistance and smooth surface to impart frictionless sewing (lubrication of thread increases abrasion resistance).
- Uniform thickness of thread free of any knots or broken pieces for smoother faster movement of thread while sewing.
- It should be a torque-free sewing thread with proper twist balance minimising tangling and missed stitches.
- Good elasticity for recovering from any type of tension on the seam.
- Low shrinkage to avoid seam puckering.
- Good colour fastness to retain its colour when exposed to different agents.
- Good chemical resistance while washing, bleaching, dry-cleaning, etc.

Classification of Sewing Thread

Wide varieties of sewing threads are available in the market for diverse textile applications. These can be made either from a single fibre type such as cotton, linen, silk, rayon, nylon, polyester, rubber, etc. or a blend of fibres such as cotton and polyester using different thread constructions imparting specific end-use properties. Fineness of the thread is an important criterion while selecting it for specific end-use. Various criteria for classifying sewing threads have been summarised in Table 4.2.

Table 4.2: Classification of sewing threads

Criteria of classification	*Types*		*Description*
Substrate, i.e. the fibre type	Natural		Cotton thread is the most commonly used due to its excellent sewability. But it is not very durable and has a tendency to shrink, and hence has limited use.
	Synthetic		Possesses all essential qualities like high tenacity, good resistance to chemicals and abrasion, etc. and is therefore most widely used. E.g. polyester, nylon.
	Combination or blend		Most commonly used combination thread is made by blending cotton and polyester which combines the sewability of cotton along with strength and abrasion resistance of polyester.
Construction process of the thread	Spun thread		Made from both natural and synthetic staple fibres, but spun polyester is preferred as it is stronger than cotton thread of the same size.
	Core-spun thread		It is a combination of staple and filament fibres where polyester filament forms the core with cotton or polyester staple fibre wrapped around it. These individual strands are then twisted together to produce multiple ply construction which has excellent sewability and durability. This type of thread is widely used in the apparel industry.
	Filament threads	Monofilament	Made from single continuous fibre having uniform thickness. Though it is strong and inexpensive, it lacks flexibility and feels scratchy and is therefore used restrictively for draperies and upholstery.
		Smooth multifilament	Made from two or more continuous strands of nylon or polyester twisted together and used for high-strength sewing in leather goods.
		Textured filament	Usually made from polyester yarn to increase extensibility and covering power, but the textured effect can be snagged during wear and wash.

(*Contd.*)

Table 4.2: (*Contd.*)

Criteria of classification	*Types*	*Description*
Finishes applied to improve sewability for specific end-use	Mercerised thread	This finish is generally given to cotton thread to impart increased lustre, dye penetration and strength when treated with caustic soda solution which causes the fibres to swell.
	Gassed thread	This finish removes fibre fuzz from the surface of the thread when it is passed through the flame at high speed.
	Glazed thread	The thread becomes highly lustrous, stronger and resistant to abrasion when it is treated with wax or any other special chemical and polished. But this glaze can be rubbed off the thread while sewing and clog the needle or machine parts. Available in limited colour and used for sewing heavy material like leather, canvas, vinyl, etc.
	Other finishes applied to the threads to increase its functional use include water repellence, anti-bacterial/anti-mildew, flame proof, etc.	

Source: Compiled by the authors.

Difference between a Thread and a Yarn

A thread is a product used to join pieces of fabrics together, i.e. for sewing, while a yarn can be used for many purposes such as knitting, weaving, embroidery and crochet, and so on. A yarn is an assembly of fibres or filaments having a substantial length and relatively small cross-section, with or without twist. Yarns which are fine, even and strong can be used as threads. Therefore, a thread is always a yarn, but a yarn may not always be a thread.

In a nutshell, it can be concluded that a wide variety of yarns are produced for different end uses by varying the method of processing, i.e. mechanical or chemical spinning and the raw materials used (staple or filament fibre). The twist inserted into the yarn defines its fineness or thickness which is designated by the yarn numbering system. This system is based on the length and weight of the yarn. Technological advances in the field of yarn manufacturing have widened the scope of producing novelty yarns imparting characteristic appearance and performance to the end product.

YARN DEFECTS

Defects in yarns are irregularities that affect the quality of the fabric. The defects are as follows:

1. **Slub**: This is caused by uneven twisting or by waste caught during the process of twisting (unwanted material like extra yarns). This would indicate a weak spot in the yarn as well as an area for possible fabric abrasion and wear. This defect can be found in both staple and filament fibre yarns.

2. **Thin places**: These are regions in the fibres that have a cross-sectional size of 30–60 per cent less than that of the average cross-section of a normal yarn. This defect can be found mainly in staple fibre yarns.
3. **Thick places**: These are regions in the fibres that have a cross-sectional size of 30–100 per cent more than that of the normal yarn. This defect can be found mainly in staple fibre yarns.
4. **Neps**: Cross sectional size of 140–400 per cent of normal yarn. They are sudden thick places and occur as small tight balls of entangled fibres.

SUMMARY

- Yarn is a long strand of fibres twisted together to hold them in place. Based on their method of construction, they can be spun yarns, filament yarns or split (tape) yarns.
- Yarn numbering system defines the fineness or thickness of yarn. Direct system of yarn numbering is based on the weight of yarn per unit length and indirect system is based on length per unit mass.
- The amount of yarn twist (number of twists per inch) is responsible for the appearance and performance of the yarn.
- Yarns can be classified as simple, complex and textured yarns.

KEY WORDS

Spun yarn: Spun yarns are composed of staple fibres twisted together.

Filament yarns: Filament yarn consists of filament fibres (very long continuous fibres) which are either twisted together or only grouped together.

Yarn twist: Yarn twist is the measure of spiral turns given to a yarn in order to hold the constituent fibres together.

Thread: Sewing thread is a flexible, small diameter yarn or strand intended to be used to stitch one or more pieces of material or an object to a material.

Yarn number: Yarn numbering systems describe the relative size or fineness of a yarn as a mathematical relationship between the length and mass of a strand of yarn.

EXERCISES

1. Define the following terms:
   - Yarn twist
   - Cotton count
   - Denier
   - Complex yarns
   - Slub yarn
   - Balance of a yarn
2. What is a yarn? Briefly discuss its classification.
3. Differentiate between the following:
   - S and Z yarn twist
   - Simple and complex yarns
4. What are textured yarns? Explain any two methods of texturising.
5. Define thread. What are the essential qualities which makes the thread appropriate for various end uses?

REFERENCES

Collier, Billie J., Martin J. Bide and Phyllis G. Tortora. 2009. *Understanding Textiles*. New Jersey: Pearson Education Inc.

Corbman, P. Bernard. 1983. 'Textiles-Fibre to Fabric'. Sixth edition. New York: McGraw-Hill Inc.

Goswami, B. C., J. G. Martindale and F. L. Scardino. 1977. *Textile Yarns: Technology, Structure and Applications*. New York: John Wiley & Sons.

Joseph, L. Marjory. 1988. *Essentials of Textiles*. Fourth edition. New York: Holt, Rinehart and Winston Inc.

Kadolph, J. Sara. 2009. *Textiles*. Tenth edition. New Jersey: Pearson Education.

Sekhri, S. 2011. *Textbook of Fabric Science: Fundamentals to Finishing*. New Delhi: PHI Learning Pvt. Ltd..

ONLINE SOURCES (accessed 12 May 2016)

http://textilelearner.blogspot.in/

http://textile2technology.com

http://textileandinfo.blogspot.in

http://textileengineerr.blogspot.in

www.textiletextbooks.com

www.tut.fi

http://www.indiantextilejournal.com/articles/FAdetails.asp?id=2247

5

BLENDS, BICOMPONENT AND BIGENERIC FIBRES

HIGHLIGHTS

- Blending
  - Objectives
  - Stages of blending
  - Types of blends
- Production of bicomponent and bigeneric fibres and their applications

A large variety of natural and man-made fibres are available today, offering a wide selection of fabrics. However, all fibres have some desirable and some undesirable characteristics. The human quest to produce perfect fabrics has resulted in the production of blended yarns and fabrics and bicomponent and bigeneric fibres. This is done to introduce special properties in the final product.

BLENDING

Blending is a technique in which two or more types of fibres may be combined to produce a new yarn. The properties of the blended yarn are a combination of the properties of its constituent fibres. The accepted definition of a blended yarn, as given by the American Society for Testing and Materials (ASTM), is 'a single yarn spun from a blend or a mixture of different fibre species.' Blending can be done at various stages in the preparation of yarns. It is important to observe that the quantities of each fibre used in a blend are measured carefully; the proportion of one fibre to the other should be consistently maintained. Two or more types of generic fibres can be combined to form any one of the following types:

- *Blend*: Two or more different generic types of fibres are spun together to form a single yarn which is then used to make fabrics.
- *Combination yarns*: Ply yarns where at least one single yarn is of a different generic fibre type.

- *Mixture/combination fabrics*: Single or ply yarns of one fibre type are used with single or ply yarns of another fibre type and made into fabrics either by weaving, knitting, knotting or braiding.

According to the ASTM definition, only the first type of blend mentioned above qualifies as a true blended yarn. Further, a fabric made from only such yarns qualifies as a blend fabric. However, many consumers have come to associate the term blend with any fabric containing fibres of two or more different generic types regardless of how they have been used in making the fabric. As a rule, woven fabrics designated as blends by fabric producers use blended yarns throughout their manufacture. These fabrics are more likely to give desired performance characteristics during use and care than fabrics made with combination yarns or single yarns of different generic types. The major problem with mixture/combination fabric occurs when yarns of one fibre type have properties and performance characteristics drastically different from the other set of yarns. Such a fabric combination can produce fabrics that do not perform adequately. They may wrinkle badly in one direction or both, may shrink in one direction and show variable absorbency and strength characteristics. These fabrics may require complicated care procedures.

Objectives of Blending

Blending of different types of fibres is done to achieve certain objectives. These objectives are as follows:

- To take advantage of the best qualities of each constituent fibre of the blend or to produce a fabric with better performance characteristics. For instance, when polyester is blended with cotton, the resultant fabric has improved absorbency and comfort due to cotton while polyester adds ease of maintenance and durability. Blending with cotton also improves the spinning efficiency of polyester. Viscose in a polyester viscose blend provides absorbency, resilience and soft texture while polyester enhances durability, shape retention and wet strength.
- To reduce the cost of fabric by blending a less expensive fibre with an expensive one. Very often fibres of lower cost are blended with the fibres that are expensive to lower the overall cost of the product. For instance, specialty hair fibres are costly because of their rarity and the processing they require, and are used only for luxury fabrics. They can be blended with wool or other fibres to enhance appearance or texture, impart softness or special effects and reduce the cost of the resultant fabric. Blending of small quantities of acrylic and nylon with wool reduces the cost of the fabric as compared to one made from 100 per cent wool. Blend of rayon with cotton also helps lower the cost of the fabric.
- To achieve decorative effects/enhanced appearance. Fibres with different dyeing affinities when used in a blend can produce differential dyeing effect, such as cross-dyeing and union dyeing (discussed in Chapter 11). Some fibres can be blended to achieve lustre or textural effects, or to improve cover and drape properties of the fabric. For instance, blending viscose with cotton improves lustre and softness, and thereby enhances the appearance of

the fabric. Small quantities of silk, vicuna and cashmere may also be blended with other fibres to enhance surface characteristics.

Thus, blends can be designed to suit the requirements of consumers. They may be designed for attractive appearance, for a combination of appearance and performance, for reducing cost or for providing easy-care properties.

Various Stages of Blending in the Processing of Fibre

Blending of staple fibres may be done at any one of the fibre processing stages: during the fibre opening stage in the blowroom, during drawing out or during the roving stage.

- During the fibre opening stage, appropriate amounts of each fibre can be weighed, cut to the length required by the particular yarn-making machinery and these can be blended in layers by spreading one on top of the other. Figure 5.1(a) shows a cross-section of yarns in which the fibres were blended at the opening stage. Blending of fibres belonging to the same generic type is done at this stage.
- Blending is very frequently done at the drawing stage. Different types of fibres are blended at the drawing stage by combining slivers of previously carded, combed and parallelised fibres of different varieties. Blending done at drawing or roving stage results in a good and thorough mixing of the different generic fibres (Figure 5.1[b]) but may not be as uniform as achieved during the opening stage (Figure 5.1[a]). If all fibres are not of similar length, long fine fibres tend to move to the centre of a yarn, whereas short coarse fibres migrate to the outer edge as shown in Figure 5.1(c). Filament fibres can be cut to staple lengths and blended with other man-made and natural fibres. The earlier the fibres are blended in processing, better the blend.

Blended man-made filament yarns can also be produced using filament fibres of different deniers or generic types. Different fibres are spun using chemical spinning methods and then blended to

Figure 5.1: Cross-section of yarns showing fibre location in blends

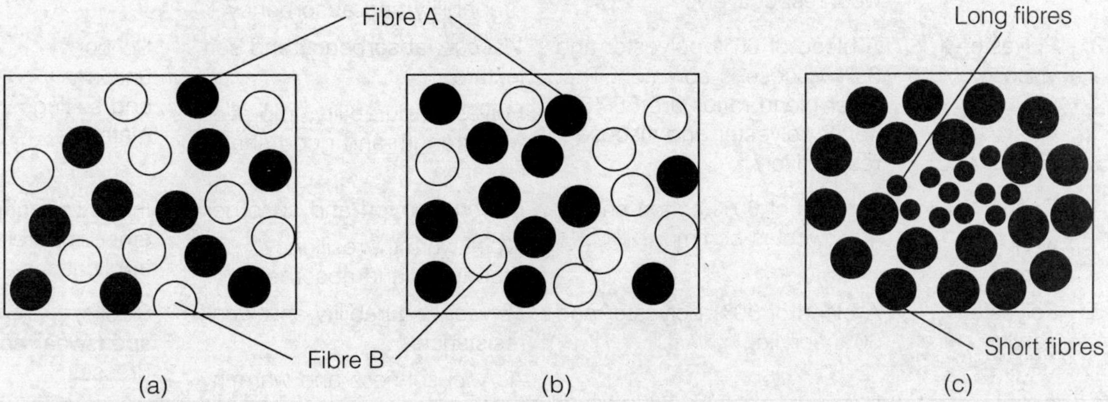

Fibre A

Long fibres

Fibre B

Short fibres

(a) (b) (c)

Source: Drawn by the authors.

form a yarn (Figure 5.2) (Kadolph 2009).

Among the various types of blends available today, the most popular fabrics are polyester-cotton (terecot), polyester-wool (terewool) and polyester-viscose blend. Polyester-cotton-viscose blends are also very common. Table 5.1 lists the various types of blends with their properties and end use. Various effects and combinations of properties can be produced from these blends depending on the fibres used and the percentage of these fibres used in each blend.

Figure 5.2: Blended-filament yarns

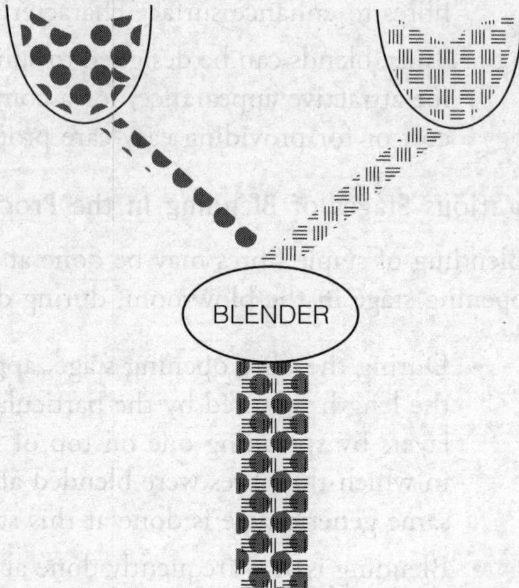

Source: Drawn by the authors.

BICOMPONENT AND BIGENERIC FIBRES

Spinneret modifications in the chemical spinning processes of man-made filament fibres can produce bicomponent and bigeneric fibres. Bicomponent

Table 5.1: Types of blends and their properties

S no.	*Name*	*Composition*	*Properties*	*End use*
1.	Polyester-cotton (terecot)	A blend of 65% polyester and 35% cotton is common. The other blend ratios are 70/30, 50/50 polyester and cotton respectively.	*Polyester*: strength, wrinkle resistance and shape retention *Cotton*: comfort and heat conduction	All types of dress material
2.	Polyester-wool (terewool)	A blend of 55% polyester and 45% wool is common. The other blend ratios are 65/35, 60/40 and 50/50 polyester and wool respectively.	*Polyester*: shape retention, wrinkle resistance, crease retention and increased strength *Wool*: warmth, resiliency, drapability and absorbency	Suiting fabrics
3.	Polyester-viscose	A blend of 65% polyester and 35% viscose is common. The other blend ratios are 55/45, 45/55 polyester and viscose respectively.	*Viscose*: absorbency and soft texture *Polyester*: durability, improved wet strength and good shape retention	School uniforms and suiting materials
4.	Cotton wool (Cotswool)	A blend of 60% cotton and 40% wool is common.	*Cotton*: comfort and absorbency *Wool*: warmth, resiliency, drapability and absorbency	Inner wear and other apparel applications
5.	Polyester acrylic	A blend of 50% polyester and 50% acrylic.	*Polyester*: durability and wrinkle resistance *Acrylic*: softness and warmth	Slacks, sportswear and dresses.

Source: Compiled by the authors.

fibres are made from two generically similar fibre solutions (e.g. the two types of nylon, viz., nylon 6 and nylon 66), whereas bigeneric fibres are made from two generically different fibre solutions (e.g. spandex and nylon).

Methods of Extrusion

Bicomponent or bigeneric fibres are produced by extruding two filaments of different composition, so that they combine as they coagulate or harden. Special spinnerets are required for all types of bicomponent or bigeneric fibres. There are three basic methods of making these fibres—the side-by-side (S/S) method, the sheath/core (S/C) method, and the matrix/fibril (M/F) method.

- *Side-by-side method*: This involves a process in which different polymers are fed to the spinneret orifice together so that they exit from the spinneret opening, alongside each other (Figure 5.3a).
- *Sheath/core method*: This involves extrusion of one component such that it forms a sheath around a second component called the core. Changing the shape of the orifice that contains the inner core can produce fibres with different behavioural characteristics (Figure 5.3b).
- *Matrix/fibril process*: This involves two polymers spun together in the same mix to form a filament in which one component forms fibrils within the other (Figure 5.3c). Fibres made

Figure 5.3: Extrusion techniques

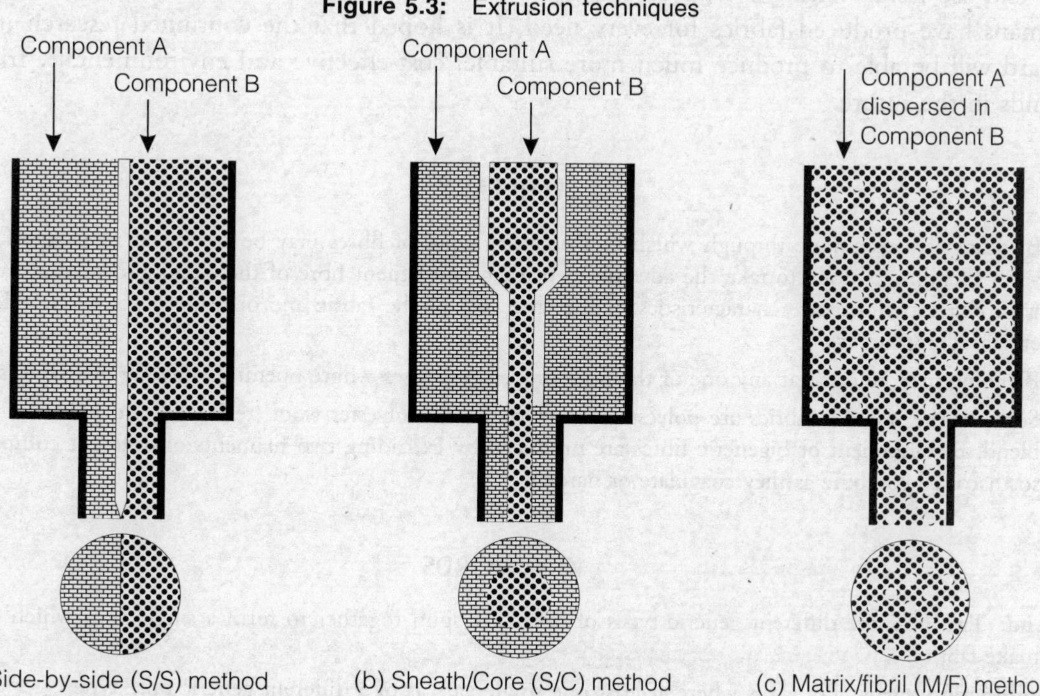

Note: The lower end of the figure shows the cross-section view.
Source: Drawn by the authors.

through this process are also called islands-in-the-sea fibres.

Advantages of Bicomponent and Bigeneric Fibres

Bicomponent and bigeneric fibres are manufactured to achieve the following effects:

- *Stretch or crimp*: Most bicomponent fibres provide stretch or crimp to the fibres. Each polymer used in the bicomponent fibre has slightly different characteristics. Often one polymer is made to shrink in heat or chemical treatment more than the other, pulling the fibre into a permanent crimp. If sufficient crimp is provided, the bicomponent fibre may also have increased stretchability.
- *Increased absorbency*: A less absorbent core fibre could be sheathed in a more absorbent fibre in order to increase comfort properties and dyeability.
- *Formation of non-wovens*: Components with different melting points could be used to bond fibres together in the construction of non-woven fabrics. On heat application, one fibre softens and acts as glue to hold the other fibre in place.

Bicomponent and bigeneric fibres exhibit properties such as varying shrinkage levels, different thermal behaviour or different shapes which result in filaments that coil. This enables yarn and fabric producers to introduce bulk, improved fit, sheerness and/or attractive appearance into the end-use item. With the help of blended yarns and bicomponent and biconstituent fibres, humans have produced fabrics for every need. It is hoped that the continued research in this regard will be able to produce much more suitable, cost-effective and environmentally-friendly blends in the future.

SUMMARY

- Blending is a technique through which two or more types of fibres may be combined to produce a new yarn. Blending is done to take the advantages of each constituent fibre of the blend and produce a fabric with better performance characteristics, reduce the cost of the fabric and/or achieve decorative effects or enhanced appearance.
- Blending may be done at any one of the fibre processing stages—fibre opening, drawing or roving stage.
- Some of the popular fabrics are polyester-cotton (terecot), polyester-wool (terewool) and polyester-viscose blend. Bicomponent or bigeneric fibres are produced by extruding two filaments of different composition so that they combine as they coagulate or harden.

KEY WORDS

Blend: Two or more different generic types of fibres are spun together to form a single yarn which is used to make fabrics.

Combination yarns: Ply yarns where at least one single yarn is of a different generic fibre type.

Mixture/Combination fabrics: Single or ply yarns of one fibre type are used with single or ply yarns of another fibre type and made into a fabric either by weaving, knitting, knotting or braiding.

Bicomponent fibres are made from two generically-similar fibre-forming polymers, e.g. the two types of nylon nylon 6 and nylon 66.

Bigeneric fibres are made from two generically different fibre-forming polymers, e.g. spandex and nylon.

EXERCISES

1. Define the following
   - Blends
   - Bigeneric fibres
   - Bicomponent fibres
2. What are various objectives of blending?
3. What are the stages at which the blending can be done?
4. Enumerate the various types of blends available in the market?
5. Differentiate between bicomponent and bigeneric fibres?
6. How are bicomponent or bigeneric fibres extruded?
7. What are the various applications of the following:
   a. Cotton-polyester blend
   b. Polyester-wool blend
   c. Polyester-acrylic blend

REFERENCES

Corbman, B. P. 1983. *Textiles: Fibre to Fabric*. Sixth edition. New York: McGraw-Hill.

Goswami, B. C., J. G. Martindale and F. L. Scardino. 1977. *Textile Yarns: Technology, Structure and Application*. New York: John Wiley & Sons.

Hollen N., Sadler J. 1973. *Textiles*. Fourth edition. Singapore: The Macmillan Company.

Joseph, M. L. 1988. *Essentials of Textiles*. Fourth edition. New York: Holt, Rinehart and Winston, Inc.

Kadolph, S. J. 2009. *Textiles*. Tenth edition. New Delhi: Dorling Kindersley (India) Pvt. Ltd.

Tortora, G. P. 1978. *Understanding Textiles*. Second edition. New York: Macmillan.

ONLINE SOURCES

http://phrontistery.info/fabric.html (accessed 11 January 2014)

http://www.textileschool.com/articles/377/blended-fabrics-textile-composites (accessed 11 January 2014)

http://www.engr.utk.edu/mse/Textiles/Dry%20Laid%20Nonwovens.htm (accessed 11 January 2014)

http://www.hsc.csu.edu.au/textiles_design/performance/4023/bicomponent_answers.htm (accessed 10 February 2014)

http://www.google.it/patents/US6811873 (accessed 10 February 2014)

UNIT II

FABRIC CONSTRUCTION METHODS

Fabrics are made from yarns or, sometimes, directly from the fibres. The second unit of this volume is dedicated to the various methods that are currently available in the manufacture of fabrics—weaving, knitting, non-woven and other methods. Chapter Six is about woven fabrics and details their various types and characteristics, and the machinery (i.e. looms) used to produce them. Knitted fabrics, which are becoming very popular these days, have been described in Chapter Seven. Here we learn about the various types, properties, use and care of these fabrics. The last chapter in this unit, Chapter Eight, deals with the manufacturing techniques, properties and uses of non-woven fabrics which are made directly from fibres. This chapter also contains a brief discussion on other methods of fabric manufacture such as nets, laces, braiding, knotting, etc.

6

WOVEN FABRICS

HIGHLIGHTS

- History of weaving
- Components and characteristics of woven fabrics
- Mechanism of weaving
  - Parts of a loom
  - Yarn preparation for weaving
  - Basic weaving operation
- Types of looms
  - Handlooms
  - Power looms
- Selvedges
- Types, construction and uses of weaves
  - Basic weaves
  - Decorative weaves

A fabric can be produced by weaving, knitting and felting, or by using other construction techniques from a framework of fibres and yarns, or combinations of these elements. Depending upon the fabric construction technique used and the type of raw material selected, fabrics vary in their texture, quality, durability and end uses (Table 6.1).

As seen in Table 6.1, many different types of fabrics can be constructed using a variety of raw materials. Weaving and knitting are the two major methods of producing textile materials. Making fabrics from interlaced yarns is the most common technique employed. Weaving is one of the most ancient art forms, where a fabric is formed by interlacing two or more sets of yarns at right angles to each other.

Woven fabrics (except triaxial fabrics) are constructed by interlacing two or more sets of yarns called warp and weft yarns at right angles. The point at which a yarn moves to the underside of the cloth from its surface and vice versa is called the point of interlacement.

Table 6.1: Various fabric types and their salient features

Raw material used	*Fabric type*	*Salient features*	*End use*
Solution	*Films*	They may be stiff, lack strength, may tear easily, waterproof, low cost, soil resistant	Shower curtains, rainwear, plastic bags, etc.
	Foam	They are too weak, hence combined with a fabric for strength and durability	Leather-like fabrics used in upholstery
Fibres	*Felt*	Does not ravel and has no grain; lacks strength	Home furnishings, clothing
	Non-woven	Their strength and durability depends on the materials and processes used in their manufacture. They may have no stretch, stretch in one direction or limited stretch in any direction	Used for both disposable and durable items in a wide variety of fields such as packaging, agriculture, sports, home textiles, medical textiles, construction, industrial textiles, protective clothing
Yarns	*Woven*	Made up of warp and weft yarns, have grain, available in different interlacing patterns	Most widely used for clothing and home textiles, tyre cord fabric, industrial filters, etc.
	Knitted	Good stretchability, porous and resilient	Clothing, blankets, etc.
	Braid	Stretchy and easily shaped	Shoelaces and belts
	Lace	Have open structure, can be hand-made or machine-made	Widely used as trims in garments
	Crochet	Made by interlocking loops of yarns, have an open structure, can be hand-made or machine-made	Used to make a variety of clothing items, laces, tablemats, etc.
	Macrame	Made by knotting of cords, no grain, hand-made	Mainly used for decoration purposes: for making wall hangings, jewellery, etc.
Composite fabrics	*Coated fabric*	Have good tensile strength, stability and water repellency. Properties vary depending on the type of coating which is used as per the end use	Soft luggage, upholstery, leather-like apparel, architectural fabrics
	Flocked fabric	They are generally patterned for a printed effect or piled all over to resemble a suede leather	Apparel, upholstery
	Embroidered fabric	Eyelet (embroidered) and applique (stitched) are examples of embroidered fabrics	Apparel, home furnishings
	Tufted fabric	Medium to high loft fabric	Carpeting, upholstery, coat linings and bed spreads

(*Contd.*)

Suede is the napped finish given to leather to impart fuzziness.

Table 6.1: (*Contd.*)

Raw material used	*Fabric type*	*Salient features*	*End use*
Multicomponent fabrics	*Bonded or laminated*	Fabrics have body, increased strength, durable, insulative	Car seats, packaging, protective and outerwear garments, flexible membranes for civil structures
	Quilted	Decorative, bulky and warm	Comforters, placemats, hot pads, baby quilts

Source: Compiled by the author.

Box 6.1: History of weaving

The art of weaving started in the early civilisations, not only to fulfil the need for clothing but shelter as well. Leaves, hair and animal fur could be assembled quickly by plaiting, twining, knotting and interlaceing to make clothes as well as sheds, walls, rugs, etc. It is believed that nature (bird's nest, spider's web, etc.) served as the inspiration for fabric construction. Woven fabrics were used during the early Egyptian civilization dating as far back as 5000 BC and high level of expertise was attained by the time of the decline of the Egyptian civilisation. Weaving was further refined and new techniques were developed in Mesopotamia, Persia, Greece, Rome, China and the Byzantine Empire. With the invention of the loom, the first improved method of holding the warp yarns taut was developed.

The art of weaving in India too originated in ancient times as fragments of woven cotton and bone needles have been discovered at Mohenjodaro and Harappa of the Indus Valley civilisation. The *Rigveda* and the epics, *Mahabharata* and *Ramayana*, dwell upon the craft of weaving at length. India was also a major exporter of textiles to most parts of the civilised world. Presently, weaving is not just limited to textiles for clothing and home furnishing, but also plays an important role in the manufacture of technical textiles—i.e. speciality textiles used for their functionality rather than aesthetic value.

COMPONENTS OF WOVEN FABRIC

Warp Yarns or Ends

Warp yarns are the yarns running along the length of the fabric (Figure 6.1). They are parallel to the selvedge and constitute the lengthwise grain of the fabric. These yarns are very strong and stable, since they must withstand great tension during weaving. And that's why fabrics do not normally stretch when pulled in the lengthwise direction.

Weft Yarns, Filling Yarns or Picks

Weft yarns are the yarns running along the width of the fabric. They are at right angles to the selvedge (Figure 6.1). They constitute the crosswise grain of the fabric. These are usually slightly weaker than the warp yarns.

Figure 6.1: Components of woven fabric

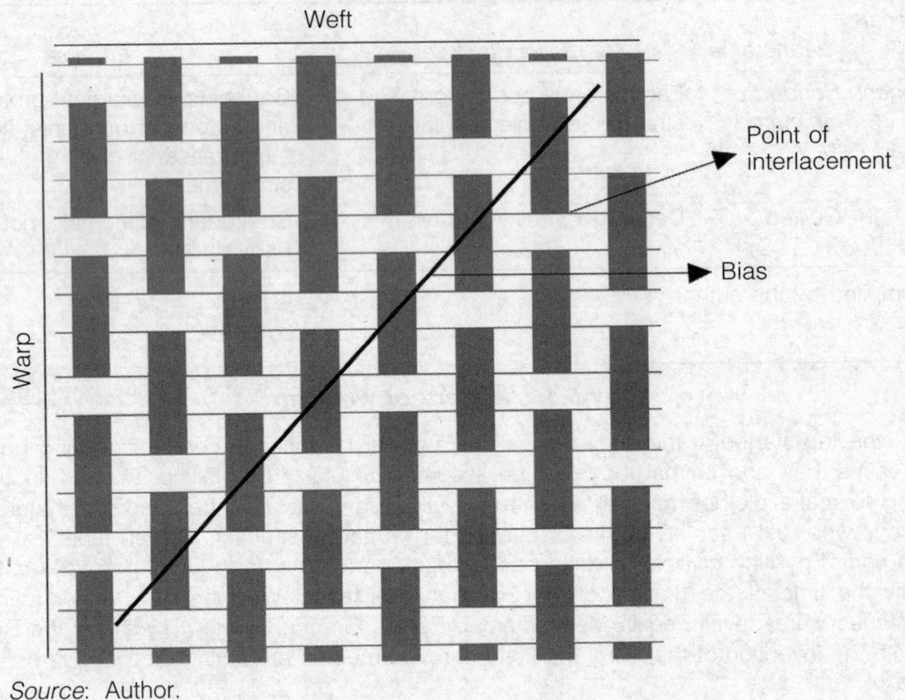

Source: Author.

Selvedge

Self-edge or selvedge is the narrow, flat woven border, or the finished edge, at both lengthwise sides of a fabric. The threads composing it are strong and densely woven. Selvedge prevents the fabric from ravelling and the edges from tearing when the fabric is under stress.

> The fraying of threads/yarns from the edges of a knitted/woven fabric is called **ravelling**.

Grain

Grain is determined by the direction of yarns in the fabric. Lengthwise grain is the direction along the length of the fabric, parallel to warp yarns and selvedge. Crosswise grain is the direction along the width of the fabric. It is parallel to the weft yarns and perpendicular to the warp yarns and selvedge. Grain is important as it decides the drape of the fabric. Garments that are made along the grain of the fabric have better drape and durability. For instance, when the fabric hangs along its lengthwise grain, it will have more body and drape.

Fabrics can be on-grain or off-grain. **On-grain fabric** has warp yarns parallel to the selvedge and perpendicular to filling yarns (Figure 6.2a). **Off-grain fabrics** do not have their lengthwise and crosswise yarns perpendicular to each other, thus creating problems in production and use (Figure 6.2b). They do not drape/hang properly and there is distortion in printed designs.

Figure 6.2: On-grain and off-grain fabric

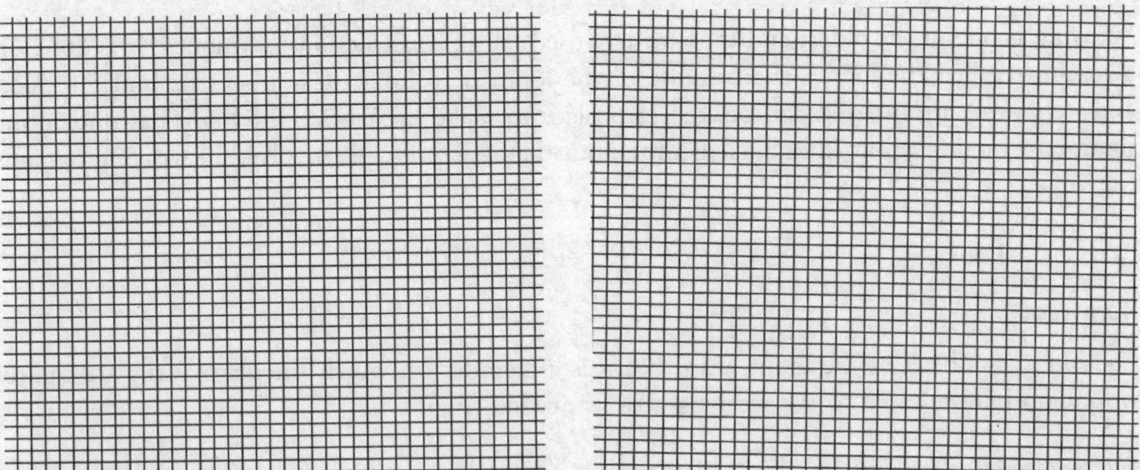

Source: Drawn by the author.

Thread Count/Cloth Count

Thread count is the number of warp and weft yarns per square inch of fabric. A closely-woven fabric has more yarns than a loosely-woven fabric, hence it is more durable. Thread count indicates the quality of fabric. The higher the count, greater will be the compactness of the fabric and better its quality.

Balance

Balance is the ratio of the warp yarns to weft yarns in a fabric. When their ratio is 1:1 (approximately), it is said to be a well-balanced fabric. For example, a muslin cloth with a thread count of 64 × 60 is considered well-balanced while a broad cloth with a thread count of 100 × 60 has poor balance. Good balance produces a fabric with good wearing qualities.

Key Characteristics of Woven Fabrics

- Woven fabrics have little stretch along warp or weft direction. These fabrics show maximum stretch along the bias direction.
- Their edges do not fray until cut, as they have a selvedge.
- They can be woven to different densities and have different weights.
- Woven fabrics are at their strongest on the grain line.
- Different interlacing patterns impart different designs and textures to the fabric.

MECHANISM OF WEAVING

Weaving is one of the oldest and most widely used fabric construction techniques. It is done on a machine called the loom. The purpose of the loom is to hold the warp yarns under tension and weft yarns are inserted and pushed into place to make the fabric. The loom has undergone significant modifications since ancient times, but the basic principle involved in weaving remains the same.

Parts of a Loom

Warp Beam

Located at the back of the loom, warp beam is a cylinder on which the warp yarns are tightly wound and are released to the weaving area as needed (Figure 6.3).

The Heddles/Healds

Heddles are wire or metal strips with an eyelet in the centre through which a warp yarn is threaded. The number of heddles is equal to the number of warp yarns in a cloth as only a single warp can be threaded through a heddle.

Figure 6.3: Parts of a loom

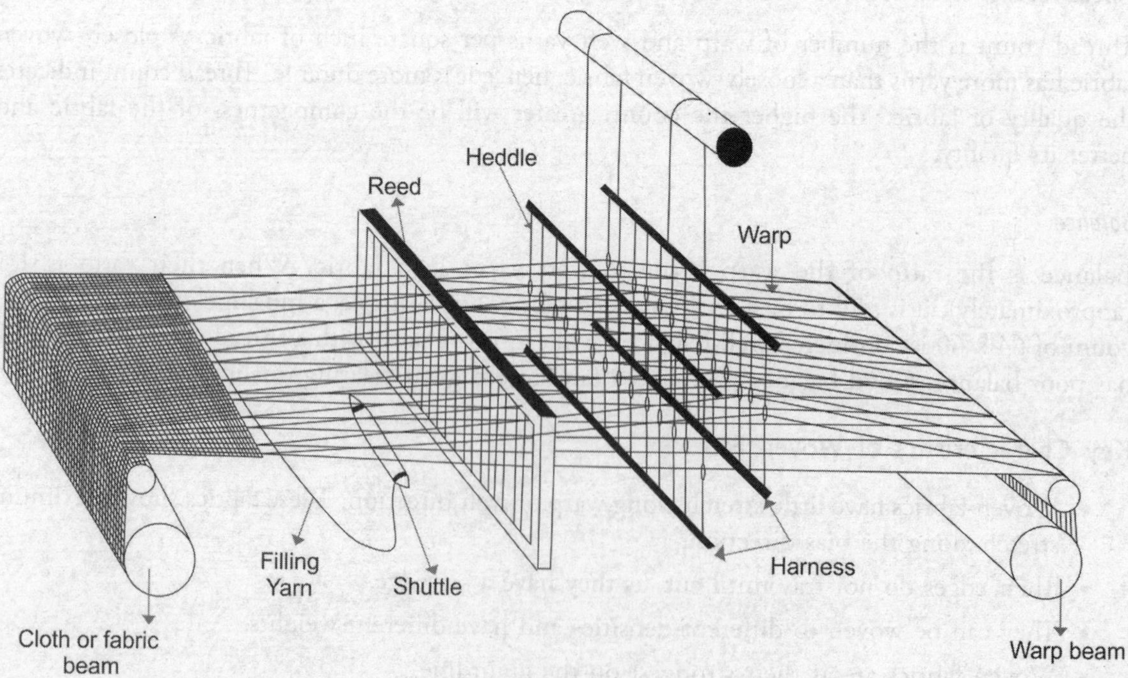

Source: Drawn by the author.

Harness

Harness is a frame to hold the heddles with the help of which a whole set of warp yarns can be raised or lowered at one time in order to produce the shed through which the filling yarn is passed (Image 6.1a). Each loom has at least two harnesses. The number of harnesses, the position of the harness and the number of heddles controlled by each harness determines the pattern of interlacement.

Image 6.1: Harness, shuttle and reed of a loom

(a) Harness

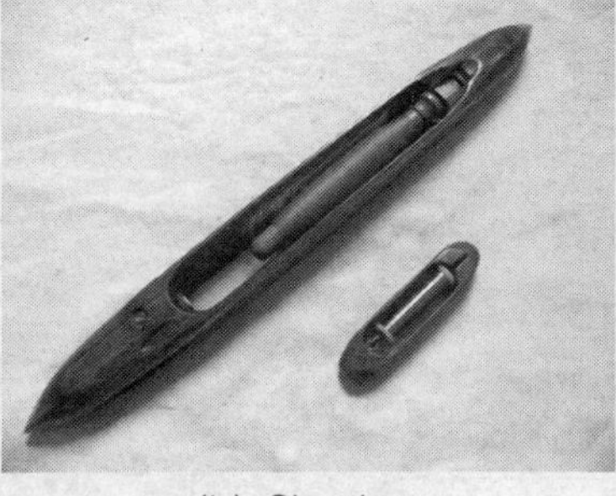

(b) Shuttle

(c) Reed

Source: 'Leeds Industrial Museum Hattersley standard loom healds 7047' by Clem Rutter (CC by 3.0 SA; Wikimedia Commons); 'Weaving shuttles' by Surya Prakash (CC by 3.0 SA; Wikimedia Commons); 'WeavingReed2' by Loggie-log (CC0 Public Domain; Wikimedia Commons).

Temple

As the selvedge is woven and the weft thread changes its direction, the tension of the weft tends to pull the warp threads together. To avoid this problem, temple is used. The temple provides an opposing tension on the cloth thus keeping the warp threads evenly spread out. A temple may be made of wood or metal.

Shuttle

Shuttle is a wooden boat-shaped device that holds the spools on which the weft yarns are wound. The weft yarn is unwound and laid across the fabric width as the shuttle travels back and forth through the shed/opening (Image 6.1b).

Reed

Reed is a comb-like device placed between the heddle and the cloth beam. It is responsible for pushing the filling yarns into place thus making the fabric firm (Image 6.1c). It also keeps the warp yarns separated on the loom and prevents them from entangling. Slots in between the wires in the reed are known as dents. After passing through the heddle, the warp yarns pass through the reed with each warp threaded through a dent/opening.

Cloth Beam

Cloth that has been woven rolls up onto a cylinder, located at the front of the loom. This cylinder is known as the cloth beam.

Yarn Preparation for Weaving

Before their use for fabric construction, the warp and filling yarns must be prepared for weaving (Figure 6.4). It is necessary to spin them to the desired size and give them the amount of twist required for the type of fabric for which they will be used. The yarns undergo the following stages before being woven into fabric.

Figure 6.4: Yarn preparation for weaving

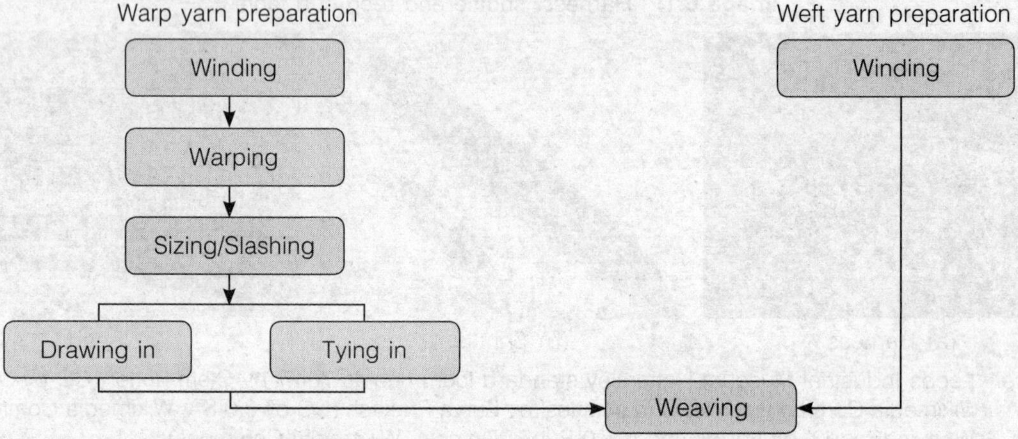

Source: Drawn by the author.

Winding

In the winding process, the yarns are rewound on large cones, tubes, etc. so that they can be used to weave a fabric on a specific loom. During this process the yarn is cleaned of any dust or fluff. This process is carried out for both warp and weft yarns. The weft cones are supplied directly to the weaving mill and the warp cones go through further processes.

Warping

The parallel winding of warp yarns from many winding packages (cheese, cones, etc.) onto a common package (warp beam) is called warping.

Sizing/Slashing

Sizing/slashing is a process where a film of thin sizing/stiffening agent is applied on the surface of the yarns to impart strength and smoothness which prevents them from breaking during the weaving process. The sizing agents generally used are PVA (polyvinyl alcohol), starch, acrylic esters, CMC, etc. This is done mainly on the warp yarns.

Drawing-in

After the sized yarns are wound on the final warp beam and placed on the loom, each warp yarn is threaded through the heddle eye and reed dents. Drawing-in is done when there is no previous warp on the loom or when the weaving pattern is different from that previously woven.

Tying-in

If the fabric to be woven has the same weaving pattern as the previous one, the new warps are tied into place by attaching them to the warps already on the loom. This process is called tying-in.

Basic Weaving Operation

Once the yarn preparation is done and both the warp and filling yarns are set in place, the loom goes through a series of motions to form the fabric. Regardless of the kind of loom or pattern to be woven, the basic weaving operation consists of the following steps:

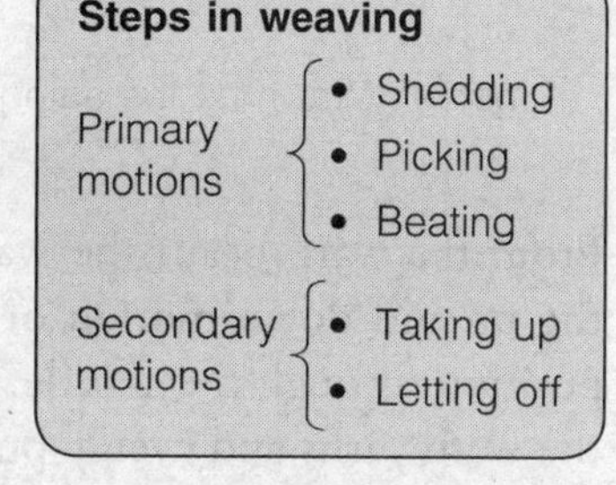

(i) *Shedding*: The raising and lowering of the harnesses creates a shed/opening between the warp yarns through which the filling yarn is inserted. This process is called shedding.

(ii) *Picking*: The insertion of the filling yarn into the shed is known as picking. A single crossing of the filling from one side of the loom to the other is known as a pick. Picking is done with the help of the shuttle or any other device which carries the weft yarn.

(iii) *Beating up (battening)*: After two to three picks have been laid by the shuttle, the newly-inserted filling yarns are pushed evenly into place against the newly-constructed cloth with the action of the reed. This makes the fabric firm and compact and the process is called beating up or battening.

(iv) *Taking up and Letting off*: As the fabric is formed, it is withdrawn from the weaving area at a constant rate and wound on or taken up on the cloth beam. At the same time, more warp yarns must be unwound or let off from the warp beam to the weaving area at the required rate. The two operations have to be co-ordinated to ensure that the weaving process proceeds smoothly without any breaks or change in tension in the yarns.

TYPES OF LOOMS

Depending on the mode of operation and method of weft insertion, there can be different types of looms. These have been briefly described below (Figure 6.5):

Handloom

The handloom (or hand shuttle loom) is the simplest type of loom used for weaving. Everything is carried out manually in this loom. The warp is wound on to the warp beam during warping.

Figure 6.5: Different types of loom

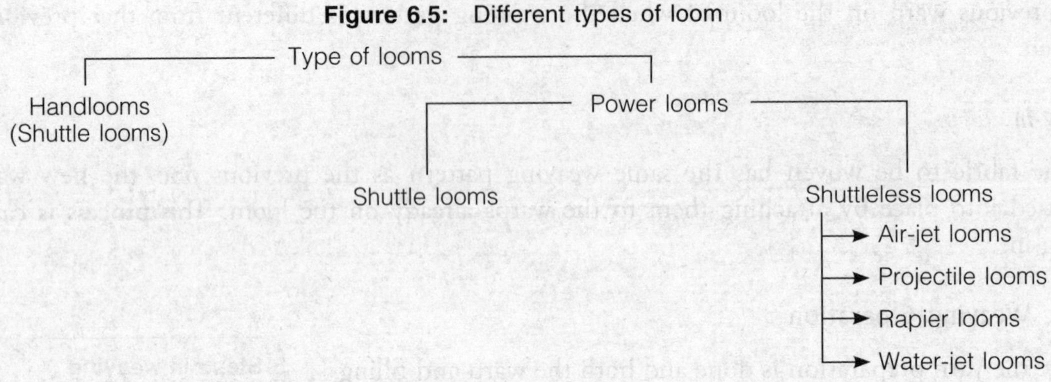

Source: Drawn by the author.

From the warp beam the warp yarn is made to pass through the heddle. Treadles activate the the up and down motion of the harnesses or shafts. The filling yarns are manually placed into position through the shuttle. The reed, which is also worked by the operator or weaver, separates the warp yarns and evenly pushes the filling yarns in place. The woven cloth is then taken up by the cloth beam. Handloom offers unparalleled flexibility and versatility, thus permitting experimentation and encouraging innovations which cannot be replicated by the power loom sector.

> **Treadles** are levers in a weaving loom that are operated by the foot to impart the reciprocating motion.

Backstrap Loom

A backstrap loom is a portable loom in which the warps are stretched from a fixed point up to the belt worn by the weaver. The device used to provide a method of separating the warp yarns into two parts is called a **rigid heddle**. Tension is created in the warp yarns by the back and forth movement of the weaver. The weaver moves the rigid heddle up, passes a weft thread through, then moves it down and again passes the weft thread through. Repetition of this process results in the formation of fabric. However, the fabric woven in this way is not too wide. This is an ancient technique mainly used by people in Northeast India, such as Manipur, Tripura, Arunachal Pradesh, Mizoram, etc.

Vertical Loom

The vertical loom, often referred to as the high-warp loom, is a permanent loom that cannot be taken apart or transported. The loom consists of two vertical posts connected by two strong, adjustable, horizontal cross beams. This type of loom is designed primarily for weaving rugs.

Power Loom

The power loom sector produces more than 60 per cent of cloth in India and accounts for more than 60 per cent of the textile exports of the country. It produces a wide variety of cloth, both

Box 6.2: Handloom industry in India

Handloom textiles constitute a timeless facet of the rich cultural heritage of India. The handloom sector is known for its excellent craftsmanship and is second only to agriculture in providing livelihood to millions. Over the centuries, handlooms have come to be associated with excellence in India's artistry due to the high quality of products like Chanderi silks of Madhya Pradesh, brocades of Varanasi, ikats of Odisha and Andhra Pradesh (Pochampalli), himroos of Hyderabad, khes of Punjab, baluchari of Bengal and jamdani saris of Uttar Pradesh, Phenek and Tongam of Assam and Manipur, maheshwari saris of Madhya Pradesh and the patola saris (double ikat) of Gujarat. Six states—West Bengal, Tamil Nadu, Uttar Pradesh, Andhra Pradesh, Assam and Manipur—produce about 75 per cent of the handloom output annually.

Advantages

1. The products are environment friendly.
2. Handlooms can be set up anywhere and anytime, with little need for specialised equipment.
3. They can be modified to produce wide and narrow pieces of cloth.
4. The strength of handloom lies in introducing designs which cannot be replicated by the power loom sector.
5. Unique and intricate designs can be produced. The latest fashion can be fused with tradition in handloom products; designer products made to specific order are possible.

Disadvantages

1. Lack of quality standardisation.
2. Unorganised structure.
3. Slow manufacturing process results in less supply in comparison to huge demand.

gray as well as processed. The major power loom weaving centres in India are Surat, Ahmedabad, Ichalkaranji, Malegaon, Sholapur, Bhiwandi, Bhilwara, Burhanpur, Salem, Kanpur and Amritsar. Production of cloth as well as generation of employment has been rapidly increasing in this sector. Power looms can be shuttle looms or shuttleless looms.

An unfinished woven fabric that is yet to be bleached, dyed or printed is known as a **gray/greige fabric** (or simply, **gray**).

Shuttle Loom

Shuttle loom is the oldest kind of loom that utilises a shuttle with a quill/bobbin to insert the filling yarns through the shed to produce the desired weave (Figure 6.7a). As the yarn is laid, the pick is battened in place, a new shed is formed and the shuttle returns to lay a second pick. This process is repeated until the fabric is complete. The following are some disadvantages of shuttle looms:

- High energy consumption
- Low speed and productivity

- More defects in fabric as the shuttle may lead to abrasion on the warp yarns and/or cause thread breaks
- More maintenance is required for the upkeep of the loom
- Deafening noise caused by the to and fro movement of the shuttle
- Small capacity of shuttle device

Shuttleless Loom

In order to overcome the inherent problems of shuttle looms, shuttleless looms have been developed which make use of entirely different methods of weft insertion. There are many kinds of shuttleless looms depending on the method employed for weft insertion: projectile looms, rapier looms, water jet looms and air jet looms (Figure 6.6).

Advantages of Shuttleless Looms

- Shuttleless looms are generally of wider widths enabling the simultaneous weaving of two or more widths.
- They are suitably designed to match the requirements of high insertion rates, shedding, beating and other auxiliary motions.
- Shuttleless weaving reduces the cost of production by almost 10 per cent in comparison to high-speed automatic shuttle looms.
- The quality of cloth obtained is far superior and is of a quality which is acceptable in the international market.
- Reduction in waste and down time.
- The machines are equipped with features like automatic pick finding and repairing.
- When patterned fabrics are woven on shuttleless looms, colours can be changed more easily.
- Lower power requirement.
- Lower sound levels and less space requirement.
- Higher speed of fabric production.

Types of Shuttleless Loom

Projectile or Missile or Gripper Loom

In the projectile loom, the weft insertion is carried out by bullet-like projectile that grips the filling yarn from the supply package and carries it across the shed (Figure 6.6b). The yarn is cut with a pair of scissors on the other side and the projectile comes back on a belt under the loom. The yarn may be inserted from one or both sides. The latter type is more common today. In a projectile machine with many projectiles, the projectiles work in sequence, i.e. they are launched

Figure 6.6: Weft insertion mechanisms in shuttle and shuttleless looms

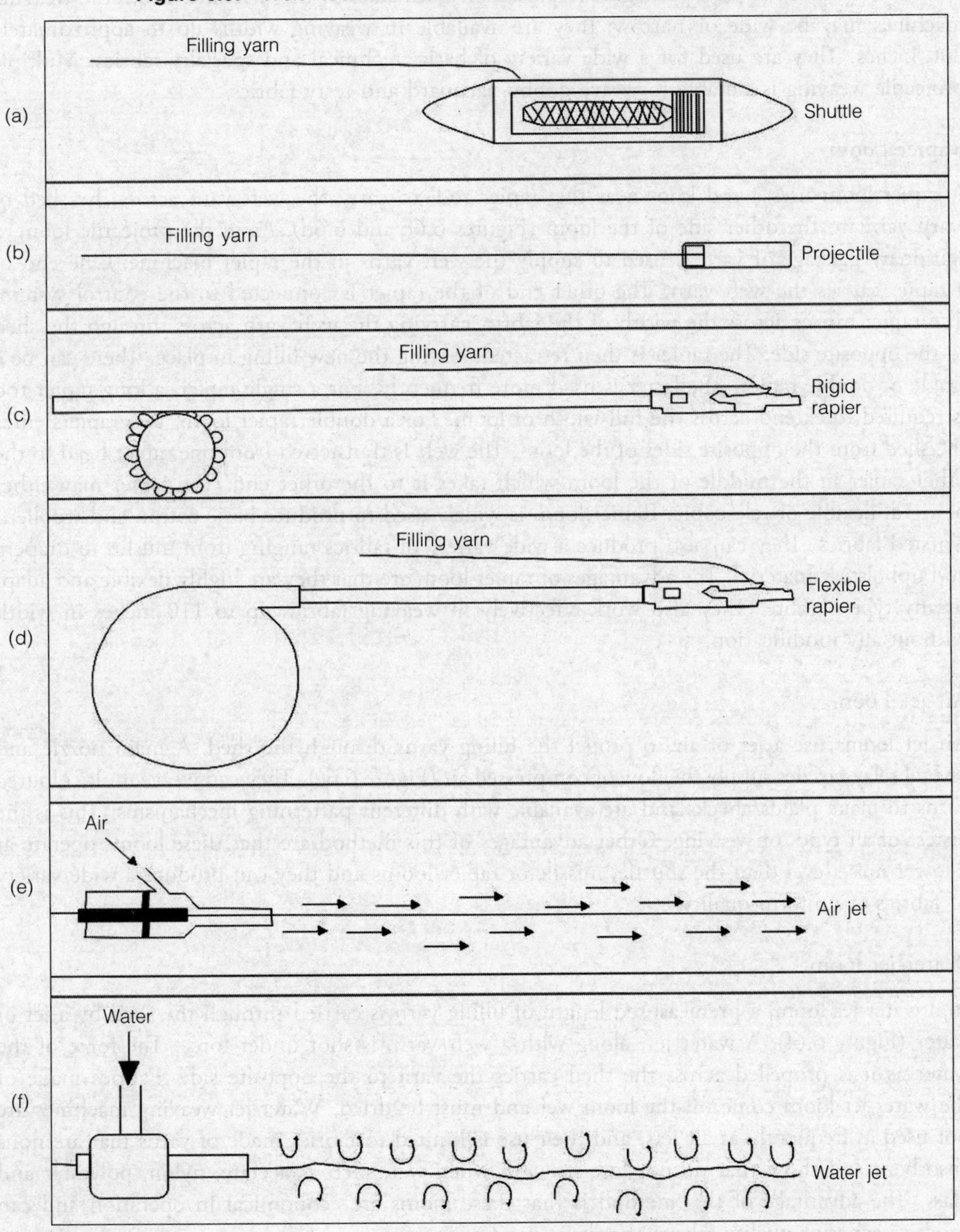

Source: Drawn by the author.

in succession and the yarn packages are placed on both sides of the machine. Projectile weaving machines may be wide or narrow; they are available in weaving widths up to approximately 200 inches. They are used for a wide variety of basic, technical and specialty fabrics. Multiple projectile weaving is suitable to weave dobby, jacquard and terry fabrics.

Rapier Loom

A rapier loom uses a rod known as the 'rapier rod' to carry the weft yarn across the shed of warp yarns to the other side of the loom (Figures 6.6c and 6.6d). As in the projectile loom, a stationary package of yarn is used to supply the weft yarns in the rapier machine. One end of a rapier carries the weft yarn. The other end of the rapier is connected to the control system. The rapier moves across the width of the fabric, carrying the weft yarn across through the shed to the opposite side. The rapier is then retracted, leaving the new filling in place. There can be a single or double rapier. The latter is used more frequently. For a single rapier, a long rapier rod is required to extend across the full width of loom. For a double rapier loom, two rapiers enter the shed from the opposite sides of the loom. The weft is transferred from one rapier head to the other rapier in the middle of the loom which takes it to the other end. The rapier may either be rigid, flexible or telescopic. Rapier loom is widely used to produce basic cotton and woollen/worsted fabrics. They can also produce a wide variety of fabrics ranging from muslin to drapery and upholstery material. The advantages of rapier loom are that they are highly flexible and adapt to any type of fibre. They also work effectively in weaving fabrics up to 110 inches in width without any modification.

Air-jet Loom

Air-jet looms use a jet of air to propel the filling yarns through the shed. A main nozzle and several relay nozzles supply the flow of compressed air (Figure 6.6e). They can weave multicoloured yarns to make plaids/checks and are available with different patterning mechanisms. This is the fastest of all types of weaving. Other advantages of this method are that these looms operate at a lower noise level than the shuttle, missile or rapier looms and they can produce a wide variety of fabrics of uniform quality.

Water-jet loom

In a water-jet loom, a premeasured length of filling yarn is carried through the shed by a jet of water (Figure 6.6f). A water jet, along with a weft yarn, is shot under force. The force of the water as it is propelled across the shed carries the yarn to the opposite side. Fabrics made of the water-jet loom come off the loom wet and must be dried. Water-jet weaving machines are not used as frequently as air jets, and their use is limited to fabrics made of yarns that are non-absorbent and those that do not lose strength when wet, such as acetate, nylon, polyester and glass. The advantage of this method is that these looms are economical in operation and can produce superior quality fabrics that have good appearance and feel.

Circular Loom

In this kind of loom a shuttle device circulates the weft in a shed formed around the machine. This method is used to weave tubular fabrics. The circular loom is primarily used for bagging material and other products include stockings, pillowcases, etc.

Triaxial Loom

In triaxial weaving, three sets of yarns (usually identical in size and twist) are woven at an angle of 60° in three directions—horizontally, vertically and on the bias. Two sets of yarns are warps and one set is weft. The fabric produced is strong, stable and doesn't crimp easily. These are mainly used for hats, outerwear apparel and in high-performance applications like air structures, sailcloth, diaphragms, truck covers, etc. The advantages of this loom are quicker production, no slippage of yarns, control of strength and stiffness in all directions and fabrics that are light in weight and isotropic.

Multiphase/Multished Loom

In all the looms that we have discussed so far in this chapter, only one shed is formed. And this shed remains open for the device which carries the filling yarn through it. The speed of the loom is limited due to the presence of only one shed. This limitation is overcome in multiphase loom in which more than one shed is formed at a time at different places along the length of the warp yarns and multiple weft yarns are inserted through them. As soon as the weft yarn is inserted, the shed closes and another shed opens for the next weft yarn. Thus the number of picks/minute (ppm) is much higher.

TYPES OF SELVEDGES

As mentioned earlier, the selvedge of the fabric is the narrow, flat woven border or the finished edge resulting at both the lengthwise sides when the crosswise threads reverse direction. There are different types of selvedges depending upon the expected use of the fabric and the type of loom used (Figure 6.7).

Plain Selvedge

Plain selvedge is made on a shuttle loom from plain weave (Figure 6.7a). The filling yarns travel back and forth creating a closed edge. The size of the warp yarns forming the selvedge is the same as the rest of the fabric, but they are more compactly packed at the edges. Such selvedges do not shrink and are durable.

Tape Selvedge

Tape selvedge is made up of plied yarns and is therefore much stronger. It is broader than the plain selvedge and is often made of a different weave for a flatter edge.

Figure 6.7: Types of selvedges

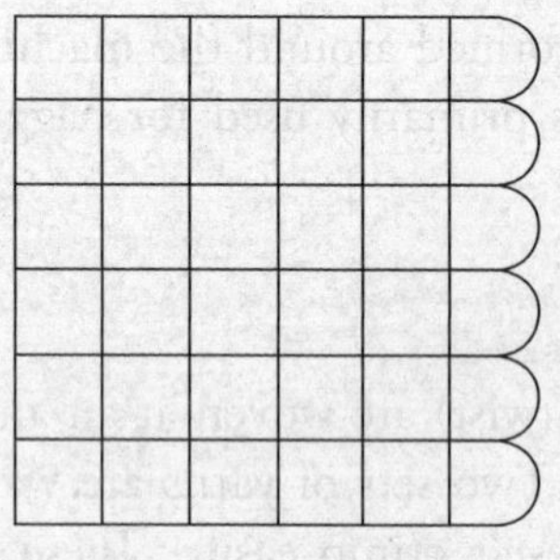

(a) Plain selvedge

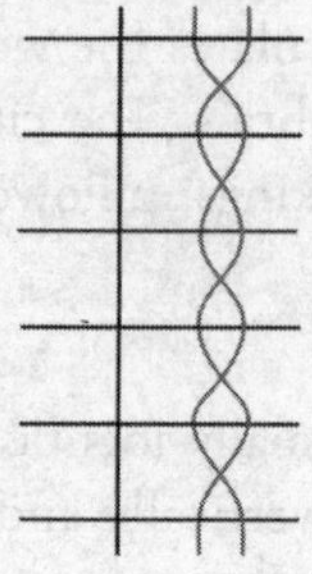

(b) Leno selvedge

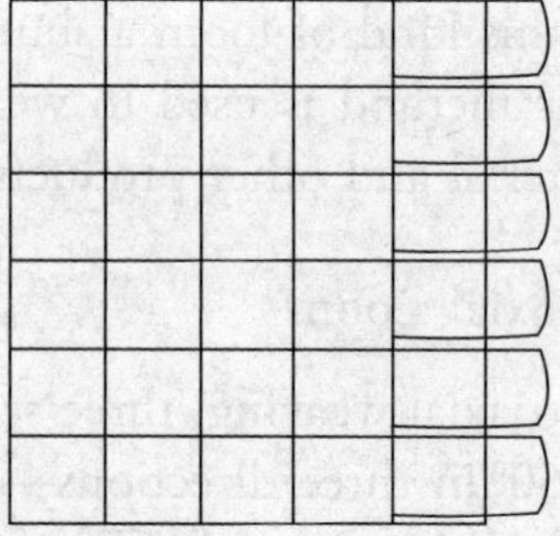

(c) Tucked selvedge

Source: Drawn by the author.

Split Selvedge

Split selvedges are produced when two fabrics are woven simultaneously. After weaving is complete, the fabrics are cut apart between the selvedges and the cut edges are finished with hemming.

In shuttleless looms, the filling yarns are cut resulting in fringed selvedges. Therefore, it is necessary to reinforce the edges to avoid fraying of fabric. This can be achieved through leno and tucked selvedges.

Leno Selvedge

Leno selvedge is formed by using the leno weave in which the warp yarns are paired and are intertwisted in such a manner that they encircle each filling yarn (Figure 6.7b).

Tucked Selvedge

For this type of selvedge, a device is used to tuck the yarns at the open edge of the fabric. Tucked selvedge is more durable (Figure 6.7c).

Fused Selvedge

Fused selvedge is used for heavy industrial fabrics made of thermoplastic fibres. The edges of the fabric are heated due to which the fibres melt and fuse together, sealing the edges.

WEAVES

Fabrics are manufactured in a wide variety of designs that are formed by different types of interlacements of the warp and filling yarns. The designs that are formed can range from simple to very complex and are known as weave. In other words, weave is the pattern in which the fabric is woven. Weaves can be broadly classified into basic and decorative weaves which can be further divided into many types (Figure 6.8). Most of the fabrics are created using one of the three basic weaves—plain, twill and satin. A combination of these with other decorative and surface figure weaves is used to create fancy fabrics. The drape of the fabric, its smoothness, stability and many other properties largely depend upon the weave style. Different types of weaves impart different characteristics to the fabric which dictate its end use.

Basic Weaves

Basic weaves use simple interlacement patterns for making fabrics. They can be one of the following three types:

Plain Weave

Plain weave is the simplest and most common of the three basic weaves. It requires a simple loom with two harnesses—one filling yarn goes alternately under and over one warp yarn at right angle.

Figure 6.8: Classification of weaves

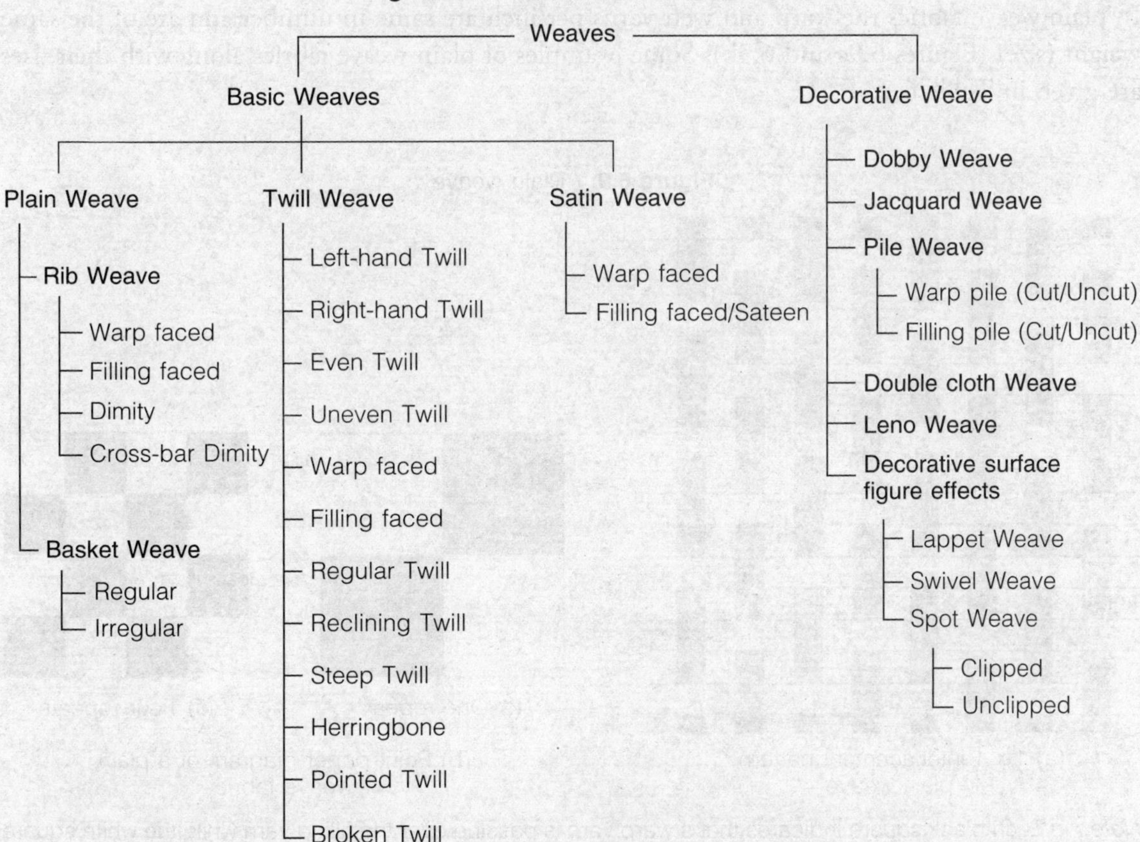

Source: Compiled by the authors.

It is the least expensive to produce and does not have a technical face or back unless printed or given a surface finish. The following are some characteristics of plain weaves:

- Plain weave is strong.
- It has a smooth surface.
- It is conducive to printing and other finishes because of the smooth surface.
- It wears well.
- It ravels less.
- It is prone to wrinkling.
- It is less absorbent than other weaves.
- It can be made with any type of yarn but the yarn type greatly influences the characteristics discussed above.

Two variations of plain weave are: Rib weave and Basket weave.

In plain weave fabrics the warp and weft yarns per inch are same in number and are of the same weight (size) (Figures 6.9a and 6.9b). Some examples of plain weave fabrics along with their uses are given in Table 6.2.

Figure 6.9: Plain weave

(a) 1 x 1 interlacement pattern in plain weave

(i) One repeat

(ii) Four repeats

(b) Point paper diagram of a plain weave fabric

Note: Each black square indicates that a warp yarn is passing over the filling yarn while the white square represents a filling yarn passing over the warp yarn.

Source: Drawn by the author.

Table 6.2: Some examples of plain weave fabrics

Light weight			*Medium weight*			*Heavy weight*		
They are constructed of fine yarns and are usually sheer			*They contain medium weight yarns and are opaque*			*They contain heavy-weight yarns, and hence are more durable and more resistant to wrinkling*		
Name of the fabric	*Characteristics*	*Uses*	*Name of the fabric*	*Characteristics*	*Uses*	*Name of the fabric*	*Characteristics*	*Uses*
Cheese cloth	Loosely woven fabric	Cheese-making, straining stocks, bundling herbs	*Calico*	Closely woven fabric with a small printed design	Sample clothing, fabric bags, curtains, pillowcases	*Butcher linen*	Linen-like plain stiff fabric made from heavy yarn	Butcher's apron
Crinoline and buckram	Heavily sized	Stiffen clothes, cover and protect books	*Chambray*	It combines coloured warp yarns with white filling yarns	Spring and summer clothing	*Crash*	Made from thick and thin yarns giving the fabric a nubby look	Toweling, dresses, caps, sport coats
Gauze	Thin fabric with loose open weave	Upholstery, theatrical costumes, medical dressings, book binding	*Chintz*	Fabric printed with large designs that is often given a polished or glazed finish	Shirts, dresses, blouses, aprons	*Homespun*	Furnishing fabric made with irregular yarns	Furnishing fabric
Chiffon	Light weight sheer fabric made from fine, highly twisted filament yarns	Eveningwear, blouses, saris	*Gingham*	Yarn-dyed fabrics with check or plaid design. They have no right or wrong side with respect to colour	Shirts and jackets	*Osnaburg*	Variable-weight fabric, bleached or unbleached, made with lower quality cotton	Industrial use, drapery lining, apparel and upholstery

(*Contd.*)

Table 6.2: (*Contd.*)

Light weight			*Medium weight*			*Heavy weight*		
They are constructed of fine yarns and are usually sheer			*They contain medium weight yarns and are opaque*			*They contain heavy-weight yarns, and hence are more durable and more resistant to wrinkling*		
Name of the fabric	*Characteristics*	*Uses*	*Name of the fabric*	*Characteristics*	*Uses*	*Name of the fabric*	*Characteristics*	*Uses*
Georgette	Made with highly-twisted yarns, it has a crinkly surface	Eveningwear, blouses, dresses, trimmings	*Taffeta*	Crisp, smooth fabric that can be yarn-dyed or piece-dyed	Ball gowns, wedding dresses, wall coverings	*Tweed*	Made of any one fibre or blend of fibres and is characterised by nubs of different colours	Informal outerwear
Organdy	It is the sheerest and crispest cotton cloth due to an acid finish imparted to it	Dresses and curtains	*Muslin*	Generally made with cotton or cotton blends	Sheets and pillowcases	*Flannel*	Soft woven fabric of varying fineness	Sleepwear, blankets, jackets, skirts
Batiste	It is the softest of the light-weight opaque fabrics	Handkerchiefs and lingerie	*Percale*	Made from both carded or combed yarns of moderate twist; it is firm and smooth	Bed linen			
Organza	Sheer fabric made of filament yarns	Bridal wear and eveningwear, sheer curtains						
Voile	Semitransparent, light-weight fabric with good drapability, made with high-twist yarns	Window dressing, lining materials						
Ninon	Filament yarn fabric	Sheer curtains						

Source: Compiled by the author.

Rib Weave

A rib weave is characterised by ribbed or corded effect on the surface of the fabric. The rib effect can be produced by using heavy yarns or groups of yarns interlacing as a unit, in the warp or filling direction (Figure 6.10). When the heavier yarns are used in the warp direction it is known as the **warp-wise rib** (Figure 6.11a); when heavier yarns run in the weft direction it is known as **weft-wise rib** (Figure 6.11b). Generally, the heavier yarns are in the weft direction as these yarns have lower twist and are therefore weaker in strength. Weft-wise rib fabrics are therefore more durable than filling-wise rib fabrics.

Figure 6.10: Weft faced and warp faced rib

(a) Warp wise (weft-faced) rib weave fabric

(b) Filling wise (warp-faced) rib weave fabric

Source: Drawn by the author.

Dimity is a variation of the Rib weave. It has a special weave pattern that is characterised by ribs of heavier yarns or a group of yarns running either warp-wise or weft-wise at regular intervals (Figures 6.11a.1 and 6.11a.2). **Cross bar dimity** is characterised by ribs in both directions at regular intervals (Figures 6.11b.1 and 6.11b.2). Both dimity and cross bar dimity can be printed or left plain.

Slippage is a problem in ribbed fabrics made with filament yarns. Also, in ribbed fabrics, an entire yarn is exposed to friction, thus making the fabric less durable. Fabrics with fine ribs are smoother and softer and have greater drapability. Those with large ribs have more body and less drapability and are good for garments where a bouffant look is desired.

Some examples of rib weave fabrics are given in Table 6.3.

Figure 6.11: Dimity

1

2

(a) Dimity in warp direction

1

2

(b) Cross-bar dimity

Source: Drawn by the author.

Basket Weave

Basket weave is a loosely-woven fabric in which two or more warp yarns are interlaced as a unit with one or more filling yarns producing a basket effect. In a basket weave the number of warp yarns interlacing as a unit are almost similar in size to the number of weft yarns interlacing as a unit. This differentiates them from the rib fabrics. When the groups of warp and weft yarns are equal in number, the basket weave is termed **regular**, otherwise it is termed **irregular**. Basket weave produces a fabric that is less firm and weaker than the regular plain weave or the rib weave. These fabrics are flexible and more resistant to wrinkles because of their loose open weave. However, garments made from these fabrics may snag easily, pill readily and stretch at points of body strain. The most common regular basket weaves are 2×2 (Figure 6.12a), 4×4, 8×8, etc., while irregular basket weaves include 2×1, 2×3, etc. (Figure 6.12b). Some examples of basket weave fabric types are given in Table 6.4.

Table 6.3: Some examples of rib weave fabrics

Light weight			*Medium weight*			*Heavy weight*		
Name of the fabric	*Characteristics*	*Uses*	*Name of the fabric*	*Characteristics*	*Uses*	*Name of the fabric*	*Characteristics*	*Uses*
Dimity	Sheer, light-weight cotton fabric with corded effect often in warp-wise direction	Curtains, blouses, dresses	*Broadcloth*	It has the finest rib because the warp and filling yarns are the same size	Shirts and blouses	*Bengaline*	It is similar to faille but has a slightly more pronounced rib	Home furnishings
Crepe de chine	It does not have a noticeable rib but drapes beautifully and has a dry, pleasant handle	Clothing items	*Faille*	It has a fine rib with filament warp yarns and spun filling yarns	Formal dresses, gowns, jackets, vests, skirts, drapes, upholstery	*Ottoman*	It has a pronounced ribbed or corded effect	Formal dress
			Poplin	It is similar to broadcloth but the ribs are heavier and more pronounced because of larger filling yarns	Dresses, upholstery	*Rep*	It is a heavy coarse fabric with a pronounced rib	Dresses, neckties, upholstery
			Taffeta	It is a fine-ribbed filament yarn fabric with a crisp handle	Evening wear	*Grosgrain*	It has very prominent ribs and is usually woven in narrow widths	Ribbons
			Shantung	It has an irregular ribbed surface produced by long irregular areas in filling yarns	Bridal gowns			

Source: Compiled by the author.

Figure 6.12: Regular and irregular basket weave

(a) 2×2 basket weave

(b) 2×1 basket weave

Source: Drawn by the author.

Table 6.4: Some examples of basket weave fabrics

Name of the fabric	*Characteristics*	*Uses*
Monk's cloth	It is an even weave fabric usually found in square counts: 2×2, 3×3, 4×4 or 6×6	Furnishings, pillows, wall hangings
Oxford cloth	It is lustrous, medium-weight, soft basket weave fabric	Casual or sporty styles of dress shirts
Shepherd's check	It has a pattern of small even black-and-white checks	Clothing items
Sailcloth, duck and canvas	Sailcloth is the lightest of the three. Canvas is the heaviest, more compact and smoother, while duck is coarser	Sail cloth, sportswear, upholstery, draperies, boat covers, awnings
Hopsacking	A loosely-woven coarse fabric of spun yarns	Upholstery, wall coverings, bags, clothing

Twill Weave

In the twill weave the filling yarn passes under or over two to four warp yarns in progressive steps right or left, thus creating a diagonal pattern on both sides of the fabric. The direction of the diagonal is opposite on the reverse side of the fabric. The diagonals formed may vary in angle, direction and can be even or uneven.

Method of Construction

Twill weave can be made on a simple loom. The smallest repeat of the twill weave uses three picks and three warp ends. At least three harnesses are required on looms for twill weave and the upper limit is not fixed but restricted as unreasonably long floats pose practical problems. Tightly-twisted yarns are used for making twill weave fabrics.

Twill weaves are denoted as fractions such as 2/1, 2/2, 3/1, 3/2 and so on where the numerator indicates the number of harnesses that are raised and the denominator indicates the number of harnesses that are lowered when a filling yarn is inserted. By adding the numerator and the denominator, we get the total number of harnesses required for the weave. Some examples of twill weave fabric types are given in Table 6.5.

Types of Twill Weave

When the diagonal moves up from the lower left hand side to the upper right of the fabric, it is a **right hand twill** (Figure 6.13a). When the diagonal moves down from the upper left hand side towards the lower right, it is a **left hand twill** (Figure 6.13b). Majority of twill weave fabrics are right hand twills.

Table 6.5: Some examples of twill weave fabrics

Name of the fabric	*Characteristics*	*Uses*
Serge	It has diagonal lines on both the face and back of the fabric	Worsted serge: military uniforms, men's suits; Silk serge: linings
Denim	It is a durable heavy-weight twill in which only the warp threads are dyed and the filling yarns remain white	Jeans, jackets, overalls, shorts, belts, handbags, upholstery
Surah	An even twill fabric, often printed, made with silk or rayon	Dresses, blouses, scarves, ties
Gabardine	It is a tightly-woven warp-faced steep or regular twill with a prominent diagonal rib on the face and smooth surface on the back	Suits, trousers, overcoats, slacks, uniforms
Drill	It is a warp-faced twill with a strong bias in the weave, durable and versatile	Sails for sailing craft, sports clothing, shirts, work clothes, uniforms
Jean	Lighter weight twill that has coloured warps and white filling yarns but it is not woven the same as denim	Jeans and jackets
Tweed	Rough unfinished woollen fabric, closely woven, usually has a herringbone pattern	Informal outerwear, men's coats, backing in musical instruments
Khaki	Closely-woven twill fabric mainly of cotton	Invariably used for uniforms
Venetian cloth	Worsted fabric in twill weave	Light-weight coats, suits, skirts and dresses

Note: Other examples of fabrics made with twill weave include canton flannel, covert cloth, coutil, outing flannel, silesia, ticking, whipcord, sharkskin, houndstooth, cashmere, cheviot, chino, cavalry twill.

Source: Compiled by the author.

Figure 6.13: Types of twill weave

(a) 2/1 right hand twill

(b) 2/1 left hand twill

Source: Drawn by the author.

In **even twill**, equal number of filling and warp yarns pass over and under each other. Thus, it can be considered reversible unless printed or finished on one side. In **uneven twill**, the filling yarn goes over either more or fewer warps than it goes under. As the number of warp and filling yarns differ, uneven twills have a right and a wrong side and are not considered reversible. When more filling yarns than warp yarns are evident on the face of the fabric, it is called a **filling-faced twill**, e.g. 1/2, 2/3 twill. When warp yarns predominate on the face of the fabric, it is called a **warp-faced twill**, e.g. 2/1, 4/3 twill. When the diagonal makes an angle which is less than 45°, it is called *reclining twill.* If the angle is greater than 45°, it is called **steep twill** and if the angle is 45°, the twill is called **regular twill** (Figure 6.14). The steeper the twill the stronger the fabric is likely to be.

Figure 6.14: Varying twill angles

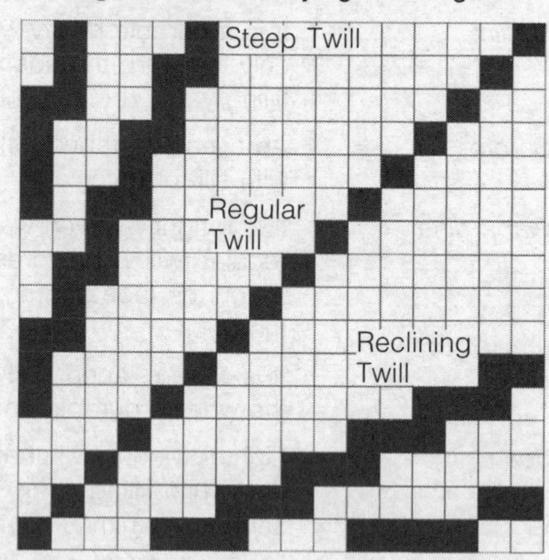

Source: Drawn by the author.

Characteristics

- Twill weave is characterised by a distinctive diagonal line on the face and often on the back of the fabric.
- Twill weave fabrics are more closely woven, heavier and have an attractive appearance.
- They are strong and durable.
- They have good drapability.
- They are pliable and resilient.

- They do not get soiled easily.
- The floats are short, and therefore snagging is not a problem.

Variations of Twill Weave

The direction of twill reverses at predetermined intervals to form a series of inverted Vs. This can make two types of designs—herringbone (Figures 6.15a.1 and 6.15a.2) and pointed twill (Figure 6.15b.1 and 6.15b.2). This design is common in suiting fabrics. Another variation of twill weave is broken twill.

> **Serge** is a type of even twill fabric. **Silk serge** is a twill silk fabric mostly used for lining parts of gentlemen's coat.

Satin Weave

In satin weave fabric there is a complex arrangement of warp and weft threads, which allows longer floats to be formed either across the warp or the weft. Due to long floats, the light falling on the yarn doesn't scatter and

> **Sateen** is usually made from spun yarns. Low- to medium-twist yarns are used to increase lustre.

Figure 6.15: Variations of twill weave

1  2

(a) Herringbone

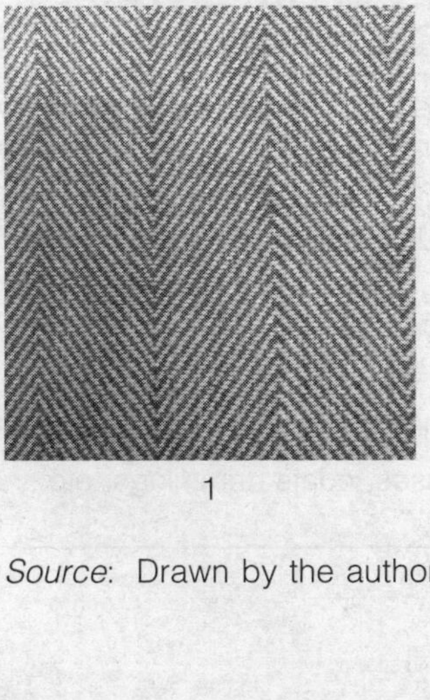

1  2

(b) Pointed twill

Source: Drawn by the author.

break up, thus giving the fabric a shiny appearance. Satin weave fabrics may be warp-faced or filling-faced. When warp ends float on the surface, it is called **warp-faced satin** (Figure 6.16a). When the filling yarns float on the surface of the fabric, it is called **filling-faced satin** or sateen (Figure 6.16b). Most satin fabrics are warp-faced. Some examples of satin weave fabrics are given in Table 6.6.

Method of Construction

Made on a simple loom, the simplest satin fabric requires at least five harnesses to form the repeat (5 shaft satin). Depending upon the complexity of the design, the number

> **Crepe-back satin**, a type of warp-faced satin, is created by combining highly-twisted yarns in the filling with loosely-twisted warp yarns. Since the back of the fabric is largely made up of filling yarns, it produces a crepe-like rougher surface.

Figure 6.16: Warp and weft satin

(a) 5 shaft warp satin

(b) 5 shaft weft sateen

Source: Drawn by the author.

Table 6.6: Some examples of satin weave fabric types

Name of the fabric	*Characteristics*	*Uses*
Damask	It is a reversible figured fabric with the pattern usually in warp-faced satin weave and the ground in sateen weave	Table linen, furnishing fabrics, clothing
Ticking	A very strong, tightly-woven fabric with more warp threads than filling	Used mostly for bedding
Antique satin	It is a reversible satin weave fabric that often uses slub-filling yarn for decorative effect	Drapery fabrics, upholstery and formal wear
Slipper satin	It is a stiff, heavy-weight fabric	Evening shoes
Peau de soie	A French term meaning 'skin of silk' is a soft closely-woven satin with a mellow lustre	Dresses, coats, trimmings, etc.

Source: Compiled by the author.

of harnesses may increase. Filament fibre yarns are generally used for making satin fabrics. The construction of a 10-shaft filling-faced satin can help us understand the making of satin weave fabrics. The number of shafts required for this purpose is 10. Numbers that will add up to the desired shaft number are paired in this manner: 1 and 9; 2 and 8; 3 and 7; 4 and 6; 5 and 5. The pair that contains number 1 and the number below the shaft number, which is 9, is eliminated. (If these pairs are used, a continuous diagonal would result, producing the conventional twill weave.) Then the pairs that have a common divisor and also those that are factors of the shaft number are eliminated (2 and 8, 4 and 6, 5 and 5). So the only interval that can be used is the pair 3 and 7. Figure 6.17a shows vertical rows of squares that represent the warp yarns and horizontal rows of squares that represent the filling yarns. The black squares indicate the point at which the filling yarns interlace the warp yarns on each successive pick. For constructing the design using the pair 3 and 7, the interlacing on the first pick will be nine squares warp yarns apart at A and B. To find the warp yarn that will interlace on the second pick, 7 is counted to the right and 3 to the left as shown in Figure 6.17b. The second interlacement occurs at point C. Similarly, the third, fourth, fifth...pick can be found and the design will start to repeat at the eleventh pick which ensures that the construction is correct.

Characteristics

- Satin fabrics are characterised by long floats on the face of the fabric which give it a shiny appearance and are selected primarily for their appearance and smoothness.
- The reverse side is invariably dull and non-shiny.
- Satin fabrics have a definite face and back.
- Satin fabrics are flat, smooth and slippery, have a shiny surface, excellent drapability and tend to reflect light easily.

Figure 6.17: Construction of a 10-shaft sateen

(a)

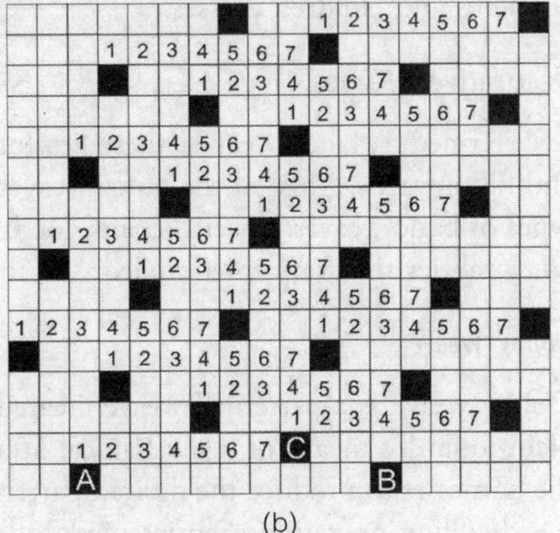

(b)

Source: Drawn by the author.

- The float length determines the strength of the fabric. The longer the float, the greater the chance that the fabric will snag and abrade easily. So, satin fabrics are not as durable as plain or twill weave fabrics.

Novelty Fabrics from Basic Weaves

Crepe fabrics: Crepe fabrics are mostly made in plain weave and some in satin weave (crepe-back satin). These are characterised by a rough and pebbly surface. Most of the crepe fabrics are made from highly-twisted or textured yarns. Crepe effect is also achieved through embossing finishes on the surface of the fabric. Properties such as drapability, durability, etc. of the crepe fabrics are largely determined by the type of yarns used. Both spun and filament yarns can be used to make crepe fabrics. Some examples of crepe fabrics are granite or momie cloth, sand crepe, moss crepe, all of which are mainly used for women's dresses and blouses.

Seersucker: Seersucker is a plain weave fabric with permanent surface effect. To make this fabric, two warp beams are used. The yarns on one beam are held at tight tension and the yarns on the other are at slack tension. Often, these yarns are of different colours in order to provide a decorative striped effect to the fabric (Image 6.2). When the filling yarns are beaten up, the warp yarns under slack tension crinkle to form the puckered stripe and the tightly tensioned yarns form the flat stripe. Seersucker is mainly used for men's, women's and children dresses.

Image 6.2: Seersucker fabric

Source: 'Seersucker02closeup' by Peteski1 (CC by 3.0 SA; Wikimedia Commons).

Decorative Weaves

Also termed as fancy weaves, these are formed by changing the interlacing pattern of the warp and filling yarns. Most of the fancy weaves are formed by a combination of at least two different types of basic weaves. Special looms, loom attachments or computer devices are used to produce fancy fabrics that vary in complexity of their designs.

Dobby Weave

Dobby weave is a patterned weave created by a combination of two or more basic weaves on a plain loom by means of a mechanical attachment, called dobby or cam, which raises or lowers the harnesses to produce the desired pattern. As many as 24–40 harnesses may be used to control the shedding operation. Fabrics made of this weave are used for shirtings for men, tie fabrics, women's apparel. The characteristics of this weave are:

- Dobby weave has small, woven designs that occur repeatedly, such as dots, geometric and floral patterns (Image 6.3).
- The weave is fairly inexpensive to produce
- The most familiar type of dobby weave is bird's eye, the small diamond pattern with a dot in the centre that gives the impression of an eye.
- Other fabric types include huck/huckaback, pique, waffle cloth, madras gingham and granite cloth.

Image 6.3: Dobby weave fabric

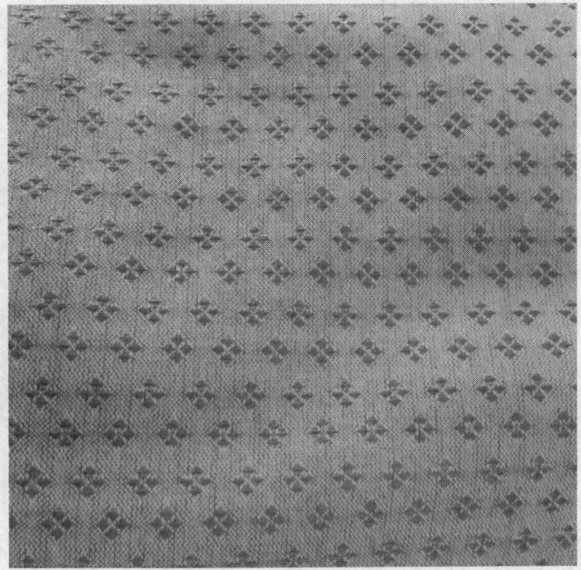

Source: Author.

Jacquard Weave

The jacquard weave is produced by the combination of plain, twill and satin weaves using the jacquard loom (Image 6.4), named after its developer Joseph Marie Jacquard. Traditionally, jacquard designs were created using a set of pattern cards. There are no harnesses in a jacquard loom; each warp is individually controlled by the needle to which it is attached. The desired design is transferred to a series of perforated cards that are laced together and placed on the jacquard attachment. The perforations on the cards allow the needles to determine which warp yarns are to be raised in forming the shed. The needles which encounter the perforations in the card raise those warps which are tied to them while the others remain down. When the shed is formed the filling yarn is inserted in place, the card moves on and a new card take its place. This process continues until all the cards are used and a repeat is formed. Many repeats are formed as per the requirement of the particular design. Of late, the only difference is that the cards have been replaced by computer tapes which determine which warp yarns are to be raised, with the other processes remaining the same. The use of various kinds of graphic software helps create

Image 6.4: Jacquard fabric

Source: Author.

complicated patterns in a much shorter duration. Both spun and filament yarns are used to produce jacquard fabrics, examples of which include damask, tapestry, brocade (Image 6.5), brocatelle, matelasse and tapestry. Jacquard fabrics are characterised by intricate designs. They are some of the most expensive forms of weaving as they require complex machinery that is costly to build.

They are used to make women's apparel such as saris, evening gowns; and home furnishings such as table cloths, drapery and upholstery.

Image 6.5: Brocade

Source: Author.

Pile Weave

The pile weave is a fancy weave that has cut or uncut loops (Images 6.6a and 6.6b) which add a third dimension to the fabric creating an effect of depth. In making of pile fabrics, an extra set of warp or filling yarns is used to form the loops. One set of warp and filling yarns interlace to form the base fabric (either a plain or a twill weave) while the extra set forms the loops or pile. Thus, at least three sets of yarns are used.

Image 6.6: Cut and uncut pile weave fabrics

(a) Corduroy: Cut pile fabric

(b) Terry cloth: Uncut pile fabric

Source: 'Corduroy fabric' by ArielGlenn (CC by 3.0 SA; Wikimedia Commons); 'Towel Stack (3249473893)' by Harald Hoyer (CC by 2.0 SA; Wikimedia Commons).

Pile fabrics can be woven entirely of either spun or filament yarns or a combination of both. They mostly range from medium to heavy weight. The pile tends to slant in one direction while pressing during finishing thus giving an up and down look as the light is reflected differently. Hence, all the pieces of a product should have the pile in one direction to have consistent colour. Pile fabrics are used in apparels, home furnishings, towels, wash clothes, bedspreads, lining of coats and jackets, bath robes, buffing and polishing cloths, etc.

There are two major kinds of pile fabrics based on the weaving method used, and they are (*a*) **Warp pile fabric**, in which pile is produced by an extra warp yarn, and (*b*) **Filling pile fabric**, in which pile is produced by an extra filling yarn.

Filling Pile Fabrics

In filling pile fabrics, the extra set of filling yarns form floats over four to six warp yarns. After the weaving is complete, these floats are cut and brushed to form the pile. The floats for filling pile fabrics are interlaced using a 'V' or 'W' method (Figures 6.18a and 6.18b). The 'W' method produces a more durable fabric as the pile is held in place by three yarns as against one yarn used in the 'V' method. Examples of filling pile fabrics are velveteen (all over pile effect) and corduroy (ribbed pile effect; Figure 6.19).

Warp Pile Fabrics

One of the following three methods can be used to form warp pile fabrics:

1. *Wire method*: In the wire method, two warp beams are used. When the extra set of warp yarns forming the pile is raised, a wire is inserted through the shed in the filling direction. The warp pile yarns are then lowered, they lie over the wire and the next filling yarn is laid in the usual manner to hold the pile yarns in place. The wire is then withdrawn. If a cut pile has to be made, a very sharp knife is attached to the end of the wire which cuts

Figure 6.18: 'V' and 'W' interlacing for filling pile fabrics

Pile

(a) V method

Pile

(b) W method

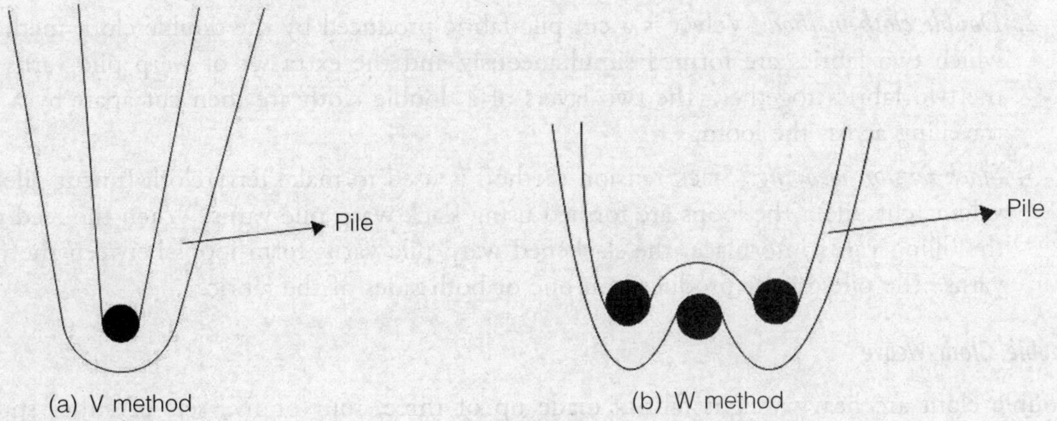

Source: Drawn by the author.

Figure 6.19: Diagram showing the making of corduroy

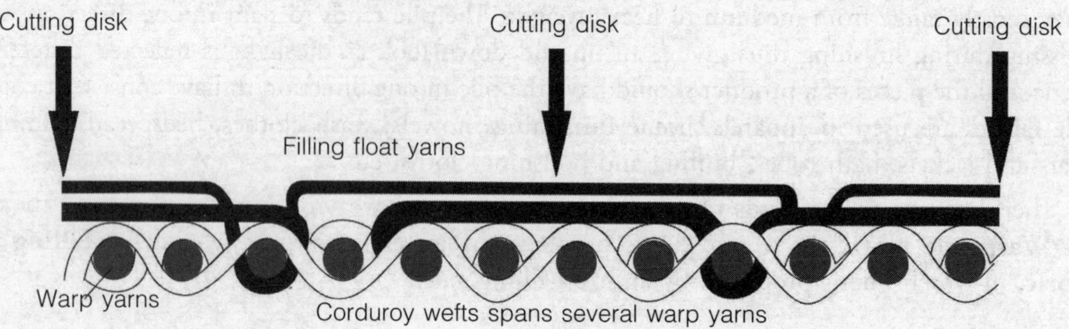

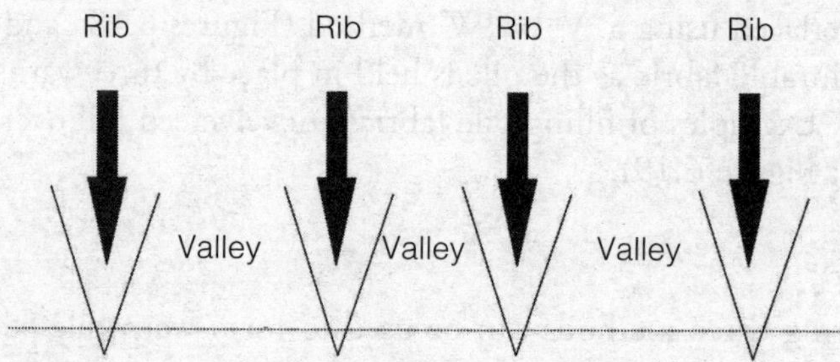

Source: Drawn by the author.

the warp pile yarns while it is withdrawn. Cut pile fabrics are soft, warm and have good absorbency but they catch lint and spot easily. If an uncut pile is to be made the wire does not have a knife. Uncut pile fabrics may wear better than cut pile; however, the loops may catch and tear. Frieze, which is often used for upholstery, is an example of uncut pile fabric produced by the wire method.

2. *Double cloth method*: Velvet is a cut pile fabric produced by the double cloth method in which two fabrics are formed simultaneously and the extra set of warp pile yarns hold the two fabrics together. The two layers of a double cloth are then cut apart by a knife travelling across the loom.
3. *Slack tension weaving*: Slack tension method is used to make terry cloth (uncut pile) and velour (cut pile). The loops are formed using slack warp pile yarns. When the reed packs the filling yarns into place, the slackened warp pile yarns form loops between the filling yarns. The pile can be produced on one or both sides of the fabric.

Double Cloth Weave

Double cloth are heavy weight fabrics made up of three, four or five sets of either spun or filament yarns. **Double-faced fabrics** are made up of three sets of yarns—two warps and one filling, or vice versa. The weave produces the same or different appearance on both sides of the fabric. Thus, these fabrics have two right sides and no wrong side; e.g. blankets and the satin ribbon. **Double-weave fabrics** are made up of four sets of yarns—two sets of warp and two sets of filling yarns. All four sets are interwoven in such a way that the fabric interlocks at some places and remains separate at some places; e.g. matelasse. The two layers of these fabrics cannot be separated without destroying the total fabric. Fabrics using five sets of yarns are woven by the double cloth method as used in the making of pile fabrics. In this, two fabrics are woven simultaneously and are combined by an extra set of the warp or filling yarns. The two sides of such fabrics may have different colours on each side and are often reversible.

Fabrics woven by the double cloth weave method are strong, heavy weight, bulky, expensive, have good durability and unique design possibilities. If fibres with good insulative properties are used, these fabrics can also provide warmth. They are used in apparels, blankets, coat materials, satin ribbons, lining and interlinings, and home furnishings, especially upholstery and draperies.

Leno Weave

Leno weave is also referred to as gauze weave. Here the warp yarns are paired and intertwisted in such a manner that they encircle each filling yarn (Figure 6.20). It is made on a loom which has a special attachment—the 'doup' or 'leno'—a needle-like device supported by two heddles. One of the two warp yarns is threaded through the eye of the needle. The needle alternately pulls the threaded warp yarn to the right or left with each pick inserted, thus causing the warps to twist around each filling yarn. Spun, filament or combination of these yarns can be used to produce fabrics in leno weave. Grenadine and marquisette are examples of leno weave fabrics.

Figure 6.20: Diagram of a leno weave

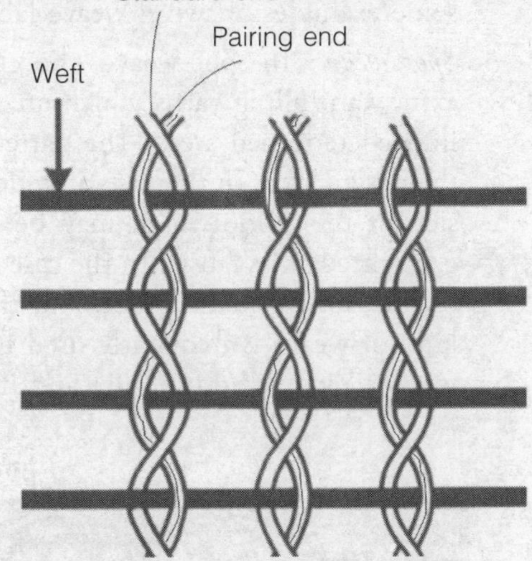

Source: Drawn by the author.

Leno weave fabrics are mostly sheer, light-weight fabrics with an open mesh effect and possess better strength than plain weave fabrics. However, snagging is a problem. They are used in making sheer curtain materials, mosquito nettings, specialised apparel, household bags, agro textiles to shade delicate plants, etc.

Surface Figure Weaves

Surface figure weaves have figures or designs on the surface of the fabric. These figures are made using one or more extra set of yarns. The base fabric is made in a basic weave using two sets of yarns and the surface design is incorporated during weaving using one or two extra sets of yarns, either extra warp or extra weft or both. They are of three types:

1. *Lappet weave*: Lappet weave is very similar to embroidery in which extra warp yarns are introduced in a way that creates a design on the fabric surface while it is being woven. The extra warp yarn that may interlace in both the warp and filling directions with the ground fabric is threaded through needles set in front of the reed. The yarns are carried back and forth to form embroidery-like design on the right side of the fabric, while the excess yarn is carried along on the wrong side from one design area to another. If long floats are formed on the backside of the fabric, they may be cut away and if the floats are short they may be left unclipped.
2. *Swivel weave*: The swivel weave uses the extra weft yarns placed in small shuttles. These shuttles are located at each design point. When the shed is made, each shuttle interweaves its design with a circular motion over very few warp yarns and the excess filling yarn floats on the underside of the fabric from one design to the next (Image 6.7). The floats may be cut, knotted and clipped off after the weaving is complete. The filling yarns that make the design do not travel the entire width of the fabric and are only confined to the design area. The advantage of swivel weave is that it allows the use of different colours in the same row as each design has its own shuttle. The design is securely fastened to the fabric and cannot be pulled out easily without damaging the fabric. Grenadine and madras are some examples of swivel weave fabrics.
3. *Spot weave*: In spot weave, the embroidery like designs can be achieved by the use of extra warp/filling yarns which are inserted along the entire length or width of the fabric in predetermined areas. The pattern is formed using dobby or jacquard attachments for the design area. In the case of widely-spaced designs, long floats are formed on the wrong side of the fabric which may be cut. This is called **clipped spot weave** (Image 6.8a); e.g. dotted swiss fabric. In the case of closely-spaced designs, the floating yarns on the back of the fabric remain uncut, and this is called **unclipped spot weave** (Image 6.8b). This kind of weave is frequently used for border designs. If the floats are cut, the design area may be removed during use and if the floats are left uncut then snagging is a problem.

Image 6.7: Swivel weave

(a) Face

(b) Back

Source: Author.

Image 6.8: Spot weave

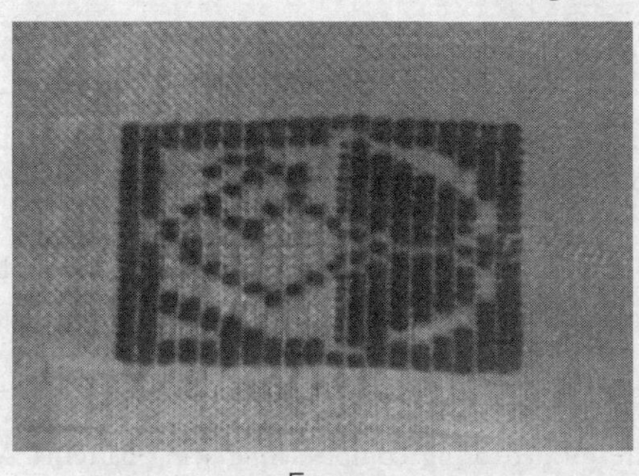

Face

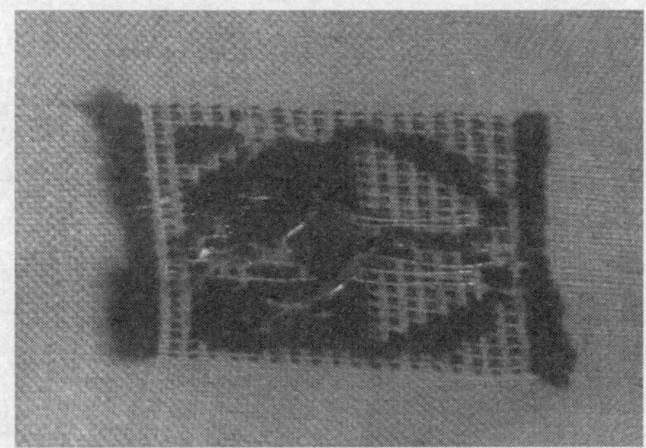

Back

(a) Clipped spot weave

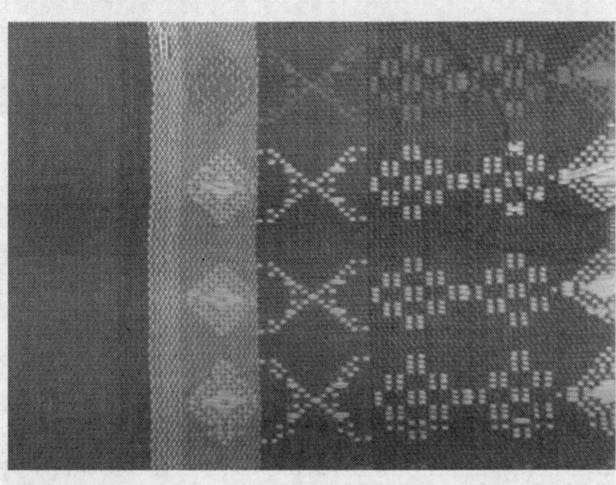

Face

Back

(b) Unclipped spot weave

Source: Drawn by the author.

SUMMARY

- Woven fabrics (except triaxial fabrics) are constructed by interlacing of two or more sets of yarns called the warp yarns and the weft yarns at right angles.
- Weaving is done on a machine called loom.
- Warp beam, heddles, harness, shuttle, reed and cloth beam are parts of the loom that carry out specific functions during the weaving process.
- Shedding, picking, battening, taking up and letting off are the basic weaving operations in fabric construction.
- Weaves can be classified as basic or decorative. Basic weaves are of three types: plain, twill and satin.
- Decorative weaves are of the following types: dobby, jacquard, pile, leno, double cloth weave and surface figure weaves (lappet, swivel and spot).

KEY WORDS

Weaving: Weaving is the art of forming the fabric by interlacing two or more sets of yarns at right angles to each other.

Grain: Grain is determined by the direction of yarns in the fabric. Lengthwise grain is the direction along the length of the fabric, parallel to warp yarns and selvedge. Crosswise grain is the direction along the width of the fabric.

Selvedge: Selvedge is the narrow, flat woven border or the finished edge at both lengthwise sides of a fabric.

Balance: Balance is the ratio of the warp yarns to weft yarns in a fabric. When their ratio is 1:1 (approximately), the fabric is said to be well-balanced.

Basic weave: Basic weaves use simple interlacement patterns for making fabrics.

Decorative weave: Also termed as fancy weaves, these are formed using complex interlacing patterns. Most fancy weaves are formed by a combination of at least two different types of the basic weaves.

EXERCISES

1. Define the following terms briefly:
   - Loom
   - Grain
   - Selvedge
   - Thread count
   - Warping
   - Shedding
   - Balanced plain weave
   - Sateen
   - Clipped spot weave
   - Uncut pile
2. Discuss the mechanism of weaving in detail.
3. Enumerate the different types of looms. Why are shuttleless looms preferred over simple looms?
4. Write a short note on pile fabrics with reference to the various methods deployed for the formation of these fabrics.
5. How are weaves classified? Give an account of the decorative surface effects in detail.
6. Differentiate between the following:
   - Dobby weave and jacquard weave
   - Projectile loom and rapier loom
   - Plain selvedge and tucked selvedge
   - Rib weave and basket weave
   - Shuttle loom and shuttleless loom

REFERENCES

Collier, Billie J., Martin J. Bide and Phyllis G. Tortora. 2009. *Understanding Textiles*. New Jersey: Pearson Education Inc.

Corbman, P. Bernard. 1983. *Textiles: Fibre to Fabric*. Sixth edition. New York: McGraw-Hill Inc.

Hollen, Norma and Jane Saddler. 1979. *Textiles*. Fifth edition. New York: Macmillan.

Joseph, L. Marjory. 1988. *Essentials of Textiles*. Fourth edition. New York: Holt, Rinehart and Winston Inc.
Kadolph, J. Sara. 2009. *Textiles*. Tenth edition. New Jersey: Pearson Education Inc.

ONLINE SOURCES (all accessed 24 May 2016)

http://textilelearner.blogspot.in/2011/08/what-is-loom-define-loom-shuttle-loom_6904.html
http://www.dsir.gov.in/reports/techreps/tsr102.pdf
http://textilelearner.blogspot.in/2012/01/projectile-loop-projectile-weaving.html
http://www.fibre2fashion.com/industry-article/10/946/weaving-weft-insertion-rapier4.asp
http://textilelearner.blogspot.in/2012/05/rapier-weaving-machine-rapier-loom.html
http://www.usti.cz/vubas/qqm/link/Weaving%20machines%20(Looms).pdf
http://textilelearner.blogspot.in/2012/05/water-jet-weaving-machine-water-jet.html
http://hexdome.com/weaving/triaxial/introduction/
http://www.fibre2fashion.com/industry-article/11/1095/different-types-of-weaves1.asp
http://belovedlinens.net/TextilePedia/dimity.html
http://textilelearner.blogspot.in/2012/05/multiphase-or-multished-weaving-machine.html
http://www.slideshare.net/zipporahthompson/history-of-weaving-14624212
http://www.tantuvi.com/history.htm
http://www.tikp.co.uk/knowledge/technology/warping-and-weaving/
http://belovedlinens.net/TextilePedia/dimity.html
http://www.ermt.net/docs/papers/Volume_2/issue_5_May2013/V2N5-108.pdf
http://planningcommission.nic.in/plans/stateplan/upsdr/vol-2/Chap_b3.pdf

7

KNITTED FABRICS

HIGHLIGHTS

- Introduction to knitting as a method of fabric construction
- Knitting industry in India
- Knitting terminology
- Knitting elements
- Types of knits
  - Manufacture of and machinery used for various types of weft and warp knits
- Defects in knitted fabrics
- Use and care of knit fabric

Knitting is the second most frequently used method of fabric construction. Knitted fabrics are constructed by interlocking a series of loops made from one or more yarns, with each row of loops coming out from the preceding row. The popularity of knitting has grown tremendously over the recent years because of the increased versatility of techniques, the adaptability of many new man-made fibres, and the growth in consumer demand for wrinkle-resistant, stretchable, snug-fitting fabrics, particularly in the vastly expanding areas of sportswear and other casual wear apparel. Today, knitted fabrics are used in hosiery, underwear, sweaters, slacks, suits, coats, rugs and other home furnishings.

Knitted fabrics are fundamentally different from their woven counterparts in terms of their construction technique, morphology and performance characteristics. A few basic points of differences between the woven and knitted fabrics are listed in Table 7.1.

KNITTING TERMINOLOGY

Knitting is the process of fabric manufacture where needles are used to form a series of interlocking loops from one or more yarns or from a set of yarns. Unlike woven fabrics—where strands usually run straight horizontally and vertically—yarn that has been knit follows a loopy path along its row, as shown in Figure 7.1. The loops of one row have all been pulled

Table 7.1: Difference between woven and knitted fabrics

Woven fabric	*Knitted fabric*
Made by interlacement of yarns	Made by interlooping of yarns
Two sets of yarns, warp and weft, are used in the making	Only one set of yarn is used
Comparatively rigid and less stretchable and hence less comfortable	Less rigid and hence more stretchable and comfortable
Has poor wrinkle recovery.	Offers better crease or wrinkle recovery.
More compact fabric, is less porous and has greater covering power	Are porous, provide breathability and comfort
Relatively stiffer and has harsh feel. Does not give warmth.	Soft in feel and provides noticeable insulation and warmth due to insulative air pockets
Offers more dimensional stability due to tighter construction and intersecting of warp and weft at right angles.	Because of loop structure and inability of yarn to return to original position, the dimensional stability is poor.
Do not have the problem of run or ladder formation.	Some knitted structures have the problem of laddering and unravelling.
Fabrics are mostly flat.	Fabrics are flat as well as tubular.
Only yardage is produced.	Both yardage and garments can be produced.
Not possible to change the designs on the loom frequently.	Designs can be changed rapidly.
Yarn may or may not be uniform. Can be done with slightly inferior quality yarn as well.	Requires more regular, uniform yarn.
Lesser yarn is required in the fabric formation.	More yarn is required due for loop formation.

Source: Compiled by the authors.

Box 7.1: *History of knitted fabrics*

The earliest known knitted fabric is a pair of thick, hand-knitted fabric socks found in an Egyptian tomb, and probably dates to the fourth century BC. But the art of knitting seems to have been perfected in Western Europe only in the fifteenth century AD. The word '*cnyttan*' was first mentioned in English literature in 1492. And within a few generations hand knitting appears to have spread rapidly throughout Europe. Initially, primitive needles of bone or wood were used to knit and they produced a coarse mesh. Then the Spaniards began using steel needles, which produced a closer mesh and a more evenly knit fabric. In 1589, William Lee invented the first knitting machine in England. By the 17th and 18th centuries the art of hand knitting was gradually replaced by guild organised cottage industry. And from 1880 to 1910 knitwear became primarily a female fashion. Later, knitted pullovers, cardigans, skirts, men's underwear, sportswear and swimwear started becoming popular. By 20th century the industry had evolved to include machines of higher production speeds and a wider choice of patterns. The following may be viewed as some of the primary reasons for the growth of the knitting industry:

1. Capital investment for starting a new knitting unit is relatively small as compared to the weaving industry.

(*Contd.*)

Box 7.1: (*Contd.*)

2. High productivity of knitting machines.
3. Time required for execution of an order is less since no preparatory processes are involved.
4. Setting up of a knitting machine is faster and simpler.
5. Knitting is more flexible than weaving. Styles and designs can be changed easily and rapidly to suit the fashion industry.
6. The knit fabric offers comfort and easy care properties like breathability, wrinkle resistance, moisture management, etc.
7. The knitting industry is less labour-intensive, and therefore much lesser incidence of problems related to workforce.

Knitting Industry in India

Knitwear forms a very important facet of the Indian textile industry. It has become an accepted form of wear because of the widespread emphasis on leisure and casualness. This industry now caters to a wide variety of markets for both the summer and winter seasons. The fact that the unit price of a knitted garment is less than that for a woven garment has also worked in favour of popularising this segment of the Indian textile industry. Knitwear has also made a good mark in exports. Indian knitwear is proving to be a substantial source of foreign revenue because of its popularity and demand abroad. While India has to compete with Hong Kong, Taiwan and Bangkok in the export of cotton knits, it has managed to carve a special niche for itself because of its innovations in styling and quality control. Another reason for the popularity of Indian knitwear is that India is able to provide reasonably priced fashion knits for the world market. While Ludhiana and Kolkata were the prime production centres in India in the early days of the knitwear revolution, today Tirupur—a small town in Tamil Nadu—has grown into the major production centre. Tirupur accounts for Rs 850 crore of knitwear exports or nearly 40 per cent of India's total knitwear outflow.

through the loops of the row below it. Since there is no single straight line of yarn anywhere in the pattern, a knit piece will be stretchy in all directions (some more than others, depending on the yarn, fibre and the specific pattern used). This elasticity is unavailable in woven fabrics, which only stretch along the bias.

The basic knit fabric has a definite 'right side' and 'wrong side'. On the right side, the visible portions of the loops are the verticals connecting two rows, arranged in a grid of V shapes. On the wrong side, the curved loop heads and the ends of the loops are visible, both the tops and bottoms, creating a much more bumpy texture. This has been illustrated in Figure 7.2.

Figure 7.1: Topology of a knit fabric

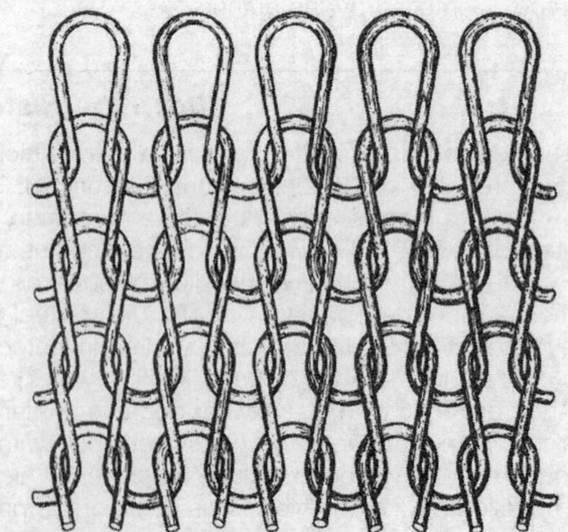

Source: 'Single jersey fabric—the most simple weft knitted structure' by Elkagye (CC by 0 Public Domain; Wikimedia Commons)

Figure 7.2: The technical front and back of a knit fabric

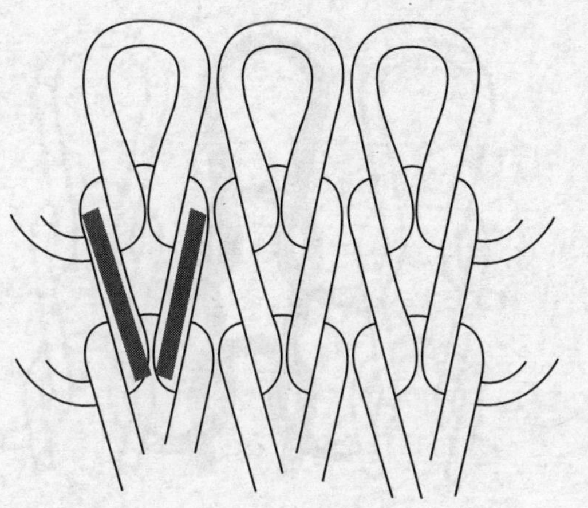

Technical Front

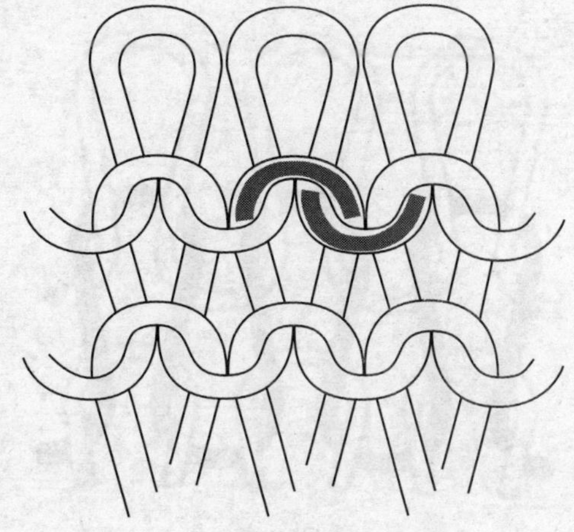

Technical Back

Source: Drawn by the authors.

Some key terms associated with knitting are:

- *Stitch*: A stitch is produced when a knitted loop is pulled through a previously formed loop. Stitches are the fundamental units in a knit fabric.
- *Courses*: A course of a knit fabric is a predominantly horizontal row of needle loops (in an upright fabric). In other words, the crosswise rows of stitches or loops are called courses. The direction of the courses corresponds to the filling of woven goods. The number of courses determines the length of the fabric (Figure 7.3a).
- *Wales*: A wale comprises a row of stitches that run in columns along the lengthwise direction of the fabric. It is a predominantly vertical column of interlaced needle loops. This corresponds to the warp direction of woven fabrics. The number of wales determines the width of the fabric (Figure 7.3b).
- *Technically upright*: A knitted fabric is technically upright when its courses run horizontally and its wales run vertically, with the heads of the needle loops facing towards the top of the fabric. The term technically upright is purely a technical description, and does not necessarily indicate orientation from the user's gaze.
- *Count*: It is the number of wales per inch and courses per inch of a knitted fabric. They are also referred to as course count or wale count.
- *Gauge*: Gauge is defined as the number of needles in 1.5 inches of a knitting machine (but this unit may be different for different knitted fabrics). It is, therefore, a measure of the closeness of the needles in the knitting machine. It determines the fineness or density of the fabric; the higher the gauge, the finer is the fabric.
- *Cut*: Cut is the number of needles per inch (NPI) in a weft knitting machine. This term is only used in weft knitting.

Figure 7.3: Course and wales

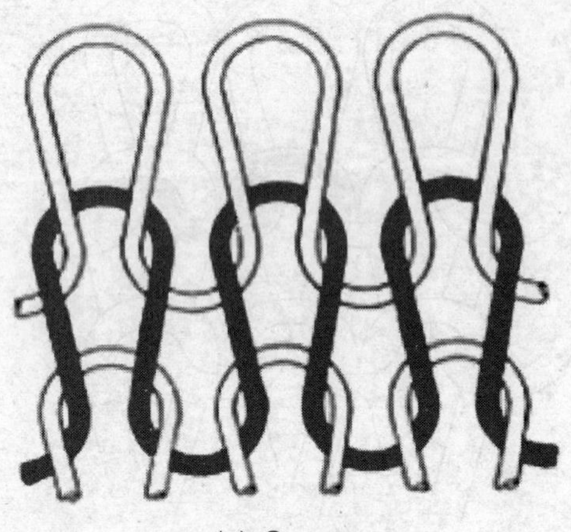

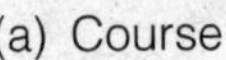

(a) Course (b) Wale

Source: Drawn by the authors.

- *Stitch density*: Stitch density refers to the total number of loops in a measured area of fabric. It is the total number of loops in a given area (such as 10 sq cm). The figure is obtained by counting the number of courses over 10 cm and the number of wales over 10 cm, then multiplying the two values.

KNITTING ELEMENTS

- Needles
- Sinkers
- Cam

The three key knitting elements are:

Knitting Needle

The needle is the main element responsible for loop formation and subsequent interlooping. The required number of needles are arranged on the needle bed at regular intervals. They are provided a vertical up-down movement or a horizontal back and forth movement to form the loops. Each needle has a hook, a closure device, and a butt which is a projection and is used to provide the needle its movement (Figure 7.4). There are three major types of needles:

(i) Latch needles
(ii) Spring beard needles
(iii) Compound needles

Latch needle is used both in warp and weft knitting constructions. The needle was developed around the 1800s. They are called so because they can be closed using a latch which operates without any extra assistance as the knitting progresses. The latch needle is robust and self-sufficient, i.e. it requires no extra element for hook closure. The latch moves up and down to close or open the hook as the needle moves to and fro. The only demerit of latch needles is that very fine gauge is not possible.

Spring beard or **bearded needle** guarantees high process reliability and precise stitch formation and are functional for both warp and weft knitting. Unlike latch needles they don't have a self-closure device. They have an additional part called the pusher to close the hook during loop formation. They are mainly used in tricot-type warp knitting. Fabrics of very fine gauge can be made using these needles.

Figure 7.4: Types of knitting needles

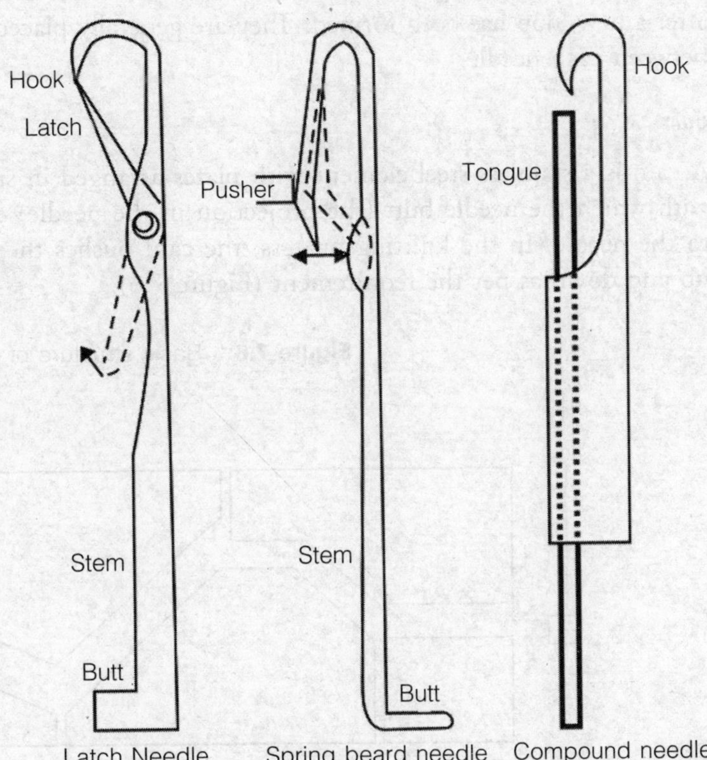

Source: Drawn by the authors.

Compound needle is the newest of the needle types. This needle is functional only in warp knit constructions. There are two distinctive elements to the compound needle: the hook and the sliding closing element (which acts as the closing device for the needle). These needles are robust and durable. The drawbacks of this type are that they have limited applications as the motions are given separately on the hook and the needle, and they cannot be made of finer variety.

Figure 7.5: Basic structure of a sinker

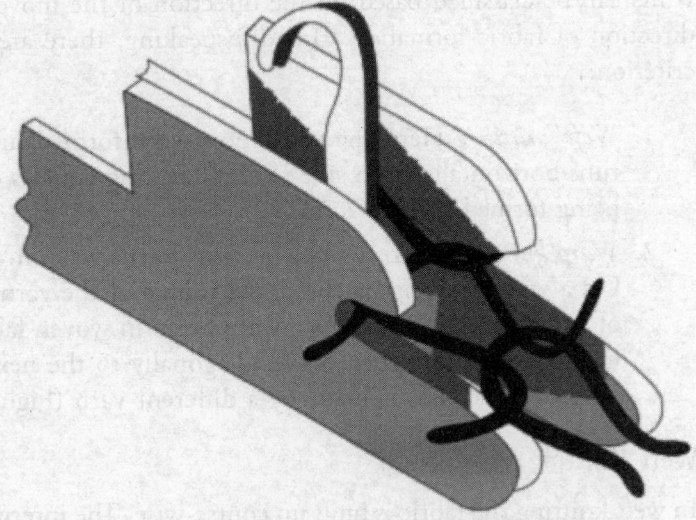

Source: Drawn by the authors.

Sinkers

Sinkers are steel elements which control the movement of the fabric during needle activation (Figure 7.5). They are responsible for holding the fabric as the needle rises, supporting the fabric as the needle descends and pushing the fabric away from needle after a new loop has been formed. They are generally placed in a horizontal position, alternately between each needle.

Cam

A cam is a stainless steel element with plates arranged in such a manner so as to create a track with which the needle butt (the projection in the needle) engages to provide necessary motion to the needle. In the knitting process, the cam pushes the needles back and forth or vertically up and down as per the requirement (Figure 7.6).

Figure 7.6: Basic structure of a cam

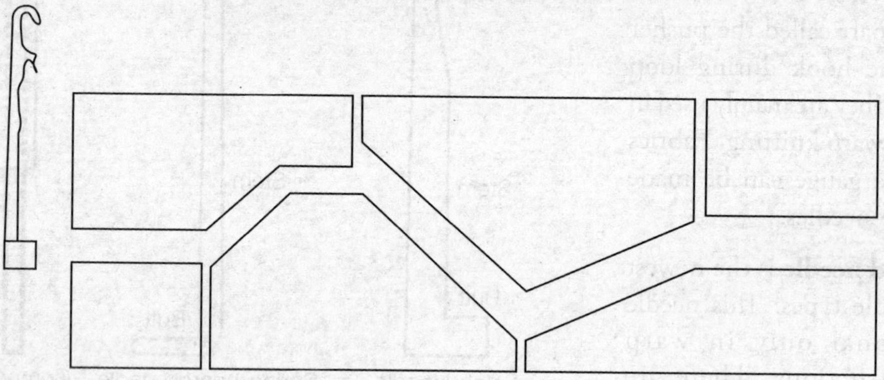

Source: Drawn by the authors.

CLASSIFICATION OF KNITS

Knits can be classified based on the direction of the movement of the yarn with respect to the direction of fabric formation. Broadly speaking, there are two types of knitting based on this criterion:

1. *Weft knitting*: Here one continuous yarn forms courses across the fabric width. The yarn runs horizontally from side to side, making courses one on top of another as the fabric is being formed (Figure 7.7a).
2. *Warp knitting*: Here a series of yarns forms wales in the lengthwise direction of the fabric. On a warp knitting machine, the source of the yarn is a warp beam which has a number of parallel yarns (similar to a warp beam in woven fabrics). Each yarn forms a vertical loop in one course and then moves diagonally to the next wale in the successive course. Each loop in the course is made by a different yarn (Figure 7.7b).

Weft Knitting

In weft knitting the fabric is built up course-wise. The intermeshing of loops is in a horizontal way from a single yarn. The loops in a course are pulled out from the loops in the previous course.

Figure 7.7: Weft knitting structures and warp knitting structures

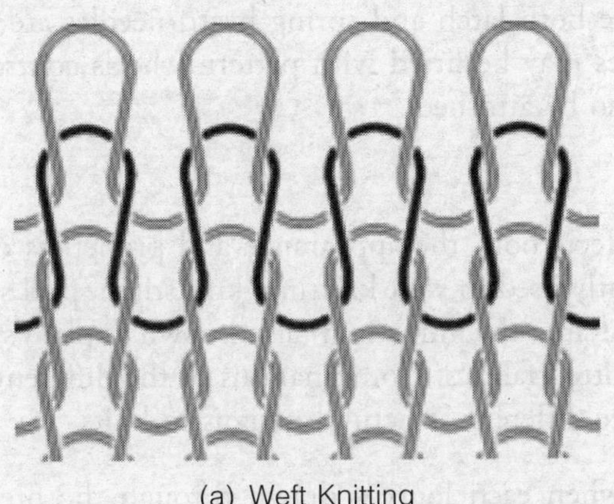

(a) Weft Knitting

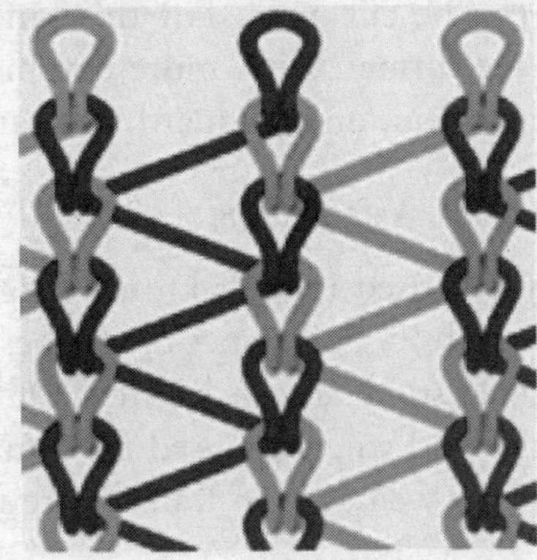

(b) Warp Knitting

Source: 'Tricot' by Martin Pley (CC0 Public Domain; Wikimedia Commons); 'Knit-schematic' by Blahedo (CC by 3.0 SA; Wikimedia Commons).

Weft knitting can be done by hand or by machine. The weft knits can be made both on flat or circular knitting machines thus producing either a flat fabric or tubular fabric. A tubular fabric can be slit open and garment pieces can be cut from these and later sewn together to construct garments. Weft knitting can even produce fully fashioned garments, i.e. those garments which are shaped while being knitted and do not require a separate cut and sew operation.

Machines for Weft Knitting

Weft knitting machines can be differentiated on the basis of their needle bed which is the platform or frame on which the needles are arranged in grooves a regular pattern. The number of needle beds in a machine may be one or two and accordingly the machines are called single bed or double bed machine. Based on the shape of the machine it can be either circular or flatbed machine.

Flat bed knitting machines have needles that are arranged parallel to each other in a flat plane. Here the knitting mechanism is in a flat horizontal plane. Knitting happens back and forth, from one side of the fabric to the other and the yarn travels to and fro across the fabric width during fabric formation. These can have one row of needles (single bed) or two rows of needles (double bed). In double bed machine, the needles are arranged in an inverted V shape. Hence, it is also known as a V bed machine. This type of machine is most often used for the construction of fully-fashioned garments. To produce such garments the width of the fabric is altered by increasing or decreasing the number of active needles, allowing production of shaped fabrics, which when sewn together make fully fashioned garments. Although flatbed machines are suited for hand operation, they are power driven in commercial use. The production time is lower for flat bed as compared to the circular machine. By selection of colour, type of stitch, cam design and jacquard device, an unlimited variety of fabrics is possible. Circular bed knitting machines are most commonly used for production of tubular fabric. The fabric is made by yarns forming loops across the width or around the circumference of the fabric. If a flat fabric is required, the tube can be cut open. For this kind of machine both latch and spring beard needles are used, with the former being more common. Machines may be fitted with pattern wheels controlling needle action, or a jacquard mechanism may also be attached.

Stitches for Weft Knitting

The stitch used in weft knitting significantly affects both the appearance and properties of the knitted fabric. Four types of stitches are prevalently used in weft knitting: knit stitch, purl stitch, miss/float stitch and tuck stitch. Of these four stitches, the knit stitch (also known as plain stitch) and the purl stitch are used for a majority of knitted fabrics. Combinations of the different knit stitches as well as loop transfer are used to create different structural designs in knits.

- *Knit/Plain stitch*: Plain stitch is formed when each loop is drawn through the previous to the same side of the fabric. This stitch has the neck of the loop formation on the face and the head of the loop at the back. The knit stitches look like 'V's stacked vertically (Images 7.1a and 7.1b).
- *Purl stitch*: In purl stitch, loops are drawn to opposite sides of the fabric, which on both sides has the appearance of the back of a plain stitch fabric with the heads of the loops showing prominently. The purl stitches look like a wavy horizontal line across the fabric (Image 7.2).
- *Tuck stitch*: Tuck stitch is formed when the needle holds the old loop while taking on one or more additional loops (Figure 7.8). Here a completed loop is not cast off from

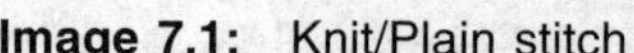

Image 7.1: Knit/Plain stitch

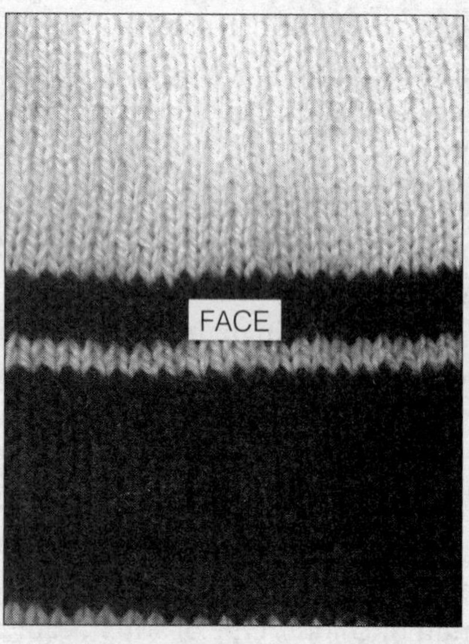

(a) Face of a plain stitch

(b) Back of a plain stitch

Source: Author; 'Stockinette example back' by Pschemp (CC by 3.0; Wikimedia Commons).

Image 7.2: Purl stitch

Source: 'Gestrick links-rechts linke Seite' by Ikiwaner (CC by 3.0 SA; Wikimedia Commons).

Figure 7.8: Tuck stitch

Source: Drawn by the authors.

some of the needles in each course, and two loops accumulating on these needles are later discharged together. The tuck stitch tends to reduce the length of the fabric being knitted and increases its width, resulting in the fabric being thicker with less extension in the width. Tuck stitch can be used to produce raised designs in the fabric to give a lofty appearance and a soft handle.

- *Miss/float stitch*: Miss stitch is formed when one or two needles are deactivated and do not move into position to accept the advancing yarn (Figure 7.9). The yarn passes by without making a loop. The missed yarns float freely on the back side of the fabric therefore sometimes it is also called a float stitch. The introduction of miss stitches results in the fabric becoming narrower in width, since the fabric wales are pulled closer together. This tends to improve fabric stability. A miss stitch can be used for coloured designs by means of passing coloured yarns on the back of the fabric.

Figure 7.9: Miss stitch

Source: Drawn by the authors.

Weft Knit Structures

Different weft knit structures can be made using the above mentioned weft stitches. All weft knit structures have the unique tendency of laddering, i.e. if the yarn breaks causing a loop to break, subsequent loops will drop vertically through the fabric causing a ladder-like formation of vertical blank streaks in the fabric. The following are the primary categories of weft knits:

Plain/Single/Jersey

The plain knit is the most basic form of weft knitting. It is a plain, flat surfaced fabric with a distinct face and back. The loops form distinctive vertical herringbone like ribs or wales on the right side (or technical face) of the fabric. On the reverse side (or technical back), the courses can be seen as interlocking rows of opposed half circles (Image 7.1). This construction gives the face of the fabric relatively smooth surface in relation to the back. It can be produced in flat-knit or tubular (or circular) form. The plain knit is also called jersey stitch because the construction is like that of the turtleneck sweaters originally worn by English sailors from the Isle of Jersey. The following are the characteristics of the plain knit:

1. The fabric is extensible in course-wise and wale-wise direction. The course-wise extension is approximately twice that of wale-wise direction.
2. Although there is a technical back and face of the fabric, both the sides can be used as the face. When the technical back is used, the fabric is known as reverse jersey.
3. Jersey fabrics have a tendency to curl at the edges. The sides will curl towards the technical back while the top and bottom of the fabric will curl towards the technical front.
4. The fabric produced by plain knits is lightweight as compared to those made by other stitches.
5. The rate of production is almost five times that of the weaving operation.
6. Pattern variations like stripes, multicoloured patterns (Intarsia patterns), textured surfaces produced by raised designs and pile effects are possible. Horizontal stripes can be knitted in plain jersey by using yarns of different colours.

Purl

The purl fabric uses the purl stitch described earlier. The construction consists of loops drawn alternately to the front in one course and to the back in the other course. This means that courses of all face loops alternate with courses of all back loops. It resembles the technical back of the plain or Jersey knit on both the sides. This type of fabric is widely used in infants and children's wear. The characteristics of fabrics made of this technique are:

1. The rate of production is slower and more expensive.
2. The fabric looks the same on both sides.
3. The yarn will run vertically up or down if the loop gets broken.
4. They do not curl at the edges.
4. The stitch exhibits crosswise and lengthwise extensibility.

Rib

Rib knits have alternating lengthwise columns of plain and purl stitches constructed in such a way that the face and back of the fabric appear alike. This can be produced on flat or circular machine.

Each course is made of alternating face and back loops. The simplest rib knit is 1 × 1, where a single column of face loops alternates with a single column of back loops. Other configurations like 2 × 2, 3 × 3, 6 × 3 are also used. The width-wise extensibility is highest for 2 × 2 rib fabric and decreases as the rib width increases. Due to its excellent width-wise elasticity, it is extensively used in apparel where snugness of fit is essential, such as wristbands, waistbands, neckbands, etc. It is also widely used for underwear and socks for men and children (Figure 7.10).

Figure 7.10: Typical rib knit structure

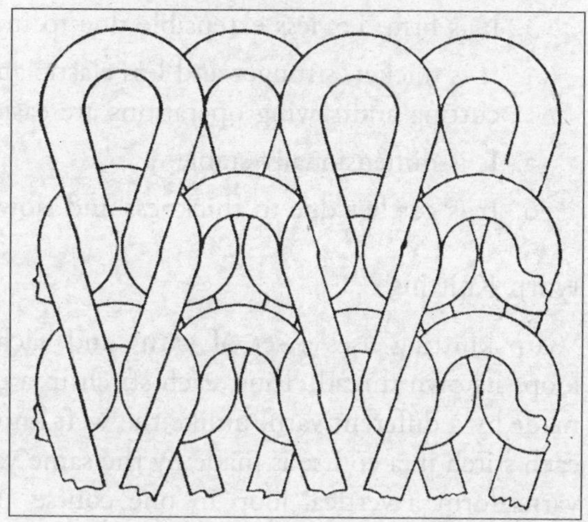

Source: 'Double knit structure (weft knitted)' by Elkagye (CC by 3.0 SA; Wikimedia Commons).

The following are some characteristics of rib knit fabrics:

1. They are double faced and balanced structures.
2. The fabric surface is virtually corrugated or ribbed.
3. It is much thicker compared to the single jersey fabric.
4. The fabric has good extensibility in length-wise direction, but the width-wise extensibility and recovery are much higher than the single jersey fabric.
5. The fabric does not curl at the edges.
6. The fabric can be easily unravelled from the end last knitted.
7. Rib construction is costlier because of the greater amount and finer yarn requirement accompanied by a slower rate of production.

Interlock

Interlock is a popular double jersey knitted structure. It is also referred to as a variation of the rib structure at times. It resembles two rib fabrics interknitted back to back. The interlocking of the two rib structures is in such a way that the face wale of one fabric is directly in front of the face wale of the other fabric (Figure 7.11).

Figure 7.11: Formation of an interlock fabric

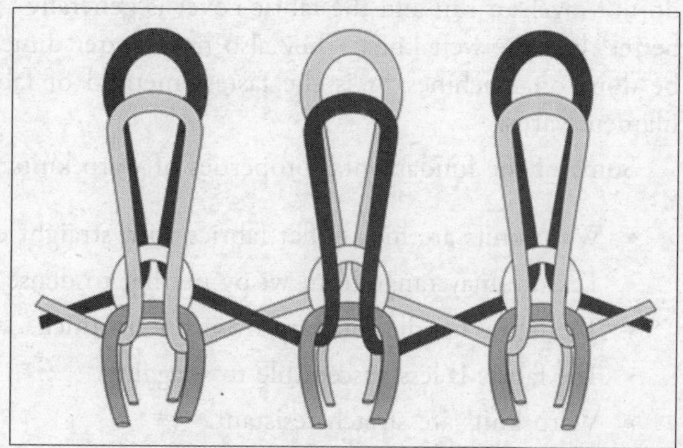

Source: Adapted by the authors.

The characteristics of interlock fabric are:

1. It is a reversible fabric.

2. It does not curl or ladder.
3. It is firm, i.e. less extensible due to interlocking structures.
4. It is thicker, stronger and less elastic and is therefore nearer to the woven structure. Hence cutting and sewing operations are easier.
5. It is dimensionally stable.
6. It is costlier due to thickness and slower production.

Warp Knitting

Warp knitting uses a set of yarns and each needle loops its own thread. Thus, each stitch in a course is made by a different yarn, unlike the weft knits where each stitch in a course is made by the same yarn. The yarns form a vertical loop in one course and then move diagonally to another wale to make the loop in the successive course. This is the action which is responsible for creating the fabric. Due to the diagonal movement of the yarn, floats called as 'underlaps' are formed at the back of the fabric. These underlaps are a distinctive feature of warp knits (Figure 7.12). The face of a warp knit fabric has quite clearly defined plain stitches generally running vertically. The back of the fabric has slightly angled horizontal floats formed by the underlaps.

Figure 7.12: Warp knit structure

Source: 'Trikot' by Ryj (CC-by-3.0 SA; Wikimedia Commons).

Consumers like warp knits because of their smoothness, possible sheerness, appreciable wrinkle and shrink resistance. As compared to weft knits, warp knits have some other advantages too. They do not ravel, or run and the fabric cover is generally better than the weft knits. They also have better dimensional stability. Warp knitting can only be done on machines. It is the fastest method of fabric production using mainly continuous filament yarns.

Some of the fundamental properties of warp knits are:

- Warp knits are mostly flat fabrics with straight edges.
- Texture may range from wispy netting to dense patterns.
- The fabric handle can vary from fine to thick.
- The fabric is less susceptible to snagging.
- Warp knits are stretch resistant.
- Warp knits are also resistant to unravelling.

Machines for Warp Knitting

A warp knitting machine consists of one or more warp beams on which yarns are wound. The yarns from the warp beam are fed to the needles on the needle bar. The needle bar comprises a row of knitting needles extending across the machine width. Each yarn is controlled by a yarn guide, which is a metal strip with a hole through which the yarn passes. The yarn guides are set on a horizontal metal bar called the guide bar, which extends across the width of the machine. Guide bars move periodically both in the forward–backward motion and side-to-side swing. Each yarn guide on the guide bar directs one yarn to the hook of one knitting needle. In this sense, the guide bars can somewhat be likened to the heald shafts of a weaving loom. The loops of one course are all made simultaneously with one guide bar movement. The guide bar then moves sideways to feed the yarn to the adjacent needles for the next course. The higher the number of guide bars, the greater is the design flexibility (Figure 7.13).

Figure 7.13: Warp knitting action

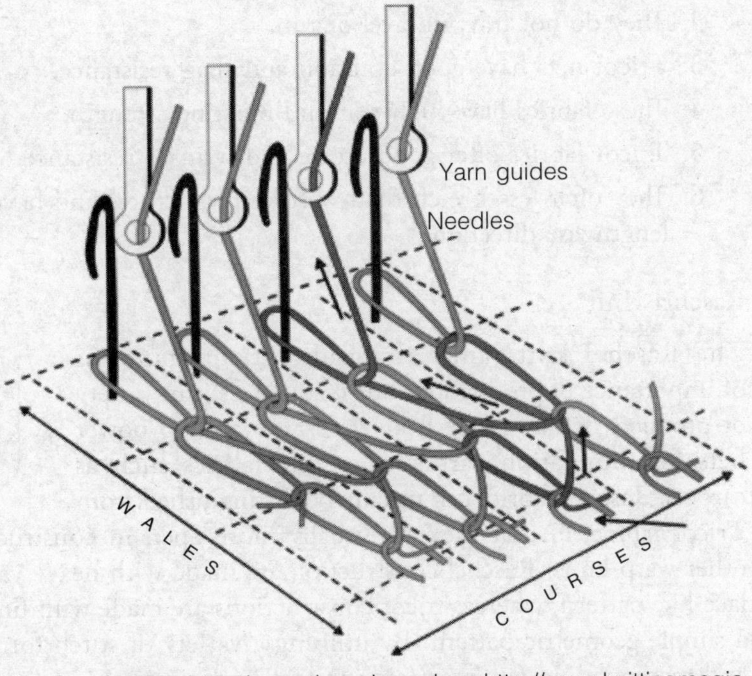

Source: Adapted by the authors based on http://www.knittingmagic.biz/practical-guide/yarn-feeding-and-loop-formation.html (accessed 18 March 2016) (CC By 3.0).

Based on the features of the warp knitting machines two machines are widely used: Tricot and Raschel. They can both be either single bed or double bed machine. The number of guide bars can vary from two to four in a Tricot machine, though three to four guide bars machines are more common. The Rascal machine can accommodate up to 24 guide bars and hence offers wide pattern capability.

Warp Knit Structures

Tricot Knit

Tricot fabrics represent the largest portion of production in warp knit structures. It produces a flat, thin textured, lightweight fabric characterised by fine vertical lengthwise ribs on the face and horizontal crosswise stitches on the back which are slightly at an angle. Variations in the tricot knits are produced by varying the size of the yarn, lightness and compactness of stitch, number of guide bars and the length of guide bar movement. Tricot knits are used for a wide variety of fabric weights and designs. They are popularly used as backing for bonded, laminated or coated fabrics. The characteristics of tricot fabrics are as follows:

1. They are soft and pliable having flowing or drapable qualities.
2. They do not fray, unravel or run.
3. Tricot nets have good abrasion and snag resistance.
4. These fabrics have high tear and bursting strength.
5. Tricot fabrics offer good crease and wrinkle resistance.
6. They offer less stretch than weft knit structures and have significantly lesser stretch in the lengthwise direction.

Raschel Knit

The Raschel knit ranks alongside Tricot in terms of importance of production, but it surpasses it in variety of products, which range from veils and laces to power nets for foundation garments to pile fabrics such as carpets. Raschel fabrics can usually be distinguished from Tricot fabrics in that they are usually more open in construction and coarser in texture than other warp knits. Raschel construction are made with heavy yarns and usually have an intricate, lace-like pattern, whereas tricot constructions are made with fine yarns and are either flat or have a simple geometric pattern. By utilising a variety of stitch formation and methods of arranging and manipulating *fed-in* or *laid-in* yarns (yarns inserted or fed through rows of loops in course-wise direction to create various design effects) in Raschel construction, a variety of fabrics across design interests, surface textures and with openwork effect can be produced. The characteristics of these fabrics:

> The fabric structure containing a pattern of openings or holes formed by pronounced inter-yarn spacing is called **openwork**.

1. They can be made in fine to heavy gauge.
2. The fabric compactness may vary from close knit to openwork crochet, laces and nets.
3. Power nets and similar elastic types for foundation garments are also possible.
4. The fabric can provide pronounced dimensional effects, ribbed and pleated effects.
5. The fabric can be double faced and reversible.

Milanese Knits

The Milanese stitch, though accomplished by a different technique, produces a fabric very similar in appearance to Tricot. It can be identified by the fine rib on the face and a diagonal pattern on the back. The stitches here are composed of two loops, one from each set of warp thread both moving diagonally from one stitch to another at each course. The crossing of several set of yarns creates a diamond effect. It is usually knitted from filament yarn into fine lightweight fabrics. Milanese fabrics are superior to Tricot in smoothness, elasticity, regularity of structure and tear resistance. Despite the apparent advantages of Milanese fabrics, production is limited due to costly low production rates and their limitations in scope of pattern.

Simplex

Simplex fabrics are a relatively small part of warp knit production. The fabric is made up of stitches that appear on both sides of the cloth and look like a double-faced Tricot. The fabrics are of fine gauge knits generally ranging from 28 to 34 needles per inch. These are generally produced within widths of 84–112 inches. Characteristics of simplex fabrics are:

1. They are made of fine yarn but are, nevertheless, relatively dense and thick.
2. The cloth is sometimes lightly napped to obtain soft, suede-like finish.
3. They are used for production of gloves, handbags, and sportswear and slip covers.
4. Eyelet and other openwork may also be produced on the simplex machine.

The differences between weft and warp knits have been summarised in Table 7.2.

Table 7.2: Difference between warp and weft knits

Weft knits	*Warp knits*
Loop formation takes place course wise in the horizontal direction	Loop formation takes place wale-wise in the vertical direction
Yarns run in horizontal or course directions during knitting	Yarns run in the vertical or wale directions during knitting
Needles knit sequentially in a knitting cycle	All the needles knit together in a knitting cycle
Only one or a few yarns are needed during knitting a fabric	A large number of yarns are needed for knitting a fabric. The number of yarns is equal to the number of needles being used.
Few preparatory processes are required before knitting	More preparatory processes are required before knitting
Lower rate of production	Higher productivity
Fabrics have good stretchability in both directions	Fabrics have low stretchability in both directions
Dimensional stability of the fabrics is lower	Dimensional stability of fabrics is higher
Machine may be flat or circular	Machines are generally flat

Source: Compiled by the authors.

DEFECTS IN KNITTED FABRICS

Bands and Streaks

There are different kinds of bands and streaks that may occur in knitting. Some of the popular band and streak defects are as follows:

1. *Barré*: A barré effect is a streak or band in the fabric caused by various factors like, lack of uniformity in yarn size, colour or lustre, varied tension on the yarns during knitting, uneven shrinkage or other finishing defects. It is horizontal in weft knits and vertical in warp knits.
2. *Bowing*: It appears as a line or a curve across the fabric. This distortion is caused by faulty take-up mechanism on the knitting machine.
3. *Streak* or *stop mark*: This straight horizontal streak or stop mark in the knitted fabric is due to the difference in tension in the yarns caused by the machine being stopped and then restarted.
4. *Skewing*: Skewing effect is seen as a line or design running at a slight angle across the cloth.
5. *Needle lines*: Needle lines or vertical lines are due to a wale that is either tighter or looser than the adjacent ones. This is caused by the needle movement due to a tight fit in its slot or a defective sinker.

Stitch Defect

The various kinds of stitch defects are as follows:

1. *Boardy*: This is caused when the stitches have been knit very tightly. In this case, the knitted fabric acquires a stiff or harsh handle.
2. *Cockled* or *puckered*: The knitted fabric is cockled or puckered due to uneven stitches or uneven yarn size and as a result does not lie flat.
3. *Dropped stitch*: This is an un-knitted stitch caused by the yarn carrier not having been set properly.
4. *Run* or *ladder*: A run or ladder indicates a row of dropped stitches in the wale.
5. *Hole*: A large hole or a press off is the result of a broken yarn at a specific needle feed so that knitting could not occur.
6. *Tucking*: This is the result of an unintentional tucking in the knitted fabric. This is also called 'bird's eye' defect.
7. *Float*: This is caused by a miss stitch which is the result of failure of one or more needles that should have been raised to catch the yarn.

USE AND CARE OF KNIT FABRICS

A large variety of knit fabrics are available in the market. The performance of one knit fabric can markedly differ from that of another. However, some common problems seem to affect the performance of the knitted fabric. These are low dimensional stability, snagging, sagging and pilling. Sagging could be avoided by taking care while drying of wet knitted garments.

These should preferably be dried flat to avoid deshaping. Pilling can be avoided by choosing gentle washing cycles and preferably washing by hand.

Dimensional Stability

Knits became popular because of the comfort they offered. The construction through formation of loops permits good drape and easy movement of the body. The stretch provides comfort, but it considerably affects the dimensional stability. At times there are problems of shrinkage, stretching and distortion of knits. To counter such stability-related problems treatments such as shrinkage control, heat setting of synthetics and special resin finishes are available. These provide good dimensional stability to the fabric. However, not all manufacturers subject their products to such treatments.

Snagging

The loop formations make the knitted fabric susceptible to rip or snag. Sharp objects can damage the knits if they are caught in the loop. The loop then pulls out from the fabric surface and a long snag of yarn may be formed. If the yarn breaks the snag will produce a hole in the fabric.

Pilling

When the fabric is subjected to abrasion during wear, the short fibre ends work their way to the fabric surface and get rolled up into small balls that hang on to the surface. Pilling is more frequent in the synthetic double knits or knits made from loosely-twisted yarns. Weaker fibres such as cotton, rayon, acetate and wool generally break off the fabric. The stronger synthetic fibres cling to the fabric and are hard to remove, and thus render an unsightly area on fabric surface.

Care

Each knit product can have different wash care requirements and they must be addressed accordingly. However, there are a few general points that have to be kept in mind while taking care of knit fabrics:

- If they are air-dried, knit fabrics shrink less as compared to those dried in the machine.
- Knits maintain their shape best if they are flat dried. The weight of the wet knit fabric hung on a line may cause the fabric to stretch out of shape.
- Hand knits like sweaters of wool or animal fibre with an open construction may need hand laundering in soft soap and blocking (stretching back into shape). Such items can be laid on a sheet of wrapping paper before washing and the outlines traced. After wash the garment must be laid on the same sheet to dry. While still wet the garment should be gently stretched to fit the outline of the original dimension.
- The broken yarn of a knit structure can be secured by a few hand stitches with the needle and a matching thread. This will prevent the hole from becoming large during handling and laundering, and eventual laddering.

SUMMARY

- Knitting is a process of fabric manufacture in which needles are used to form a series of interlocking loops from one or more yarns or from a set of yarns.
- A knitting machine is an apparatus for applying mechanical movement in order to convert yarn into knitted loop structures.
- Gauge relates to the number of needles per inch in a knitting machine, which accounts for the number of stitches or loops produced per square inch of fabric.
- Based on the direction of movement of yarn with respect to the direction of fabric formation, the two main types of knit fabrics are warp knit and weft knit.
- Weft knitting is a process in which one yarn or yarn set is carried back and forth or around in the course-wise or horizontal direction to form the fabric. Four types of weft knitting stitches are: knit stitch, purl stitch, miss stitch and tuck stitch.
- Warp knitting is a method of forming a fabric in which the loops are made in a vertical way along the length of the fabric from each warp yarn. The yarns are interlooped in the wale-wise direction in adjacent courses to form the fabric. Tricot, Raschel, Simplex and Milanese are four basic warp knit structures.
- The three major types of knitting needles are latch needles, spring beard needles and compound needles
- Knitting machines could be either flat bed or circular bed and could have either a single needle bed or a double needle bed.
- Defects in knitted fabrics may occur as bands and streaks or as stitch defects.
- The common problems that seem to affect the performance of the knitted fabric are in the areas of stability, snagging and pilling.

KEY WORDS

Course: A course of a knit fabric is the horizontal row of needle loops. The direction of the courses corresponds to the filling of woven goods. The number of courses determines the length of the fabric.

Wale: A wale comprises a row of stitches that run in columns along the lengthwise direction of the fabric. This corresponds to the warp direction of woven fabrics. The number of wales determines the width of the fabric.

Knitting needle: This is the main element responsible for loop formation and subsequent interlooping of the yarn.

Sinkers: Sinkers are the steel elements which control the fabric movement during needle activation.

Cam: A cam is a stainless steel element with plates arranged in a manner so as to create a track along which the needle moves during the knitting process.

Knit/plain stitch: The plain stitch is formed when each loop is drawn through the previous to the same side of the fabric. It has the neck of the loop formation on the face and the head of loop on the back.

Purl stitch: The purl stitch is formed when loops are drawn alternately to the front in one course and to the back in the other course.

Rib knits: Rib knits have alternating lengthwise columns of plain and purl stitches constructed so that the face and back of the fabric appear alike.

Interlock knit: Interlock is a popular double jersey knitted structure. It resembles two rib fabrics interknitted back to back.

EXERCISES

1. Briefly describe the process of knitting and bring out the differences between knitting and weaving.
2. Why are knitted fabrics more suitable for end use such as leisure wear, sportswear, etc.?
3. Give a detailed classification of knitted fabrics.
4. Enlist the general properties of warp and weft knit fabrics.
5. Give diagrammatical representations of the various types of machine knitting needles and list their salient features.
6. Define the following terms:
   - Wales
   - Courses
   - NPI
   - Knitting machine
   - Gauge
   - Laddering
   - Guide bar
   - Sinker
   - Cam
7. Enlist the common problems that affect the performance of knitted fabric and the measures that must be adopted to overcome these problems.

REFERENCES

Brackenbury, T. 2005. *Knitting Clothing Technology*. New York: Blackwell Science Publishers.

Corbman, P. B. 1989. *Textiles: Fibre to Fabric*. Sixth edition. New York: McGraw-Hill.

Joseph, M. L. 1988. *Essentials of Textiles*. Sixth edition. Florida: Holt, Rinehart and Winston Inc.

Kadolph S. J. and A. L. Langford. *Textiles*. Tenth edition. New Jersey: Pearson Education.

Spencer, D. J. 2005. *Knitting Technology: A Comprehensive Handbook and Practical Guide*. Fourth edition. Cambridge: Woodhead.

8

NON-WOVENS AND OTHER FABRIC CONSTRUCTION TECHNIQUES

HIGHLIGHTS

- Production, properties and uses of non-woven fabric
  - Selection, web formation, web bonding, finishing
  - Bonded fabrics
  - Felts
- Other methods of fabric construction
  - Nets
  - Laces
  - Braids
  - Macramé
  - Stitch-bonded fabrics
  - Quilted fabrics
  - Laminated fabric
  - Tufted fabric
  - Crochet

Most fabrics we normally use are made by the two major methods we have already discussed—weaving and knitting (Figure 8.1). However, these are not the only methods of fabric construction. There are other methods which, though constitute a small percentage of total textile production, form critical components of the garment and other textile applications. These include non-wovens such as bonded fabrics and felts (woollen felts and needle felts). Fabrics produced by other methods of construction are nets, laces, braids, macramé, stitch-bonded, quilted, laminated, tufted and crochet fabrics.

Figure 8.1: Schematic of the production of woven and non-woven fabrics

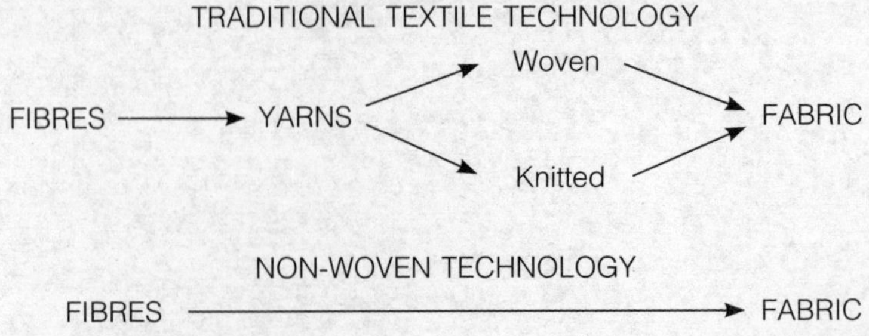

TRADITIONAL TEXTILE TECHNOLOGY

FIBRES → YARNS → Woven / Knitted → FABRIC

NON-WOVEN TECHNOLOGY

FIBRES → FABRIC

Source: Drawn by the authors.

NON-WOVENS

The American Society for Testing and Materials defines non-woven fabrics as textile structures constructed of fibres held together 'by bonding or interlocking or both accomplished by mechanical, chemical, thermal, or solvent means, and the combination thereof' (ASTM 1976). In other words, non-wovens are made from fibrous webs bonded together by either mechanical entanglement, thermal fusion, by forming chemical complexes or by using resins/adhesives. Recent technological advances have resulted in a wide variety of non-woven fabrics with a very diverse group of end uses such as apparel (shoe linings, interfacings, bra paddings, gloves), household (wipes, bags for vacuum cleaner, tea bags, garment bags), health care (bandages, surgical gowns, beddings, shoe covers, curtains), furnishings (carpet backings, mattress cover, bedspreads, upholstery backing, floor coverings), packaging (carry bags, gift wrappings), air and liquid filters, and geotextiles.

Non-woven fabrics are typically not strong. Their flexibility is generally less than that of a similar woven or knitted substrate. However, the primary advantage in non-woven fabrics is the speed with which the final fabric is produced. Since yarn-making does not happen here, all yarn preparation steps are eliminated (the initial stage only involves the basic steps in fibre preparation). The fabric production itself is faster than conventional methods. These are the steps involved in making non-woven fabrics:

1. *Selection and preparation of fibres*: Opening, cleaning and blending
2. *Web formation*: Dry-laid, wet-laid, air-laid and other special techniques
3. *Web bonding*: Mechanical, thermal, chemical/adhesive and other methods
4. *Finishing*: Dyeing, printing, imparting suitable finishes

Selection and Preparation of Fibres

The first step in the production of a non-woven is the selection of fibre(s) with suitable specifications depending on the end use of the fabric. Both filament and staple fibres can be used and they can be natural, man-made or of mineral origin. The fibres then need to be opened, cleaned and blended for further use. These steps have been explained in detail in Chapter Four.

Web Formation

The next step is the arrangement of fibres/filaments in a sheet or web. While the dry-laid and wet-laid techniques are generally used for short staple length fibres, spun-laid and melt-blown are used for continuous filament fibres (Figure 8.2).

Dry-Laid

There are two types of techniques to form dry-laid webs: carding and air-laying.

Carding

Carding is a mechanical process of forming a web. After opening and blending, the fibres are conveyed to the carding machine by air transport. The fibres are then combed into a web by the carding machine. The precise configuration of cards will depend on the fabric weight and fibre orientation required. Relative speeds of the cards and web composition can be varied to produce a wide range of properties.

Figure 8.2: Different techniques of making non-woven fabric web

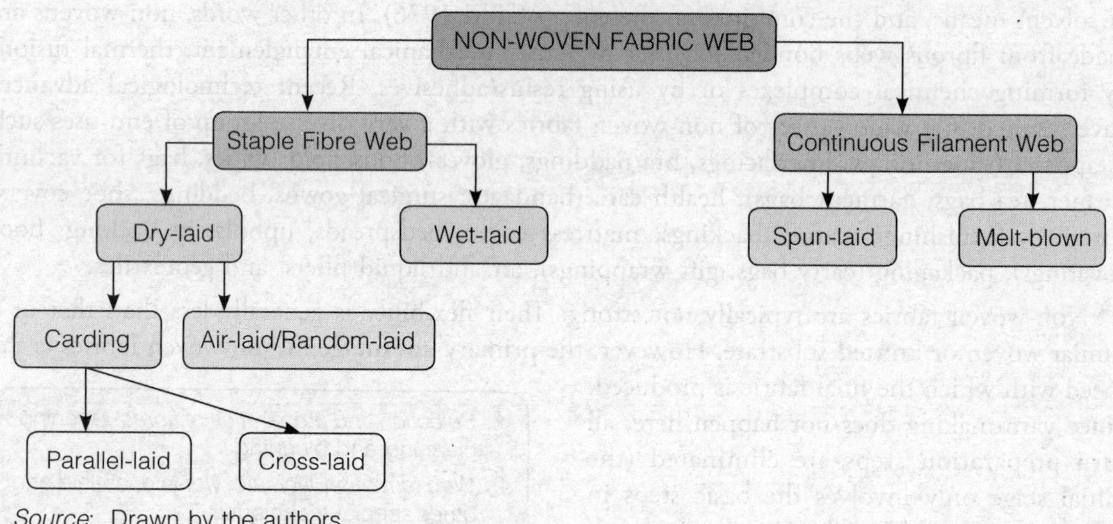

Source: Drawn by the authors.

Carded webs can be of two types:

1. *Parallel-laid web*: In this web, the fibres are aligned more or less parallel to each other and to the direction in which the card produces the web, i.e. in the machine direction (Figure 8.3a). The web is stronger when pulled lengthwise than crosswise because there is more friction between the fibres in lengthwise direction. They are the cheapest to produce and are used in applications including interlinings, headrest covers, cable insulation, wipes, etc.
1. *Cross-laid web*: To increase the strength of web in both lengthwise and crosswise directions, cross-laid web is used. To achieve this, the fibres which make up the web will be orientated equally in both lengthwise and crosswise directions (Figure 8.3b). This technique is more costly to produce, but the fabric has greater crosswise strength. Applications include CD liners, dusters, table covers, table napkins, etc.

Figure 8.3: Parallel-laid and cross-laid web

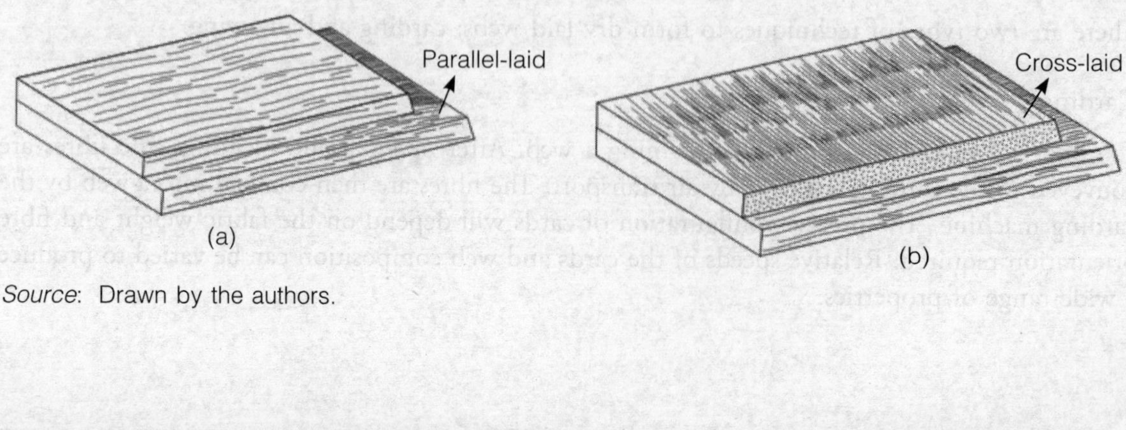

Source: Drawn by the authors.

Air-laid/Random-laid

This process produces a remarkably uniform web from staple fibres with equal strength in all directions. In air-laying technique, the fibres which can be very short are fed into an air stream and from there to a moving belt or perforated drum where they can form a randomly-oriented web. In comparison to parallel-laid and cross-laid webs, air-laid web is expensive to produce. However, they have lower density and greater softness and offer great versatility in terms of the blends that can be used and are capable of producing products for a wide range of end use applications.

Wet-laid

In this method, dilute slurry (mixture) of water and fibres is first deposited on a moving wire screen and then drained to form the web. The web is then dewatered, consolidated by pressing between rollers and dried. Impregnation with a binder is often included in a later stage of the process. Formation of wet-laid web allows a wide range of fibre orientations, ranging from near random to near parallel. A wide range of natural, mineral and man-made fibres of varying lengths can be used. Wet-laid webs are the cheapest to produce and are used in disposable end products such as handkerchiefs, napkins, aprons, gloves, tea bags and surgical gauzes. More durable applications include interlinings, filter cloths and carpet underlaying.

Spun-laid

In this process polymer granules are melted and molten polymer is extruded through the spinneret. After the extrusion of the filaments from the spinneret they are laid on the conveyor belt to form the web. A high temperature is maintained at this stage so that the fibres do not solidify completely, leading to sticking of the filaments with one another as they criss-cross on the belt.

However, it has to be noted that spun-laid process is not the principle method of bonding of fibres in the web. Also referred to as spun-bonded at times, the process has the advantage of giving greater strength to the non-woven fabric but flexibility of the raw materials is more restricted. As spun-laid non-wovens tend to have low bulk and high tensile and tearing strength, they are used for numerous industrial applications such as protective clothing, filters, packaging and geotextiles.

Melt–blown

In melt-blown web formation, low viscosity polymers are extruded in to a high-velocity airstream through the spinneret. This scatters the melt, solidifies it and breaks it up into a fibrous web. Surgical face masks, surgical towels, operating room drapes and gowns, industrial wipes, instrument wrappings, etc. are some end uses of fabrics made by this technique.

Web Bonding

Once the web of fibres is made, it has to be consolidated to form the fabric. This is carried out during bonding. The choice of method of bonding is as important to the ultimate functional properties as is the type of fibre in the web formation process. There are three basic types of bonding (Figure 8.4):

- Mechanical bonding
- Chemical bonding
- Thermal bonding

Figure 8.4: Methods of web bonding

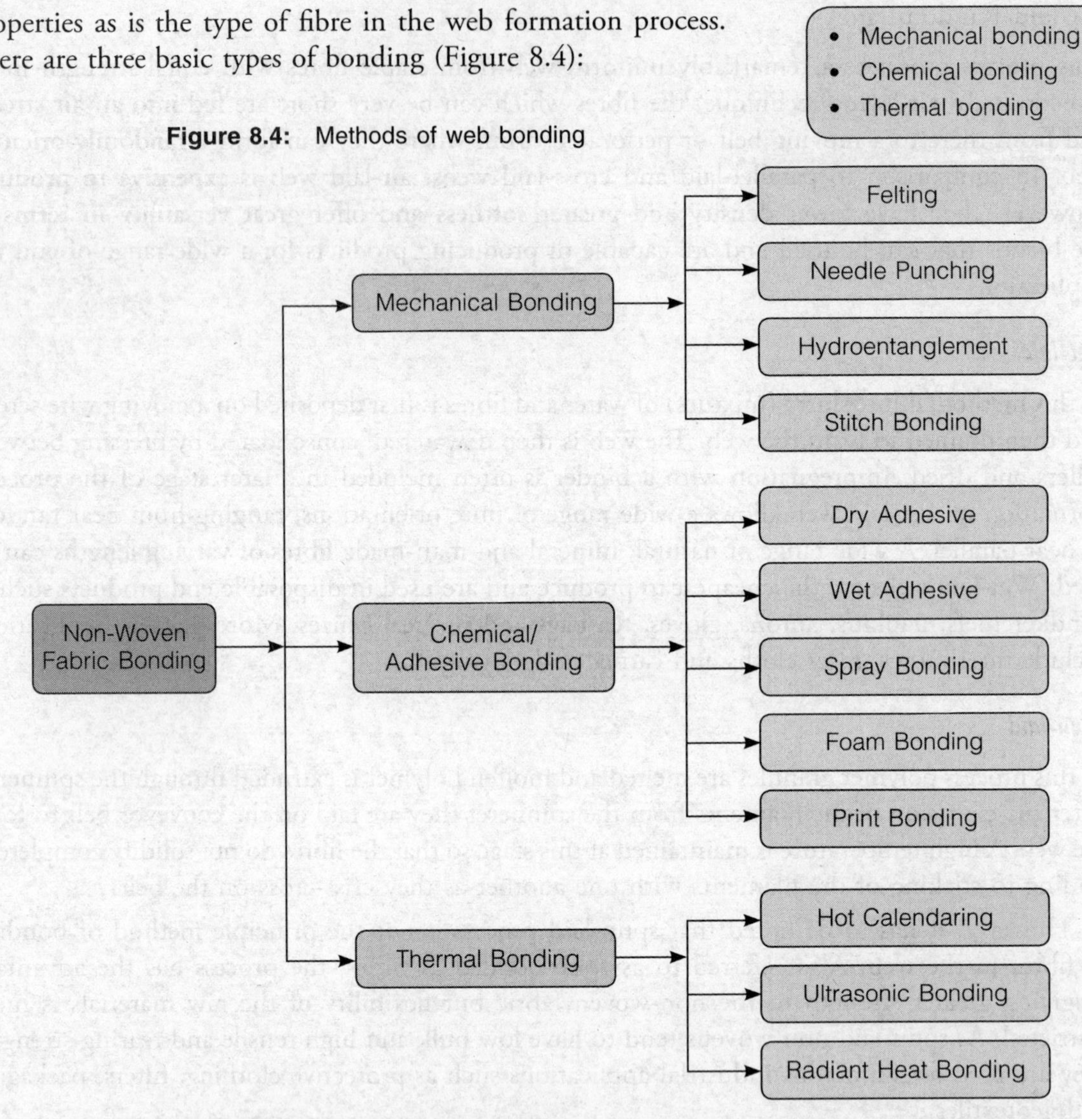

Source: Drawn by the authors.

In **mechanical bonding** the strengthening of the web is achieved by inter-fibre friction as a result of the physical entanglement of the fibres. There are four types of mechanical bonding: felting, stitch bonding, needle punching and hydro-entanglement. Felting process utilises the scaly structure of wool for entanglement. Needle punching can be used on most of the fibres in which specially-designed needles are pushed and pulled through the web to entangle the fibres. Hydro-entanglement is mainly applied to carded webs and uses fine, high pressure jets of water to cause the fibres to interlace. A stitch-bonded non-woven fabric is made on a weaving machine that bonds the web, or holds the web in place, with longitudinal yarns. The stitch-bonding method gives non-woven fabric a texture as soft as that of the original web.

Chemical bonding mainly refers to the application of a bonding agent in liquid or dry form to the web. There are many ways of applying the binder. It can be impregnated, coated, sprayed or applied intermittently as in print bonding. **Impregnation** or **saturation bonding** can be done between screens or rollers. In **print bonding**, the print paste contains the adhesive. Print bonding is used when specific patterns are required and where it is necessary to have the majority of fibres free of binders for functional reasons. In **spray bonding** the adhesive is sprayed on both sides of the web. Bonding can be brought about by either curing/heating or drying depending upon the type of adhesive used. The bonding agents that are commonly used are acrylate polymers, styrene butadiene copolymers and vinyl acetate polymers. Water-based systems are the most widely used, but powdered adhesives, foam and, in some cases, organic solvent solutions are also used.

Thermal bonding method uses the thermoplastic properties of certain synthetic fibres to form bonds under controlled heating. In some cases the web fibres themselves are used, but more often a low-melt fibre is introduced at the web-formation stage to perform the binding function later in the process. On application of heat, the thermoplastic fibres soften and fuse.

Finishing

The non-woven fabrics can be given suitable finishing treatments to impart aesthetic as well as performance-enhancing properties. This can be achieved by dyeing, printing or application of finishes which we will see in subsequent chapters.

TYPES OF NON-WOVEN FABRICS

Non-woven fabrics are broadly divided into two main types on the basis of the raw materials used and the process of manufacture. These are:

- Bonded fabrics
- Felts
    - Woollen felts
    - Needle punched felts

Bonded Fabrics

Bonded fabrics are generally made by either the thermal or chemical methods of fusing fibres together. The processing and manufacturing of bonded fabrics vary with the fibres, method of laying the fibres and the bonding agents used. The general method of constructing these fabrics is as follows: The fibres of staple length, ranging from about 1½″ to 6″ in length are selected. The fibres are processed through a series of opening, cleaning and blending operations. Layers of webs of fibres are then formed. Dry-laid, wet-laid, spun-laid and melt-blown techniques could be employed to form the web. Adhesion of fibres is accomplished by thermal or chemical bonding. Dyeing is usually done after the fabric has been formed. Fabrics may be calendared for smoothness or embossed for textured effects. Softeners may be added to improve the handle. Cotton, rayon, wool, acetate, nylon, polyester, acrylic, modacrylic and polyethylene fibres are used separately or in combinations to make bonded fabrics.

Characteristics of Bonded Fabrics

- The appearance of these fabrics may be paper-like, felt-like or similar to that of woven fabrics.
- They are easy to sew. They have no grain and they do not ravel.
- They may be as thin as tissue paper or as bulky as the thickest padding.
- They may have a soft, resilient handle or may be hard, stiff with little pliability.
- They may be translucent or opaque.
- Porosity may range from high with free air flow to totally impermeable
- Drapability varies from good to none at all.
- Some fabrics have excellent launderability and some have none.
- In appearance, texture and strength they may be similar to a woven fabric.

Uses of Bonded Fabrics

A spectrum of products with diverse properties can be produced with bonded fabrics. They may be used to form disposable as well as durable end products. Disposable uses include dust cloth, sanitary napkins, diapers, industrial masks, towels, under pads, surgical pads, masks and bandages, while durable uses include home furnishings, such as bedding, coated fabrics; backing for quilting, draperies, upholstery, mattress padding, carpet backing; and apparel such as caps, interlinings and interfacings. They are also used in industries in the making of geotextiles and filters, and for insulation and road bed stabilisation sheeting.

Felts

Woollen Felts

Felt was probably the first kind of fabric used by primitive humans when they discovered that applying heat and water to wool fibres and pounding it with rocks resulted in a cloth that could be worn to keep warm or used to sit on. This formation of a fabric is due to a unique property of wool or other hair fibres—when subjected to moisture, heat and agitation they tend to coil, interlock and shrink, a process that is known as felting. And this art of producing felted fabrics directly from fibres matted together began before spinning and weaving were invented. Many cultures have legends about the origins of felt-making. One such story is that of St. Clement and St. Christopher. As the legend goes, while fleeing from persecution, the men wrapped their feet in wool to prevent blisters. At the end of their journey, the vigorous movement and sweat had turned the wool into felt socks. Today felt is made from wool with or without the admixture of another animal fibre, vegetable fibre or man-made fibre.

Construction

Felting is based upon the physical characteristics of the crimp and scaly surface of the wool fibres that cause them to cling and intermesh and allow them to be pressed into a compact fabric. The fibre content of felt ranges from 50 per cent to 100 per cent wool, with other fibres generally being rayon or cotton. The steps in the making of wool felt are as follows:

- The selected fibres are subjected to blending, opening and cleaning.
- Web formation is achieved by carding process. Two carding operations make the fibres parallel and of even thickness in the form of a fine web. This fibre web is also known as **mass** or **batt**. The mass or batt (layers) may then be cut, and the edges trimmed to the desired width. A number of batts may be superimposed on one another (usually perpendicular to the adjacent layers) to get the final felt of desired thickness. Finished felt may vary in thickness from a few thousandths of an inch to 3 inches.
- The bonding of the web is achieved by a mechanical process. The batts are evenly sprinkled with warm water, passed over a steam box to thoroughly warm the fibres, and then pressed between two rollers. The top roller rests on batt and with an oscillating motion exerts the pressure that, combined with moisture and heat, produces the final felting action. After the batts have been processed, they are allowed to drain and cool off for about 24 hours.
- The felt/web is further strengthened by the **fulling** process. In this process, the felt is dampened with a suitable lubricant, such as a combination of soap and soda (sodium carbonate), and then subjected to pounding action with hammers. The longer the felt is fulled, the firmer it will be. But if this operation is continued for too long, the felt will be spoiled.
- Finishing of felt is done by washing, stiffening, ironing and brushing to raise the nap, and shearing to produce a smooth, even surface.
- Finally, if desired, the felt may be made water-repellent, flame-retardant or moth-proof.

Properties of Woollen Felts

- Felts have no warp, filling or selvedge which simplifies their use in garment construction.
- They do not fray or ravel as they do not have a system of threads.
- They absorb sound and shocks.
- They are easy to shape.
- They have good to excellent resilience and will retain their shape unless subjected to undue tension.
- They have high thermal insulating properties and provide warmth.
- They are more impervious to water than untreated woven or knitted fabrics.
- Felts show little tensile strength as fibres are not twisted or interlaced.

- They lack pliability and elasticity.
- They have poor drapability.
- They need to be dry-cleaned as they shrink.

Uses of Woollen Felts

- They are used industrially for padding, sound-proofing, insulation, filtering, polishing, and wicking.
- They have wide use in manufacturing hats, house slippers, clothing decorations, crafts, home furnishing items such as rugs, floor matting, under liners for upholstered furniture, table pad and pillow backing.
- Flexible felts are used for apparel such as skirts and jacket, and thick felts fabric are used in insulation.
- The **Namda** is a traditional floor mat used popularly in Kashmir (Image 8.1). It is made of wool felt.

Image 8.1: Namda rug from Kashmir made from wool felt

Photo credits: 'Bilal Bahadur/Kashmir Life (http://www.kashmirlife.net/woolen-treasure1004-53807/)'

Needle Felts

Needle felts resemble woollen felt in appearance, but they are made primarily from fibres other than wool. The fabric is produced by inserting needles in fibre webs and is therefore called needle-punched felt. These fabrics have high density but retain the same bulk. They are available in different weights and thicknesses. These fabrics are characterised by 3-D fibre entanglement produced by the mechanical action of barbed needles without any application of heat, moisture and pressure.

Construction

The fibres, which may vary from short to long lengths, are cleaned and blended and taken for web formation. The fibre web or matt is prepared by carding or air-laying technique. The web is then fed into the machine and moved between two sets of plates—a metal plate and a stripper plate (Figure 8.5). The needle-punching machine has specially-designed barbed needles which are essentially hooks mounted on a needle board that moves up and down. The movement of the needles through the perforated plates punches the web and brings about mechanical interlocking among the fibres. Filament fibres may be inserted into the web to improve the strength and structural integrity of the needle-punched fabric. They are finished by pressing, steaming, calendaring, dyeing and embossing. Pile fabrics/surfaces can be created if fibres are pulled above the web surface.

The strength of needle-punched fabrics is affected by the arrangement of fibres within the web. If fibres are placed parallel to each other, the finished fabric will have good strength in that direction but will tend to be weak in the opposite direction. If fibres are in a random arrangement, the strength is equal in all directions. The number of needles, the type of barbs on them, their angularity and depth of penetration, and the kind of fibre used determine the characteristics of the fabric produced.

Figure 8.5: Schematic of manufacture of needle-punched felts

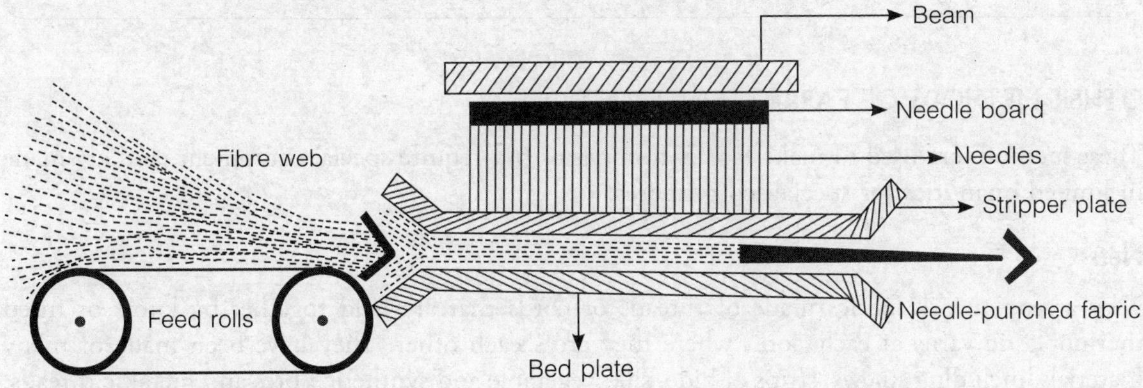

Source: Drawn by the authors.

Properties of Needle-punched Felts

- The fabric is impervious to moisture.
- When it is produced from appropriate fibres it has low flammability characteristics.
- Blankets produced by this method may be lofty and soft.

Uses of Needle-punched Felts

- Attractive blankets, carpets, floor coverings, wall coverings, padding material, insulation material, industrial fabrics and fabric for various types of vehicles are made by needle punching.
- Indoor/outdoor needle-punched carpeting made of olefin fibres are used extensively.
- The Indian army has developed a ballistic protective clothing vest for combat use from needle-punched fabrics.

Care of Non–wovens

Care and maintenance of non-woven fabrics depends on several factors. These are: fibre used, web formation in terms of thickness and direction of fibre lay, method of bonding, adhesive used, finishes and colour applied. If a care label is present, it should provide the directions for care. Because of the lack of yarn structure, wrinkling or twisting of thin non-woven fabric should be kept to a minimum.

Box 8.1: INDA: International Non-wovens and Disposables Association

INDA is the association of the non-woven fabrics industry. It serves hundreds of member companies in the non-woven/engineered fabrics industry to connect, innovate and develop their businesses. INDA educational courses, market data, test methods and consultancy help members by providing them the information they need to plan and execute their business strategies. When it was started in 1968, the association was known as 'The Disposables Association'. In July 1972, the name was changed to reflect the scope of its interests and the membership it serves more accurately. INDA is well-known for its conferences, expositions and training courses as well as for monitoring government policies affecting the industry.

OTHER METHODS OF FABRIC CONSTRUCTION

These methods are used to make fabrics from yarns but require special equipment, and are made in limited quantities for specialised purposes.

Nets

Nets are openwork fabrics made of threads or cords that are held together by knots or fused thermoplastic yarns at each point where they cross each other. They have been made of many materials, including sinews, strips of hide, silk, vegetable and synthetic fibres and metallic threads. Meshes could be square, hexagonal or octagonal.

Nets were originally made by knotting at the points of intersection. Before the 1800s all nets were knotted by hand, but the process was later mechanised and machine nets were made. A large proportion of nets are presently being made by Tricot and Raschel warp knitting machines. Instead of knots at the intersections they have loops and are not as durable as the knotted nets.

Nets are generally fragile and require care in cleaning and handling. Care depends on the type of fibre and yarn and the size of the opening. They have to be protected against snagging and abrasion damage (which can be prevented by ensuring minimal rubbing of the net against another surface). Applications of this fabric include nets for capturing fish, birds and insects, a mesh for holding the hair in place, sports products like rackets in tennis and badminton, goal in soccer, hockey and lacrosse. Nettings are also used for safety nets and for making hammocks.

Types of Nets

- *Bobbinet*: It is a thin to medium-weight hexagonal net; typical use is for bridal veil.
- *Tulle*: It is a fine, stiff netting of hexagonal mesh and is generally used as trimming for dresses.
- *Malines*: It is a very thin, diaphanous diamond-shaped net.

Laces

Lace is another basic fabric made from yarns. Lace, which is a derivative of netting, has its origins in antiquity. It is an openwork fabric with complex patterns or figures, is either hand-made or made on special lace machines. It consists of a decorative design created with threads on a net-like open background. The technique of lace-making involves looping, knotting, braiding, twisting or stitching thread into decorative, openwork patterns. Yarns may be twisted around each other to create open areas. Laces may be full fabric width, generally used to make formal dresses or narrow width to be used as a trim. The early 1800s saw the first machine-made laces. Leavers machine produced fine and intricate Leavers lace. Nottingham lace was a heavier rough-textured lace made on the Nottingham machine. The better-known knotted laces are **tatting** and **macramé**. These are made of heavy yarn knotted into geometric patterns. Raschel knitting machines make laces that look similar to Leavers lace but are produced at much higher speeds and thus less expensive. Coarser laces are generally made from filament yarns and used for tablecloths, draperies, etc.

Laces need special care since they can easily snag and tear. They should be hand-washed without rubbing or agitation. They can also be machine-washed by putting in a protective bag, or dry-cleaned.

Braids

Braids are narrow fabrics in which yarns are interlaced lengthwise and diagonally. They are made by plaiting or braiding three or more yarns or fabrics. Narrow braids can be joined together to form wide fabrics. They have good elongation characteristics and considerable stretch in the lengthwise direction. They are very pliable and curve around the edges nicely. Three-dimensional braids are made with two or more sets of yarns. Braids are of two types:

- Flat braids such as those used for trimming.
- Round braids used as shoe laces, ropes and insulation for wires.

Braided fabric has to be given care appropriate for the fibres used and the method of construction. Loose stretchy braids need careful handling while compact braids can be cleaned and laundered using normal care procedures.

They are used to make straw hats, small rugs, as a narrow fabric for trimmings, as cords for shoe lace, parachutes and gliders, as covering cord for tyres, tubing, hose, wires and cables, and as ropes, tapes and elastic of various types.

Crochet

Crochet is a process of creating fabric by interlocking loops of yarn, thread, or strands of other materials using a crochet hook. The name is derived from the French term 'crochet', meaning small hook. The important difference between crochet and knitting is that each stitch in a crochet is completed before proceeding to the next one, while knitting keeps a large number of stitches open at a time. Stitches are made by pulling one or more loops through each loop of the chain. The crochet hook comes in many sizes and materials, such as bone, bamboo, aluminium, plastic, and steel. Like knitting, crochet can be worked either flat or circular.

Macramé

Macramé or macrame is a technique of making textiles using the technique of knotting. Materials used in macramé include cords made of cotton twine, linen, hemp, jute or leather. Macramé can produce self-supporting three-dimensional textile structures, as well as flat fabrics, and is often used for ornamentation. Leather or fabric belts, friendship bracelets, jewellery, bags, and other decorative articles and wall hangings are often created through the macramé technique.

> A **knot** is a method of fastening or securing linear material such as rope by tying or interweaving. While many textiles use knots to repair damage, they can also be applied in combination to produce complex objects such as netting or for making macramé textiles.

Stitch-bonded Fabrics

Stitch-bonding uses needles and threads/yarns to stitch yarns or fibre webs into a fabric. Four different types of fabrics can be made by this method: malimo, schusspol, maliwatt and malipol.

Malimo fabrics are made using two groups of yarns: one lengthwise and the other widthwise. These layers of yarns are stitched together using warp-knit stitch tricot, which looks very similar to chain stitch. When the stitching threads are finer than the base yarns, the fabric resembles a woven structure since the stitching threads disappear into the surface. When the stitching threads are thicker than the base yarns it resembles a knit fabric. These fabrics are used for upholstery, draperies, conveyor belts, packing materials, etc.

Schusspol fabrics need a sheet of weft yarns, threads for stitching and another set of yarns to form the pile surface. The pile yarns are held together by the stitching yarn and form the raised surface only on one side. These fabrics are used for floor coverings, upholstery, wall coverings, etc.

Malliwatt fabrics are made by stitching a web of fibres with yarns/threads. The base web could be either cross-laid or random-laid. These are used for interlinings for apparel, padding for carpets, insulation and temperature-control fabrics.

Malipol uses preconstructed fabric as the base to which yarns are stitched to form a pile surface. Pile can be left cut or uncut and looks like terry cloth or velveteen structures. They are used as imitation furs, floor coverings, towels, etc. and are less expensive than their woven counterparts.

Quilted Fabrics

Quilted fabrics consist of three layers: (*a*) top/face fabric; (*b*) fibre in a mat form (cotton, down, fibrefill or foam) for the middle layer; and (*c*) a backing layer of fabric. Most quilted fabrics are made by stitching these three layers with a thread. Polyester fibrefill is most widely used in quilted fabric and provides excellent resiliency combined with good insulation. Foam results in a stiffer fabric with poor drape. The quality and durability of the finished fabric is largely dependent on the type of thread and stitch used for quilting. Quilting threads must be durable so that they can hold the structure together. Patterned fabric can be used for the face of the quilt, but for reversible quilts patterned fabrics should be used on the front and the back too. Quilting can be done in straight and curved lines and it may outline printed figures/designs. The stitching should be close enough to prevent slippage and movement of the middle layer. Size of the stitch should be small so that there is less breakage during use and care. For all end uses, fabric beauty and fashion appeal are the prime considerations. All the three layers of the quilted fabric must be compatible in terms of care so that they retain the shape and size and exhibit similar dimensional stability. Some quilted products can be easily laundered but others require cleaning or dry-cleaning by professionals. Quilted products are used for both home furnishings and apparel.

Laminated Fabric

A laminated fabric is a layered structure where a base fabric is joined to a continuous sheet material such as polyurethane foam. When only one layer of fabric is attached to a layer of foam it is termed as foam-backed fabric. When two layers of fabric are used with the layer of foam in between it is called a sandwich construction. Laminated fabrics may also be made by a thin layer of film bonded to a woven or knitted fabric. Foam may or may not be used between them. These fabrics are called **supported films**. An additional method for laminating uses a thermoplastic material as an adhesive between the outer and lining fabric. The middle layer can be a thin micro porous film that allows the material to breathe.

> Quilted and laminated fabrics are two kinds of **layered fabrics**.

Tufted Fabric

Tufted fabrics are another type of pile fabrics but, unlike the pile produced by weaving and knitting processes, the base fabric could be a non-woven fabric also. The structure is produced by inserting extra yarns into an already woven fabric of relatively open weave. A row of needles placed along the width of the fabric are provided with a yarn from a spool. The needles come down together and go through the fabric. A hook holds the loop as the needle is withdrawn. Yarns that form the tuft or the pile effect may be of any fibre. Tufts can be cut or left uncut (Figure 8.6). Tufted fabrics are less expensive than the woven and knitted pile fabrics. End uses include blankets, rugs, carpets and bedspreads.

Figure 8.6: Tufted fabric

Source: Drawn by the authors.

SUMMARY

- Non-woven fabrics are textile structures made directly from fibres bonded together by mechanical entanglement, chemical or thermal means.
- Two types of non-woven structures are bonded fabrics and felts.
- Two types of felts are woollen felts and needle punched felts.
- Stitch-bonded fabrics use needles and threads/yarns to stitch yarns or fibre webs into a fabric.
- Quilted fabrics consist of three layers: face fabric on the top, fibre in a mat form as the middle layer and a backing layer of fabric.
- A laminated fabric is a layered structure where a base fabric is joined to a continuous sheet material such as polyurethane foam.
- Nets, laces, braids and macramé are other methods of making fabrics from yarns. The unique characteristic of these fabrics is their open-mesh structure.
- Structures similar to pile fabrics can also be made using the process of tufting. Tufted fabrics are produced by inserting extra yarns into an already woven fabric of relatively open weave. The base fabric could also be a non-woven fabric.

KEY WORDS

Non-woven: Non woven fabrics consist of one or more layers of fibres that are bonded together by chemical, thermal or mechanical processes to be made into textile products.

Felt: Felt is made from wool with or without the admixture of another animal fibre, vegetable fibre or man-made fibre. The felting property of wool is exploited to effect the bonding of fibres in the web.

Nets: These are openwork fabrics made of threads or cords that are held together by knots or fused thermoplastic yarns at each point where they cross each other.

Laces: Lace is an openwork fabric with complex patterns or figures, hand-made or machine-made on special lace machines.

Braids: Braids are narrow fabrics in which yarns are interlaced lengthwise and diagonally. They are made by plaiting or braiding three or more yarns or fabrics.

Crochet: It is a process of creating fabric by interlocking loops of yarn, thread, or strands of other materials using a crochet hook.

Macramé: It is a technique of making textiles using knotting rather than weaving or knitting.

EXERCISES

1. Enumerate various methods of fabric-making besides conventional weaving and knitting.
2. What are the various kinds of non-woven fabrics?
3. Enumerate the various methods of web making and web bonding.
4. What are the major uses of non-woven fabrics?
5. Distinguish between the following:
   - Mechanical and chemical methods of web bonding

- Needle-punched felts and woollen felts
- Stitch-bonded fabrics and non-woven fabrics
- Laminated fabrics and bonded non-woven fabrics

REFERENCES

American Society of Testing Materials (ASTM). 1976. *Compilation of ASTM Definitions*. Third edition. Philadelphia: ASTM.

Cohen, C. Allen and Johnson Ingrid. 2010. *J. J. Pizzuto's Fabric Science*, Ninth edition, New York: Fairchild Books.

Corbman, P. Bernard. 1983. *Textiles: Fibre to Fabric*. Sixth edition. New York: McGraw-Hill Inc.

Joseph, L. Marjory. 1988. *Essentials of Textiles*. Fourth edition. New York: Holt, Rinehart and Winston Inc.

Kadolph, J. Sara. 2009. *Textiles*. Tenth edition. New Jersey: Pearson Education.

Sekhri, Seema. 2011. *Textbook of Fabric Science: Fundamentals to Finishing*. New Delhi: PHI Learning Pvt. Ltd.

ONLINE SOURCES (all accessed 12 May 2016)

www.textileworld.com

www.edana.org

www.engr.utk.edu

https://iheartindian.wordpress.com/2016/02/04/namda-the-amazing-felted-rugs-from-kashmir/

UNIT III

FINISHING, DYEING AND PRINTING OF TEXTILE FABRICS

Depending on our end-usage, we may require fabrics with different functionalities and aesthetics. These functionalities and aesthetics can be added to the fabrics by applying certain types of finishes and by dyeing and printing. Five chapters of this last unit are devoted to these various mechanical and chemical processes of preparing the fabric, while the last chapter details the methods of appropriately labelling textiles for the consumer. Various routine and special finishing processes have been discussed in Chapters Nine and Ten, respectively. These finishes significantly alter the performance and various characteristics of the fabrics. Colour application or dyeing of fabrics is dealt with in two chapters: Chapter Eleven which gives an overview of the basics of colouration, the stages of dyeing and fastness properties of dyed fabrics; and Chapter Twelve which discusses in detail the various types of dyes and their dyeing mechanisms on various textile substrates. The aesthetic appeal of textiles can be further enhanced by various methods and styles of printing which are dealt in Chapter Thirteen. Finally, labels and certifications that help us make the right selection of textile products are discussed in Chapter Fourteen. This chapter also includes information on various international and Indian organisations working in the area of developing standards and specifications related to textiles on one hand and consumer protection and redressal on the other.

9

ROUTINE FINISHES

HIGHLIGHTS

- Classification of finishes
- Purpose of routine finishes
- Types and application of routine finishes
  - Singeing
  - Desizing
  - Scouring
  - Bleaching
  - Mercerisation
  - Degumming
  - Weighting
  - Carbonising
  - Crabbing
  - Decating
  - Fulling
  - Optical brightening
  - Heat setting
  - Tentering
  - Calendering
  - Sizing
  - Softening
  - Stabilisation

The term **finish** is used to refer to the treatments that change the appearance, handle and performance of the fibre, yarn or fabric. The main objective of finishes is to impart desired end-use properties to the textile materials. **Greige** (or grey) fabric is an unfinished woven or knitted fabric which may contain fatty, waxy and resinous impurities. It may also contain the sizing agents employed during fabric manufacture. These impurities render the fabric non-absorbent and makes subsequent processing, dyeing and printing difficult. Moreover, at this stage, the fabric may not have all the properties to fulfil the intended end use. Therefore, these impurities must be removed by applying various kinds of finishing treatments. The technique of finishing is capable of wide variation, but it basically depends on: (*a*) the physical and chemical properties of the fibre/fibres; and (*b*) the type of yarn and fabric construction. Finishes have a wide variety of functions which makes the fabric more suitable for its intended use such as:

> **Sizing agents** are compounds that are applied to warp yarns during weaving to impart strength, reduce hairiness and minimise breakage on the loom; e.g. starch, polyvinyl alcohol.

- Accentuate or inhibit some natural characteristic of the fibre.
- Impart new characteristics or properties to the fabric.

- Increase life and durability of the fabric.
- Set the fabric so that it maintains its shape and structure.

CLASSIFICATION OF FINISHES

Textile finishing is a broad term and the various finishes can be classified as below:

Preparatory or Basic or Routine Finishes

Preparatory finishes are customarily given to fabrics before dyeing or final processing. These can also be referred to as basic or routine finishes as these are applied in routine processing of fabrics. They impart some basic properties to the fabrics as well as prepare them for further processing, dyeing or printing. Natural fibres have various kinds of impurities, such as grease and vegetable matter (in wool), gum (in silk) or vegetable matter and waxes (in cotton). Preparatory finishes help remove these impurities.

Special Finishes

Special finishes include those finishes that are not applied in routine processing of fabrics but are used to obtain a specific effect or property depending on the end use. These finishes are applied to limited lengths of fabric as the properties imparted are very specific/special. The finishes can be further characterised into aesthetic and functional finishes.

1. *Aesthetic finishes*: These finishes have the primary function to change the appearance and/or handle of the fabric and modify the texture of the material.
2. *Functional finishes*: Functional finishes, also known as performance finishes, are those which improve the performance or characteristics of the fabric and impart certain functionality to the fabric. Functional finishes are either applied on the surface of fibres, yarns and fabrics (as film-forming finishes) or they enter the fibre structure and bond chemically. Functional finishes do not generally alter the appearance of the fabric but may modify the handle to some extent. Therefore, the parameters for the application of these finishes should be carefully monitored.

Finishes can also be classified on the basis of the length of the time they remain on the fabric. From this perspective, finishes can be classified as temporary, durable and permanent. A **temporary finish**, also known as renewable finish, lasts until the product is washed or dry-cleaned. Some of these finishes can be renewed or reapplied after washing/dry-cleaning. **Durable finish** is a finish that lasts longer than a temporary finish but not for the life of the product. A **permanent finish** is one that bonds with the fibre structure or brings about permanent change and hence lasts the life of the product. Table 9.1 lists all the finishes which are either defined as preparatory or special (aesthetic and functional).

Table 9.1: Classification of finishes

Preparatory/basic/routine finishes	*Special finishes*	
	Aesthetic finishes	*Functional finishes*
Singeing	Schreinering	Flame retardant
Desizing	Emerising	Antimicrobial
Scouring	Embossing	Insect and mite resistant
Bleaching	Moireing	Ultraviolet protection
Mercerisation	Raising (gigging, napping, flocking)	Anticrease/durable press
Degumming	Polished surfaces (chintz, cire)	Abrasion resistant
Weighting	Acid finish	Water repellent and water proof
Carbonising	Basic finish	Antislip
Crabbing	Shearing	Antistatic
Decating/decatising	Beetling	Stabilisation finish
Fulling/milling	Glazing	Stain- and soil-resistant
Optical brightening		Soil release
Heat setting		
Tentering		
Calendering		
Sizing/stiffening		
Softening		
Mechanical stabilisation		

Source: Compiled by the authors.

PREPARATORY/BASIC/ROUTINE FINISHES

Singeing

The term 'singe' means 'to burn'. Fabrics manufactured with spun yarns have certain protruding ends which interfere with finishing. Therefore, it becomes important to remove these protruding ends. Singeing is a finish applied to burn away the 'fuzz' produced due to the protruding fibre ends on the yarn or fabric surface. It is a chemical finish since it involves oxidation reaction. It is the first step in finishing. Singeing is carried out on cotton and wool yarn and fabric to obtain smooth surface. Singed fabrics soil less easily and have reduced risk of pilling. Singed fabrics allow printing of fine intricate patterns with high clarity and detail.

During the singeing process, the fabric is opened in full width. It is then brushed lightly to raise the fibre ends on the fabric surface. Fabric is then moved rapidly in full width over heated copper plates or a direct flame to destroy the fibre ends. Singeing is immediately followed by quenching in water to extinguish the smouldering fibres. If singeing is followed by desizing, the singed fabric is immersed in the desizing bath. During singeing, the fabric moves at the speed of 200–250 metres per minute. The passage of the cloth can be arranged in such a manner that one or both sides of the fabric may pass over and be in contact with the heated plate(s), in order to accomplish singeing of one or both sides of the fabric in a single passage. This process is carried out on a plate singeing machine. The drawback of this process is that the protruding ends present in the interstices of the fabric may not get singed completely, thus reducing the efficiency of the process.

Apart from this, there are two more types of singeing machines: rotary-cylinder singeing machine and gas singeing machine. In rotary-cylinder singeing machine, the cloth passes over and is in contact with a heated rotary cylinder made of copper or cast iron. The direction of rotation of the cylinder is opposite to the direction of the fabric so that the protruding fibres or nap of the fabric is raised. This type of machine is particularly suitable for the singeing of velvets and other pile fabrics. If singeing of both sides of the fabric is required, then two cylinders are employed, one for each side of the fabric. In gas singeing machine, the fabric passes over a burning gas flame at such a speed that only the protruding fibres burn and the main body of the fabric is not damaged by the flame. This is the most common type of machine used for singeing fabrics as well as yarns.

The fibres that protrude on certain fabrics, giving them a raised surface is known as **nap**. Such fabric surfaces are said to be napped fabrics.

During singeing, the burning characteristics of the fibres must also be considered. Heat sensitive fibres melt and form tiny balls on the surface of the fabric. These balls interfere with dye absorption. Filament yarns do not require singeing as there are no short fibre ends projecting onto the surface of the fabric. Fabrics that are to be napped are not singed. Yarns, sewing threads, felts and carpet backing can be singed. Yarn singeing is usually called **gassing**.

Desizing

During weaving, sizing agents such as starch or polyvinyl alcohol are applied on the yarns in order to impart strength to the yarns. However, these sizing agents impair the wettability of the fabric and result in the non-uniformity of the dyeing and finishing processes. Therefore, once the fabric is made, these materials need to be removed to improve the wettability of the fabric and make it receptive to subsequent chemical treatments. Desizing is a chemical finish as it involves chemical action on the sizing substances present on the fabric.

Water soluble sizes are removed with the help of surfactants by immersing the fabric in hot water. Starch can be removed either by steeping the fabric in warm water or a mild acid solution for a certain period of time or by treating the fabric with enzymes. The enzymes break up the starch polymers into smaller molecules that can be washed away. Starches can also be removed by the oxidative method in which oxidising agents degrade and depolymerise the synthetic sizes. Polyvinyl alcohol can be removed by hot alkaline wash.

Scouring

Scouring is a preparatory process to remove natural and acquired impurities by wet treatments and is therefore a washing treatment. The scouring treatment will depend on the amount and types of impurities present in a material which will further depend on the type of fibre in the material. Grey cotton contains waxes, pectins and vegetable matter which must be removed in scouring. For scouring of cotton, alkali (sodium hydroxide) and surfactant are common scouring agents. Alkali emulsifies waxes and fat on the cotton fibres to form soap. This soap then emulsifies the remaining wax and washes away any dirt or other impurities. Scouring of cotton was traditionally done in the kier vat or J-box. Nowadays, the enzyme pectinase is used to scour cotton fabric. Protein fibres are damaged in strong alkaline conditions and so they are scoured using neutral synthetic detergents. The treatment temperature is also lower than that for cotton so as to avoid damage to the protein fibres. Man-made fibres have very small amounts of impurities in the form of size or spin oils, or some dirt which may be acquired during their storage or transportation. They therefore require very mild scouring which is usually carried out using neutral detergents. Scouring can be applied as a separate step or in combination with other treatments (usually bleaching or desizing).

Bleaching

The main purpose of bleaching is to decolourise the impurities that mask the natural whiteness of fibres. Cotton and other natural fibres contain natural pigments that absorb light causing the fibres to have a creamy, yellowish or dull appearance. Bleaching is therefore required unless the material will be dyed very dark or dull shades. Bleaching is a chemical finish and may be durable depending on the process used and subsequent treatment and usage conditions of the product. The classification of different bleaching agents is given in Figure 9.1.

Reducing bleaching agents destroy the colouring matter by reductive reaction. Commonly used reducing agents are sodium dithionite, popularly known as sodium hydrosulphite or Hydros, sodium bisulphite and sodium thiosulphate. Reducing agents are not commonly used for bleaching as the white produced is not permanent. The pigments in fibres may reoxidise and return to the original colour. One of the oldest methods of bleaching by reducing action is **stoving** of wool and silk. In this method, moist wool or silk yarns are exposed overnight to the fumes of SO_2 produced by burning sulphur. This method gives adequate bleach for low-quality goods, is cheap and requires simple equipments. However, this method is no longer in use.

Nowadays, oxidising agents are used to bleach textile fibres. Oxidising agents produce active oxygen which is responsible for the bleaching action, by destroying the colouring matter present in textile materials and leaving them white. The commonly used oxidising bleaching agents are hydrogen peroxide, sodium hypochlorite and sodium chlorite. Hydrogen peroxide is the most widely used bleaching agent. It is also known as universal bleaching agent as it can be used on most textile fibres. The bleaching processes must be closely controlled so that the colour in the fibres is destroyed while damage due to oxidation of the fibrous material is minimised.

While cotton is commonly bleached using chlorine bleaches (as in our homes), commercially, cotton is bleached with hydrogen peroxide. Chlorine-based bleaches are not used for wool and silk as they may severely damage the protein fibres. Hence, hydrogen peroxide, under mild alkaline conditions, is used for these fibres. Chlorine-based bleaches react to generate chlorinated

Figure 9.1: Classification of bleaching agents

Bleaching agents		
Oxidising		**Reducing**
Peroxide	Chlorine	Sulphurdioxide
Hydrogen peroxide	Sodium hypochlorite	Sodium hydrosulphite
Sodium peroxide	Sodium chlorite	Sulphoxylates
Sodium perborate	Lithium hypochlorite	Sodium bisulphite
Potassium permanganate		Sodium thiosulphate
Peracetic acid		

Source: Drawn by the authors.

volatile organic compounds (AOX), which are toxic to fish and other aquatic organisms when released with the effluent. Considering the effluent problems (AOX) created by hypochlorite bleaching, their use is being discouraged and hydrogen peroxide has gained importance.

> **AOX** is the abbreviation for Adsorbable Organic Halides such as chlorides and bromides.

Manufactured fibres do not contain any natural pigment, hence they require very mild or no bleaching unless they have been stained during manufacturing and/or storage and transport.

Mercerisation

Mercerisation is the treatment of cellulosic textiles, primarily cotton, with a concentrated solution of sodium hydroxide. The process is named after John Mercer who discovered it in 1850. Mercerisation is an optional step in the finishing of cotton and cotton-blends. During this process, cotton yarns or fabrics are immersed in 18–30% solution of sodium hydroxide (NaOH) at about 20°C under controlled conditions. The excess liquor is squeezed and a short time (about 30–60 seconds) is allowed for the solution to penetrate in the fibre. The alkali is then washed off and material is neutralised with acetic acid. At normal mercerising temperatures of 20°C or below, penetration of mercerising liquor into yarns and fabrics may be slow and incomplete and may result only in surface swelling. Therefore, hot mercerisation is being practised in which NaOH is applied at 60°C. Hot mercerisation results in fast and uniform swelling.

During mercerisation, a number of physical, chemical and structural changes take place in cotton. Under the action of concentrated NaOH, native cellulose (cellulose I) forms alkali cellulose which is converted to cellulose II upon washing and neutralisation.

$$\underset{\text{Cellulose I}}{C_6H_7O_2(OH)_3 + NaOH} \longrightarrow \underset{\text{Alkali cellulose}}{C_6H_7O_2(OH)_2(ONa)} \xrightarrow[\downarrow H_2O]{} \underset{\text{Cellulose II}}{C_6H_7O_2(OH)_3 + NaOH}$$

The penetration of sodium ions from the solution into the fibre leads to the breaking of internal hydrogen bonds pushing the cellulose chains apart and causing swelling of the fibre. This results in an increase in the amorphous content and the number of OH groups. These changes are responsible for enhanced moisture absorbency, dye uptake and increased reactivity. Mercerised cotton fabrics have increased lustre which is attributed to their rounder cross-section and reduced convolutions. The polymer chains are pushed apart and there is better distribution of stress resulting in increased strength.

Mercerisation can be of two types: (*a*) mercerising under tension and (*b*) slack mercerisation. Mercerisation under tension results in better lustre and increased strength due to increased orientation of the polymer chains. Slack mercerisation is the mercerising of the fabric in the absence of tension or under reduced tension. In the slack method, moisture and dye absorption increases but not much effect is observed on lustre. After washing-off, the fabric remains in a shrunken state and consequently a high degree of crimp is obtained. Hence, the fabric becomes more extensible. Slack mercerisation is used to produce stretchable fabric which finds its end use in casual wear and sportswear.

Mercerisation can be carried out at both the yarn and fabric stages. Mercerised yarns are used as sewing and embroidery threads and in expensive knitted and woven fabrics, where a high degree of lustre, strength and smoothness is required.

Liquid ammonia treatment has been used as an alternative to mercerisation and is often done prior to the application of functional finishes. The ammonia treatment swells cotton fibres and imparts lustre and improves dyeability just like mercerisation. However, the depth of colour attained is less as compared to mercerised cotton.

Mercerisation of blends of cotton with other fibres depends on the effect of concentrated NaOH on the other fibre. Polyester and polyamide fibres are resistant to normal mercerising conditions but acrylic fibres are adversely affected at these conditions. In cotton-viscose blend, viscose fibre is slightly affected by NaOH at mercerising strength. Linen/flax fibre has natural lustre, therefore, mercerisation is not a frequent process for this fibre.

Degumming

Silk filaments are encased in a gum called sericin which is responsible for the brittleness, harsh handle, matte appearance and yellowish colour of silk. Therefore, silk yarns and fabrics have to be degummed as a preparatory process so as to make silk fabric lustrous, soft and absorbent.

The weight of sericin could be as much as 30 per cent of the total weight of the silk material. A classical degumming method is to treat the fabric with 'Marseilles soap' (an alkali-free olive oil soap) in a bath at 90–98°C for 2–4 hours. Degumming silk in an alkaline solution of sodium carbonate and sodium bicarbonate is cheaper and is therefore commonly practised.

Other methods such as enzymatic or extraction with water are also being used for degumming. In enzymatic degumming, silk fabric is treated with enzymes such as papain and alcalase which act on the peptide bonds causing the removal of sericin from the fibres. The extraction with water involves the boiling of the fabric under pressure at 120°C.

The degree of degumming can be evaluated quantitatively by determining the weight loss or by means of a dyeing test with a Direct Red dye and then measuring the colour value. The presence of sericin will result in a deeper shade as compared to the shade obtained on degummed silk. The extent of degumming can also be assessed qualitatively by viewing the samples under a scanning electron microscope. The existence of lumps on the surface of fibre will show the presence of residual sericin.

Weighting of Silk

Silk is conventionally sold by weight. Degumming of silk causes 20–30 per cent decrease in its weight. In order to make up for this loss in weight, and sometimes to further increase the weight so as to maximise profits, silk is weighted. Therefore, weighting is a kind of sizing of silk which is done after degumming. It is carried out by the addition of metallic salts (tin phosphate-silicate weighting), plant material (tanning agents) or synthetic substances (methacrylamide) onto silk fibroin. Weighting of yarn is done in rope form before it is dyed.

Tin phosphate-silicate weighting process is the most commonly used. The other weighting process using tanning agents is easy and causes less damage but results in only 10 per cent increase in weight and large quantities cannot be weighted. The weighting of degummed silk can also be done by graft copolymerisation process using monomeric methacrylamide but large variation in the degree of weighting limits its usage.

> Tanning agents can be extracted from the following plant sources: *Acacia catechu* (Khair), *Quebracho colorado, Acacia nilotica* (Babul), *Cassia fistula* (Konnoi), *Tamarindus indica* (Tamarind), *Zizyphus xylopyrus* (Kath ber) and *Terminalia chebula* (Myrobalan).

> **Graft copolymerisation** is a polymer modification process where two or more monomers are attached as a side chain or as a branch to the main polymer; thereby altering its structure and properties.

Sericin has to be removed initially during the processing of silk because it obstructs the silk fibre from achieving the necessary handle, lustre and dye uptake. However, sericin can be applied later as a weighting agent where it will function as a finish. In this case, sericin on silk will not only increase the weight of the fabric but also provide antioxidant properties.

Weighting results in increased stiffness and fullness of the fabric. It also imparts improved sheen and handle to the fabric. However there are many disadvantages as well—high cost, ecological impact and damage to the fibre. Weighted silk is prone to water spotting, cracking and splitting. Also, weighted silk is more sensitive to sunlight, air and perspiration as compared to silk with no weighting. Therefore, nowadays, weighting is done only for specific articles such as ties, etc.

A **water spot** is an area of dried mineral deposits left on the fabric surface after it has been allowed to air dry.

Carbonising

Carbonising is a finishing step in the preparatory processing of wool. It is a chemical finish which is used to remove vegetable impurities which may remain on scoured wool. In this process, woollen yarns, fabrics and blends are immersed in 6–9 per cent sulphuric or hydrochloric acid, centrifuged at a constant speed to remove extra liquid and then subjected to high temperature of around 140°C to convert the vegetable matter to carbon particles. The carbonised particles are then removed by beating, brushing and washing the fabric. The fabric is then neutralised to eliminate the presence of any acid and then rinsed again. The process should be carefully controlled to prevent the weakening of the fabric. Irregular carbonising becomes evident in piece dyeing in the form of skitteriness. Woollen textiles (fibre, yarn or fabric) can be carbonised before or after dyeing.

Crabbing

Crabbing is a mechanical finish used to set woollen and worsted fabrics in a smooth flat state so that they will not pucker or wrinkle during subsequent wet processing. In this preparatory finish, wool fabric is treated in the flat, open width form in hot water. The fabric is immersed for 10–30 minutes and then cooled down in open width form by passing through cold water. The exposure to hot water releases the latent stress in the fabric and cooling sets the flat fabric. The term flat setting is particularly used in the finishing of woven wool fabrics, where setting is usually achieved by setting under pressure.

Decating / Decatising

Decating is a mechanical finish which produces a smooth wrinkle-free surface with lustre and soft handle on the woollen fabric and sets the grain of the woven fabric. It is also used to soften the handle of rayon, silk and blends. Decating can be dry or wet. In dry decating, the fabric is wound under tension on a perforated cylinder and steam is forced through the cylinder. The moisture and heat causes the wool fabric to become plastic; tensions relax and wrinkles are removed. The yarns become set in the shape of the weave and are fixed in this position by the cooling off, which is done with cold air. In wet decating, steam and cold air are replaced with hot and cold water, respectively.

Fulling / Milling

When a woollen fabric is taken off the loom, it is loose and hard in texture. To achieve the fabric which will fulfil consumer's satisfaction, in terms of warmth and softness, the fabric is given fulling finish. During this process, appropriate amount of heat, moisture and friction are applied. As a result, the yarns shrink together, thus making the fabric soft and compact. Therefore, fulling makes use of the felting property of wool; the process involves partial felting of the fabric under controlled conditions. Since it is done to close the interstices and open areas in the fabric after weaving, so as to produce a compact, bulkier or fuller fabric the process is termed fulling.

Optical Brightening

Generally, textile materials are not completely white and efforts have been made since ancient times to remove this yellowish tinge. Bleaching in the sun, blueing and later chemical bleaching of the textile increase the brightness of the products and eliminate the yellowish tinge to a certain extent. Application of an optical brightening agent (OBA), also known as Fluorescent brightening agent (FBA), further adds brightness to the fabric. These organic products are based on Diaminostilbene sulphonic acid derivatives (Figure 9.2).

Figure 9.2: Diaminostilbene sulphonic acid derivative

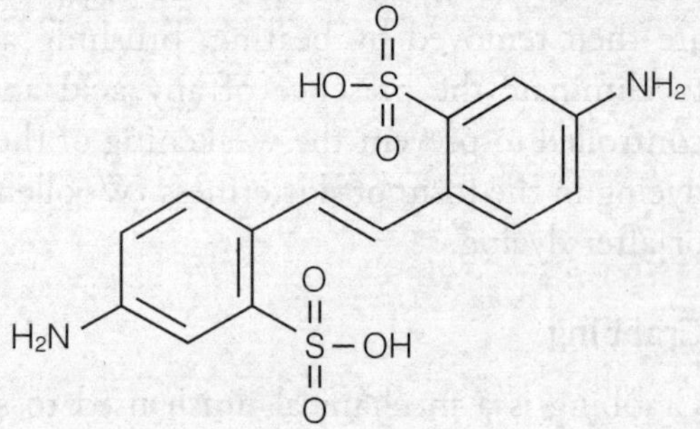

The optical brightening agents absorb invisible ultraviolet light and emit it again within the visible range at the blue–violet end of the spectrum. Thus the total light reflected from the textile surface increases and the material looks brighter. Optical brightening agents can be broadly classified into two groups: (*a*) direct (substantive) brighteners and (*b*) disperse brighteners. Direct optical brightening agents are predominantly water-soluble substances used for brightening natural fibres and occasionally for synthetic materials such as polyamides. Disperse optical brightening agents are mainly water-insoluble and, as with disperse dyes, they are applied either from an aqueous dispersion or added to the spinning solution/dope. They are used for synthetic materials such as polyamide, polyester, acetate and occasionally on paper.

Heat Setting

Heat setting is a mechanical finish that imparts dimensional stability to synthetic (thermoplastic) fibres with the aim to achieve a stable fabric or to produce a special shape. Heat setting is used to produce a smooth, flat shape, pleated or creased fabric. Thermoplastic fibres, such as nylon, polyester and polyolefins, have a melting temperature (T_m) as well as a glass transition (T_g) temperature. At temperature above glass transition, the polymer chains in the amorphous regions start vibrating. The fibres can be shaped and set at this temperature. Therefore, heat setting is carried out at a suitable temperature above T_g. The heat changes the physical characteristics of the polymer. The temperature, period of exposure and the amount of force/tension used to hold the fabric in the desired shape and size, during heat setting affect the dimensional stability of the fabric. Heat setting also affects resiliency of the fabric, which subsequently affects wrinkle resistance and elastic recovery. The temperature and duration of treatment is governed by the conditions that the fabric will be exposed to during subsequent processing, usage and care. Heat set fabrics should be laundered and treated at temperatures below the stabilisation temperatures otherwise the fabrics will tend to shrink, wrinkle and loose shape.

Tentering

Tentering is a mechanical finish that sets the warp and weft of woven fabrics at right angles to each other, stretches and fabric to its final dimensions and dries it. The fabric is stretched by the use of a tenter frame. It consists of chains fitted with pins or clips which travel on tracks and hold the selvedges of the fabric. As the wet fabric passes through a heated chamber, creases and wrinkles are removed, the weave is straightened, and the fabric is dried to its final size. The fabric is then released and rolled on take-up rolls. If the fabric is picked up by the tenter frame in such a way that the filling yarns are not absolutely perpendicular to the warp yarns, the fabric is off-grain and exhibits skew. Due to fabric skewness, the fabric will not hang properly and patterns may become distorted. The grain of the fabric cannot be corrected after heat-setting if the fabric is thermoplastic. However, there are ways by which the fabric can be set on a true grain. A tenter frame with a variable chain drive enables the operator to slow down one chain and keep the filling threads in their proper location. The same variable mechanism can be controlled by electronic controls that adjust the speed of the chains and bring the filling and warp into proper relationship to ensure that warp and filling yarns are at right angle to each other. The marks of the clips or pins used to hold fabric on the tenter chains are often visible in the fabric selvedge. Many fabrics have selvedges heavier than the rest of the fabric in order to minimise the damage from tenter clips or pins.

Calendering

When a fabric is dried after being subjected to scouring, bleaching, mercerising, dyeing and printing, it loses its lustre. This happens because the threads are left highly crimped or wavy so that the fabric surface is not smooth and flat. The two essential conditions required for a lustrous fabric is the flat fabric surface and the individual fibres should be as parallel to each other as possible. Thus, to increase its lustre the fabric must be flattened and smoothened. This is achieved by a process known as calendering. Calendering produces a certain amount of flattening of the fabric surface and results in a moderate lustre. It also alters the handle, texture and appearance of the material. This finish is generally applied to cotton, linen and rayon. The type of calender used depends on the type of cloth to be run and the effect desired.

This type of mechanical finish is basically ironing, but done at a commercial scale. The moist fabric is passed between calenders under controlled pressure and temperature for a particular amount of time. The calenders are an assembly of hard and soft rollers (known as 'bowls' in this process), of iron and paper or cotton, respectively, that are normally mounted in vertical frames. The metal rollers have a very smooth surface and are heated and they smoothen out the fabric. The soft rollers provide support to the fabric and protect creases and thicker areas of the fabric from potential damage. This arrangement is similar to the arrangement during ironing, where the fabric is placed on the padded soft ironing board to absorb the pressure exerted by the heavy and heated metal iron. The calendering arrangement can have two, three or a maximum of seven rollers in a vertical arrangement. The various parameters affecting the calendering process include fibre content, cloth construction, finish applied, moisture content, bowl speed, temperature and pressure applied, bowl composition and number of bowls (Schindler and Hauser 2004).

Sizing/Stiffening

Sizing/stiffening finish is applied to keep the fabrics stiff, crisp and attractive. It also prevents snagging and slipping of yarns. Cellulose fabrics are generally sized with starch or resins. Starch or other sizes add weight to a fabric and enhance the appearance of the inferior product until laundered. They also help prevent the fabric from soiling. During weaving, yarns are sized to increase the strength of the fabric. Sizing can be temporary, permanent or durable as per the requirement. The permanence of the finish is dependent on the type of stiffening agent and method of application. Temporary stiffening agents include starch, naturally-derived gums, carboxy methyl cellulose (CMC) and polyvinyl alcohol. These get removed during washing and need to be reapplied after each wash. A permanent effect is produced if the finish is resin based and is heat set. Acrylic binders, polyvinyl chloride, and polyvinyl acetate emulsions act as permanent stiffening agents. A metal salt such as stannic chloride ($SnCl_4$) is used to stiffen silk as it has affinity for silk and no binder is needed. Nowadays, many stiffening agents are available in the market that are easy to use and apply.

Softening

The prime interface between a user and a product is the sense of touch. Softness of a fabric is a subjective sensation felt by the skin when a textile fabric is touched and gently compressed. Often, exposure to various processes and chemicals—for instance, heat setting or crease resistant finish—imparts stiffness to the fabrics and adversely affect their handle. Therefore, softening of textile forms a very important part of the chemical processing. Softening is the process of modifying the handle or feel of textiles for better comfort, better wear and performance under defined conditions of use. It is the process associated with modification of surface properties of textiles described in terms of handle, volume, softness and drape. However, some disadvantages are also associated with these softening finishes, and they are: reduced crock fastness, yellowing of white goods, change in the hue of dyed goods and fabric structure slippage. Softeners are almost invariably applied on knit fabrics in the final rinse after the wet treatments to restore their handle. Softeners consist of molecules with both a hydrophobic and hydrophilic part. Typically, they contain long alkyl groups with 16–22 carbon atoms. Most softeners have low water solubility. Glycerine and Turkey Red Oil (TRO) are commonly used chemicals which are used to impart softness.

> **Crock fastness** or rub fastness is the resistance of the dye to be transferred from one surface to another during rubbing.

Softeners are classified according to their ionic nature into cationic, anionic, non-ionic, amphoteric and silicone softeners. **Cationic softeners** are the most preferred class of softeners for industrial applications. They have several advantages such as high softening efficiency, substantivity to most fibres and good lubricant properties. The major disadvantage is that they render a hydrophobic surface and poor rewetting properties because hydrophobic groups are oriented away from the surface. Examples of cationic softeners include quarternary ammonium salts. **Anionic softeners** are compounds which have anionic groups oriented towards outside. They are stable to heat during textile processing and are compatible with other components of dyes and bleaching baths. They provide good antistatic and rewetting properties and can be used in combination with anionic flourescent brightening agents. Turkey Red Oil is an example of anionic softening agent. **Non-ionic softeners** are generally less efficient than anionic and cationic softeners, but they can withstand the effect of hard water and extreme pH conditions. Examples are ethers and polyglycol esters, oxiethylates products, paraffins and fats. **Amphoteric softeners** have strong ionic character and hence provide good softening effect, low permanence to washing and high antistatic effect. Examples are compounds based on betaine and amine oxide. **Silicone-based softeners** are generally polysiloxane derivatives of low molecular weight. They are insoluble in water, and therefore must be applied on fabrics after dissolution in organic solvents or in the form of disperse products. They feature quite good fastness to washing. They create a lubricating and moderately waterproof film on the surface, and give fabrics a velvety silky handle. They can be used for velvets, upholstery fabrics and emerised fabrics. Enzymes such as cellulases have also been used in the softening of cotton. Popularly known as **bio-polishing**, this treatment removes protruding fibres and slubs from fabrics, reducing the yarn diameter and leading to softening of the handle and smoothing of the surface to counteract the harshness produced in cellulosic textiles by pretreatment methods like alkaline mercerisation. The process is eco-friendly and precise, but it is expensive and requires strict control of the parameters.

Stabilisation Finishes

Manufacturing processes create stresses and tension in fibres, yarns and fabrics. When such fabrics are given any wet treatment, the fibres swell because of water and heat and this causes the rearrangement of internal forces. This results in the relaxation of internal stresses, bringing the fabric to a tensionless state. This type of shrinkage is known as **relaxation shrinkage** as it is caused due to the relaxation or release of tension build-up during fabric manufacture. The shrinkage may be **progressive**, i.e. the shrinkage may be observed over a number of wet treatments or laundering. Apart from relaxation shrinkage, another type of shrinkage is due to felting of wool and other hair fibres. These undesirable shrinkages in fabrics can be controlled by various stabilisation treatments. These treatments may vary for different types of fibres and fabrics. A few of them have been described here.

Shrinkage Control in Cellulosic Fabrics

1. *Drying in tensionless state*: A simple method to control shrinkage is to wet the fabric thoroughly and dry it in completely tensionless state. The fabric can then be smoothened out by calendering or ironing. Alternately, wet fabric can be dried on the tenter frame in a slack condition by over-feeding the frame.
2. *Compressive shrinkage*: Since shrinkage in woven fabrics is mostly lengthwise rather than widthwise, compressing the filling yarns closer together stabilises the fabric against further shrinkage. During this process, the yarns are pushed mechanically in the fabric closer to

one another, thereby preventing shrinkage. Compressive shrinkage reduces shrinkage to less than 2 per cent. **Sanforising** is the best known compressive shrinkage process. The compressive shrinkage machine consists of a roller over which a thick blanket moves (Figure 9.3). As the blanket bends over the roller, its outer surface extends. The damp fabric is fed over the roller at this point. It comes in contact with the stretched outer surface of the blanket. A shoe plate over this arrangement ensures that the fabric is in close contact with the blanket surface. The blanket and the fabric leave the guide roll and press against a heated cylinder. At this stage the blanket retracts due to change in curvature and the fabric also compresses along with it.

Figure 9.3: Compressive shrinkage

Electrically heated shoe
Palmer drying cylinder
Feed roller
Fabric
Fabric and blanket moves in this direction
Felt blanket
Place of occurrence of compressive shrinkage in fabric

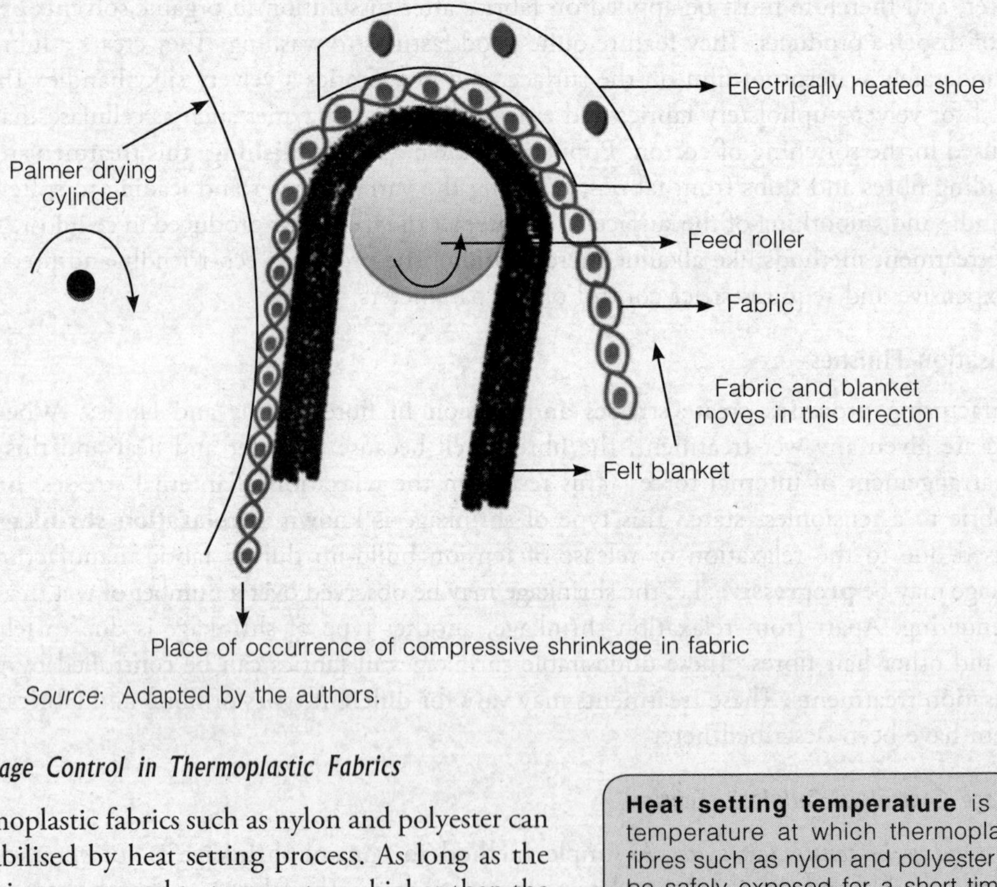

Source: Adapted by the authors.

Shrinkage Control in Thermoplastic Fabrics

Thermoplastic fabrics such as nylon and polyester can be stabilised by heat setting process. As long as the fabric is not exposed to temperatures higher than the heat setting temperatures, the fabric remains stable.

> **Heat setting temperature** is the temperature at which thermoplastic fibres such as nylon and polyester can be safely exposed for a short time to impart dimensional stability.

Shrinkage Control in Wool Fabrics

Chlorination is the chemical treatment of wool fabric with a dilute solution of calcium or sodium hypochlorite so as to make the fabric shrink-proof. This treatment partially removes the scales of the fibres, thus felting is reduced and the fabric is made shrink-proof. Due to the removal of scales, the affinity for dyes and lustre of the fabric also increase. Chlorinated wool is generally used in woollen underwear, socks and sweaters. Chlorination is also done on wool fabrics which are to be printed subsequently.

SUMMARY

- Preparatory finishes are customarily given to fabrics before dyeing or final processing.
- Special finishes are further classified into aesthetic (finishes that change the appearance) and functional finishes (finishes that improve the performance).
- Finishes are also classified as temporary, durable and permanent, on the basis of the length of the time they remain on the fabric.
- Singeing is a finish applied to burn away the 'fuzz' due to the protruding fibre ends on the yarn or fabric surface.
- Desizing is a chemical finish to remove sizing substances present on the fabric.
- Bleaching is done to decolourise the impurities that mask the natural whiteness of fibres.
- Weighting is a kind of sizing of silk which compensates for the loss in weight of the silk fabric or yarns caused by degumming.
- Carbonising is a chemical finish which is used to remove vegetable impurities which may remain on scoured wool after mechanical treatment.
- Decating is a mechanical finish which produces smooth wrinkle-free surface with lustre and soft handle on the woollen fabric and sets the grain of the woven fabric. It is also used to soften the handle of rayon, silk and blends.
- The optical brightening agents absorb invisible ultraviolet light and emit it again within the visible range at the blue–violet end of the spectrum, thus making the fabric appear whiter and brighter.
- Tentering is a mechanical finish that sets the warp and weft of woven fabrics at right angles to each other, stretches and dries the fabric to its final dimensions.
- Calendering is a mechanical finish which smoothens out the fabric and results in a moderate lustre.
- Compressive shrinkage stabilises the fabric and prevents shrinkage of fabrics when they are laundered or exposed to heat and water.

KEY WORDS

Finish: The treatments that change the appearance, handle and the performance of the fibre, yarn or fabric.

Scouring: A preparatory process to remove natural and acquired impurities by wet treatments so that the impurities do not interfere with dyeing and finish applications.

Mercerisation: The treatment of cellulosic textiles, primarily cotton, with a concentrated solution of sodium hydroxide to enhance moisture absorbency, dye uptake and lustre.

Degumming: The process of removal of sericin from silk.

Sanforising: A compressive shrinkage process which is applied on woven fabric to achieve shrinkage before making the garments.

Sizing: A process where chemical substances such as starch and resins are added to a fabric to produce stiffness.

Fulling: Controlled felting of wool primarily carried out to impart softness and fullness to woollen products.

Thermoplastic: This is a plastic material, polymer, that becomes pliable or moldable above a specific temperature and solidifies upon cooling.

EXERCISES

1. Enumerate the various types of finishes. Give one example of each.
2. Describe briefly the various preparatory processes of cotton.
3. Explain the process of calendering.
4. Differentiate between:
   (i) Slack mercerisation and mercerisation under tension
   (ii) Oxidising and reducing bleaching agents

REFERENCES

Hall, J. A. 1966. *Textile Finishing.* Third edition. New York: American Elsevier Publishing Co. Inc..

Heywood, D. 2003. *Textile Finishing.* London: Society of Dyers and Colourists.

Joseph, L. Marjory. 1988. *Essentials of Textiles.* Fourth edition. New York: Holt, Rinehart and Winston Inc.

Marsh, T. J. 1957. *An Introduction to Textile Finishing.* Sixth edition. London: Chapman and Hall Ltd.

Schindler, D. W. and J. P. Hauser. 2004. *Chemical Finishing of Textiles.* London: Woodhead Publishing.

10

SPECIAL FINISHES

HIGHLIGHTS

- Chemistry and process of aesthetic finishes
  - Special calendering
  - Basic finish/Plissé effect
  - Raised surfaces
  - Shearing
  - Flocking
  - Beetling
  - Acid finish
- Chemistry and process of functional finishes
  - Flame retardant
  - Abrasion-resistant
  - Antimicrobial
  - Water-repellent
  - Insect and mite resistant
  - Stain- and soil-repellant
  - Ultraviolet protection
  - Antistatic
  - Crease resistant/durable press
  - Soil-release

Varied end uses of textiles may require special functionality and performance properties. To achieve these, fabrics are given specific finishing treatments generally categorised under special finishes. These finishes are generally given in addition to the basic/routine finishes and are applied on comparatively smaller lots of fabrics. They are applied to a limited length of fabric as the properties imparted are very specific/special. The finishes can be further characterised into aesthetic and functional finishes.

AESTHETIC FINISHES

Special Calendering

Calendering is a mechanical finish in which fabric is passed between rollers under controlled pressure and temperature for a particular amount of time, producing a certain amount of flattening of the fabric surface and resulting in a moderate lustre. However, special effects can also be imparted onto the fabric with the help of certain special calendering techniques that are schreinering, embossing, moireing and friction calendering.

Schreinering

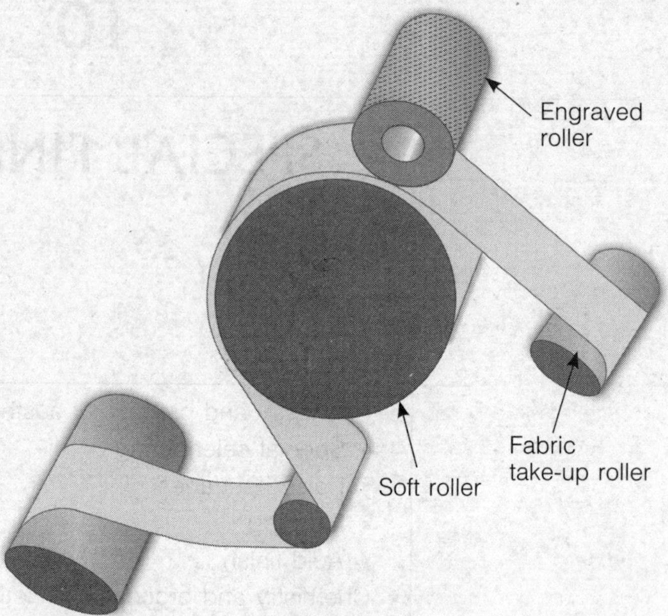

Figure 10.1: Schreiner calender machine

Source: Drawn by the authors.

Schreinering is a form of calendering in which fine lines are engraved at an angle of around 26° to the construction of the fabric on the metal roller. The roller has almost 250–350 lines per inch and the angle of the engraved lines corresponds with the twist in the warp yarns. When the fabric is passed through the nip between the heated engraved roller and a central roller, the lines on the metal roller are transferred to the fabric (Figure 10.1). This causes yarn flattening, reduces the gap between the yarns and thus increases the opacity of the fabric. This finish also provides a soft lustre and handle.

When this finish is applied on thermoplastic fibres, the heat produced by the rollers sets the fibres giving a permanent effect. On non-thermoplastic textile substrates a resin treatment is given before the process to produce a durable finish. There are chances of lint getting stuck on the fine engravings of these calenders, thus spoiling the optical effect. Another disadvantage is that all schreiner finishes reduce tensile strength.

Sateen and Damask fabrics are mainly finished by schreinering. The fabric finished with schreiner roller cannot be refinished with the same effect, as this may cause a 'moire' or 'watered effect'.

Embossing

Embossing is a finish to produce three-dimensional designs on fabrics, thus this finish is also known as figured or sculptured calendering. A 'bulky but thin' product can be made in any pleasing or functional construction, depending on the design embossed on the rollers. During the process of embossing, the pattern is impressed on smooth surfaces under high pressure and temperature. The embossing calenders consist of two or three bowls. The two-roll method consists of a soft roll made of paper or cotton and a heated embossed metal roll, whereas the three-roll system consists of a metal roll in between two soft rolls. In this process, the engraved metal roller first presses the design on the surface of dampened soft roll. Once the design is impressed deeply on the soft roll, then the fabric is fed through the calender. The engraved and the shaped soft rollers then work together to mould the design on the surface of the fabric.

The designs produced by this finish are permanent if they are produced on thermoplastic fibres and durable when they are applied on resin-treated fabrics. Embossing is used to obtain special effects such as leather graining, simulated weave, brush strokes and mock tiling.

Friction Calendering

Friction calendering is used to produce glazed surface by applying friction on the face of the fabric. A typical arrangement of friction calender consists of three bowls: the upper highly polished chromium plated steel bowl, cotton or paper bowl and a lower steel bowl. The peripheral speed of the steel bowl is almost three times faster than the soft bowl so that the fabric receives a lower or higher degree of glaze depending on the friction produced. In the process, pressure is applied hydraulically. The steel bowls are heated by gas, hot oil or electricity. Apart from the lustrous effect, the cloth becomes papery, thin and transparent. This may be because all the internal air spaces and hence light reflecting surfaces have been eliminated during the process. To enhance the surface glazed effect, the fabric can be dipped in a suitable stiffening agent before passing through the friction calenders. In this process, if starch is used to produce the glaze, then the finish is temporary and if resins are used to produce the glaze, then the finish is durable.

Chintz or Everglaze finish is produced by padding the fabric with glazing agents such as starch, glue or shellac, followed by friction calendering. Heavy calendering causes the surface to become highly glazed. Ciré is another special finish produced by friction calendering on silk, rayon or nylon and blends. In this process, fabric is impregnated with wax or other appropriate substance and then passed through the rollers to produce a durable finish.

> **Ciré** is a shiny surface produced only by hot calendering on a fabric made of thermoplastic fibre such as acetate.

Moiréing

The term originates from the French word *moiré*, which refers to a textile material with a rippled or 'watered' appearance. Traditionally, this effect was produced on silk but now it is also done on cotton and synthetic fabrics too. Rib fabrics such as failles, taffetas and bengalines produce effective moiré effects. The moiré finish is characterised by a soft lustre and a design created by differences in light reflection. This finish can be produced in two ways. In the first method, two lengths of plain fabric are placed face to face and are then fed between the rollers, where one is a large soft roller and the other is a heated metal roller which carries the rib design. The rollers with rib design flatten or distort the layers more at some places than the other, thus causing light to reflect differently across the surface of the fabric and produce the moiré effect. Patterns at regular intervals can be produced by this method. In the second method, two layers of the ribbed fabric are put under pressure by the smooth plain moiré rollers, thus ribs of each layer flattens the yarns of the other layer producing the moiré effect. This method is used to produce a random pattern. Moiré finish produces an optical effect which is caused by the differential reflection of light from the surface of the fabric. The finish can be temporary, durable or permanent as achieved in other calendering processes.

Raised Surfaces

There are two types of raised surface finishes: napping and emerising.

1. *Napping*: Napping is a finish in which raised fibre ends are produced on the fabric surface. In this process, cotton, rayon and wool fabrics with loose twist are passed through rollers

with fine metal hooks on its surface. The hooks pull the fibre ends to the surface and thus create the nap. The hooks on the adjacent rollers are in opposite directions known as pile and counter pile rollers. The pile roller creates long nap while the counter pile creates short nap (Figure 10.2). The rollers must travel faster than the cloth in order to achieve napping. Plain and twill weave fabrics are generally preferred for napping in which the warp yarns provide the strength and the loosely-twisted filling yarns are pulled to create the effect. The loose fibre ends on the fabric also entrap air, thus providing warmth and insulation. This property of napped fabrics is utilised in blankets, coating fabrics, sweaters and sleepwear. Other examples of napped fabrics are flannels and flannelettes.

Excessive napping weakens the fabric structure by pulling up too many fibres. Also, napped fabrics are difficult to work while cutting before garment construction since nap leads to varied light reflection and alters the appearance of the fabric. Therefore, when a garment is made from these napped fabrics, the patterns should be cut in the same direction to ensure uniform appearance.

Figure 10.2: Napping

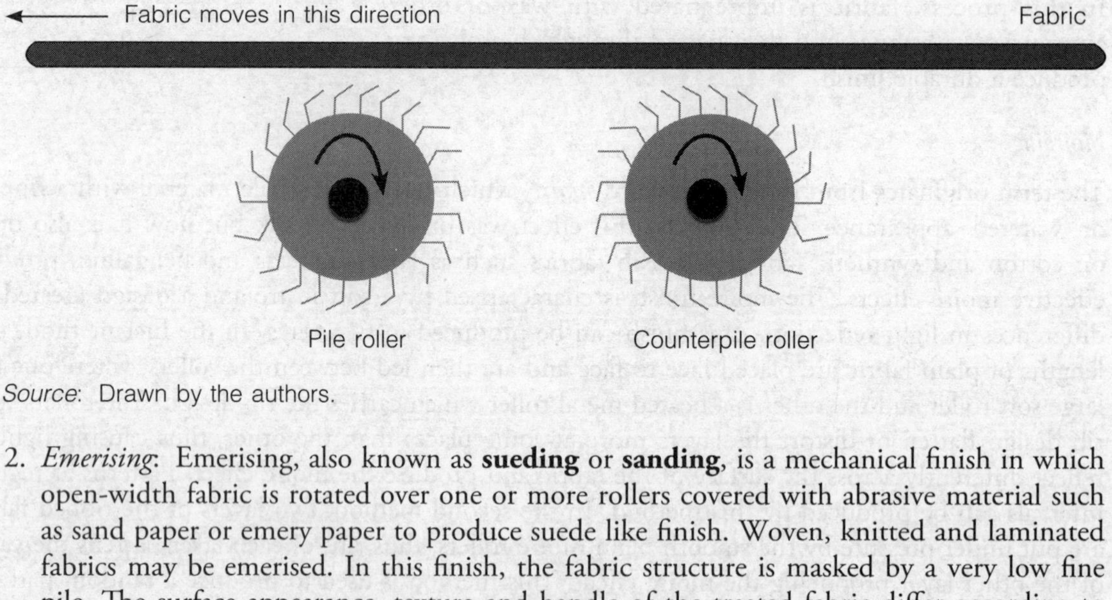

Source: Drawn by the authors.

2. *Emerising*: Emerising, also known as **sueding** or **sanding**, is a mechanical finish in which open-width fabric is rotated over one or more rollers covered with abrasive material such as sand paper or emery paper to produce suede like finish. Woven, knitted and laminated fabrics may be emerised. In this finish, the fabric structure is masked by a very low fine pile. The surface appearance, texture and handle of the treated fabric differ according to the type of fibre present, the fibre linear density and the intensity of emerising action on the fabric. The emerised fabric has a much softer handle. This gives the fabric a 'peach-skin finish' because the short fibres protruding from the fabric surface simulate the soft handle and appearance of the natural peach-skin. Emerising can be carried out after bleaching and before dyeing. Emerising a dyed fabric can result in very slight unevenness. Another problem arises because the coloured dust generated by emerising the fabric settles on the fabric and must be removed. When grey fabric is emerised, the subsequent wet processing like dyeing and rinsing removes the dust.

Flocking

Flocking is a finishing process that produces a pile appearance. Short fibres, usually nylon or rayon, are applied onto the substrate which has been previously coated with an adhesive. The length of the flock ranges from 0.25 to 5 mm. Adhesives that capture the fibres must have the same flexibility and resistance to wear as the substrate. This method of creating a raised surface is different from that of napping, gigging and sueding. A flocked surface is created by first applying the adhesive in the desired pattern, after which flock can be applied by either of the following two methods:

1. *Mechanical/Vibration method*: In this process, adhesive-coated fabric is passed over a conveyor belt. Above the belt is a container which contains the flock. As the fabric vibrates, static build-up takes place that attracts the flock. The flock adheres to the areas where the adhesive has been applied. The fabric then passes through a heated chamber where the adhesive dries with the flock embedded in it. The flock which didn't adhere to the surface is then removed by brushing (Figure 10.3a).
2. *Electrostatic flocking*: In this electrical charge is generated by the use of two electrodes. The electrostatic charge on the fabric surface propels the fibres onto the adhesive-coated substrate. Flocking fibres penetrate and embed in the adhesive at right angles to the substrate. This forms a high-density uniform flock coating or layer. Controlling the electrical field by increasing or decreasing either the applied voltage or the distance between the electrodes and the substrate controls the speed and thickness of the flocking. The fabric is then dried and the adhesive holds the flock in place (Figure 10.3b). Flocking is a permanent finish but high temperatures and dry cleaning can damage this finish by dissolving the adhesive.

Figure 10.3a: Flock application by mechanical method

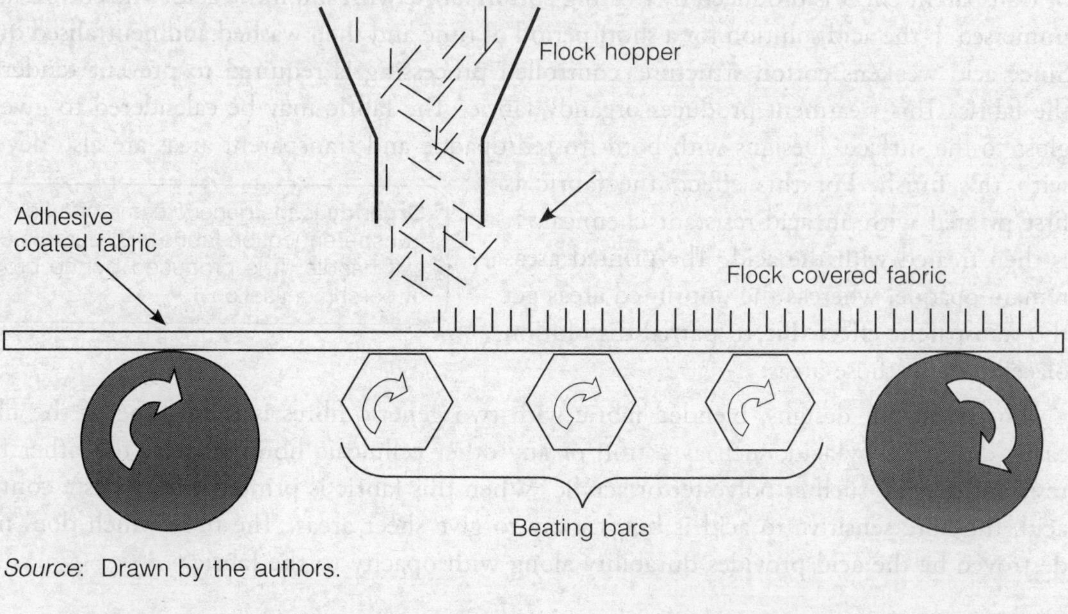

Source: Drawn by the authors.

Figure 10.3b: Flock application by electrostatic method

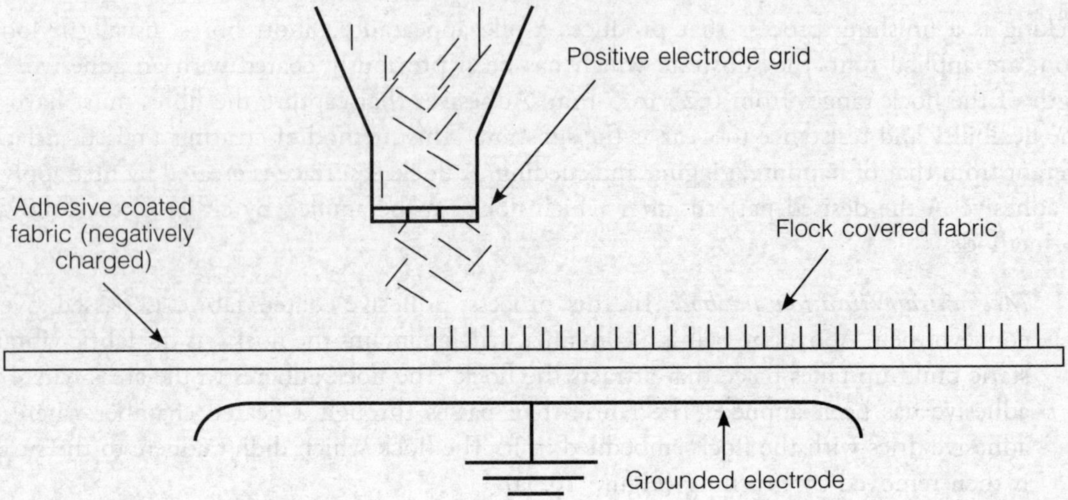

Source: Drawn by the authors.

Flocked fabrics have been used for products such as t-shirts, wallpaper, gift/jewellery boxes or upholstery. Flocked surfaces reduce water condensation, act as good thermal insulators and are used in the automotive industry for glove compartment boxes, door mouldings and window trim.

Acid Finish

Acid finish can produce two kinds of effects, parchmentisation and burnt out designs. Parchmentised or transparent effect is produced by treating cotton fabric with sulphuric acid. The cotton fabric is immersed in the acid solution for a short period of time and then washed and neutralised quickly. Since acid weakens cotton structure, controlled processing is required to prevent tendering of the fabric. This treatment produces organdy fabric. The fabric may be calendered to give more gloss to the surface. Designs with both frosted/opaque and transparent areas are also developed with this finish. For this effect, the fabric is first printed with an acid-resistant chemical. It is then treated with the acid. The printed areas remain opaque, whereas the unprinted areas get the transparent effect due to partial dissolution of cellulose in those areas.

> **Organdy** is the generic name of a fine, thin, transparent cotton fabric with durable crisp/stiff handle. It is produced by the process of parchmentisation.

For burnt-out designs, blended fabric with two generic fibres is used. One of the fibres is easily destroyed by acid, such as cotton or any other cellulosic fibre, whereas the other fibre is resistant to acid, such as polyester or acrylic. When this fabric is printed with a paste containing acid, the fibre sensitive to acid is burnt away to give sheer areas. The fibre which does not get destroyed by the acid provides durability along with opacity to the fabric.

Basic Finish/Plissé Effect

A three-dimensional effect called plissé is produced by imprinting the cotton fabric with sodium hydroxide paste. The printed areas of the fabric shrink while other areas remain unaffected. This causes untreated areas to puff up between treated areas thus giving a three-dimensional effect. Sodium hydroxide produces this effect on cotton and rayon whereas acid is used to achieve this effect on nylon. However, these fabrics cannot be ironed as it may flatten the fabric.

Shearing

Shearing is the mechanical finish to cut or shear the undesirable staple fibres ends protruding from the fabric surface. The shearing machine consists of a wide cylinder, brushes and cutting blades. When fabric is passed over the cylinder, brushes raise the loose fibre ends and cutting blades shear off the ends to give a uniform surface. The irregular fleecy layer formed during raising treatment can also be sheared to a uniform height. The extent to which shearing is done varies according to the requirement:

1. Light shearing, e.g. pretreatment of grey cotton fabric.
2. Normal shearing, e.g. close cutting of light fabrics.
3. Heavy shearing, e.g. shearing of velour pile or knitted loops

> **Velour** is a type of knitted fabric that resembles velvet.

Beetling

Beetling is a mechanical finish to increase the lustre of cotton and linen fabrics. During the process, the fabric is fed over the rollers rotating in the machine. Continuous pounding with large hammers on the surface of the fabric flattens the yarns and closes the weave, thus providing more surface area for light reflection. The beetled-finished fabric should be laundered and ironed carefully. This finish is used on damask and other linen-like fabrics. Matka silk is beetled to add lustre to the fabric. A new approach to produce a beetled fabric made of thermoplastic fibres is by using high pressure and resin. The high pressure flattens the yarns resulting in lustre similar to that provided by beetling finish. Heat and resin result in a permanent finish.

FUNCTIONAL FINISHES

Flame-retardant Finish

When heat is applied, the temperature of the substrate increases and it undergoes irreversible chemical changes, producing non-flammable gases (carbon-dioxide, water vapour and the higher oxides of nitrogen and sulphur), carbonaceous char, tars and flammable gases. When the substrate attains combustion temperature, the flammable gases combine with oxygen in the process called **combustion** and lead to formation of flames.

Most of the textile fibres are flammable. As a result, in case of a fire accident, they may not only get damaged themselves, but may also be a potential threat to human life and property. Therefore, flame retardancy is an important and desirable performance characteristic of textile products for safety. Flammability of textiles is influenced by factors such as fibre type, fabric weight and construction, method of ignition, extent of heat and absence or presence of flame retardants. The flame resistance of textiles is generally evaluated by determining the **LOI (Limiting Oxygen Index)**, which is 'the minimum concentration of oxygen, expressed as a percentage, that will support combustion of a polymer' (McCrum, Buckley and Bucknall 1998). LOI can be experimentally measured by passing a mixture of oxygen and nitrogen over a sample that is burning, and reducing the oxygen level until the critical level is reached (ibid.).Since the content of oxygen in air is 20 per cent by volume, all textiles with LOI values equal to or lower than 20 will burn quite easily in air while those with LOI values higher than 20 will not burn. Therefore, cotton and viscose, with LOI of 18.4 and 18.6, respectively, burn readily. Polyester and nylon with LOI of 20–21, acrylic with LOI of 18.2 are also flammable. However, modacrylic with LOI of 29–30 and wool with LOI of 25, are highly flame resistant.

Three types of flame resistant textile products can be produced:

1. *Use of inherently flame resistant fibres*: Companies like DuPont have created inherently flame resistant fibres—i.e. those materials that have flame resistance built into their chemical structures—such as, Kevlar® and Nomex®. Modacrylic is another fibre with excellent permanent flame retardancy. In the case of these fibres, protection against combustion can never be worn away or washed out.
2. *Use of fibres modified by adding flame retardants to the dope*: Flame resistant fibres can also be produced by incorporating flame-retarding chemicals in the dope at the time of fibre production. Examples include HEIM Polyester, Extra FR Polyester. Various phosphorous- and bromine-based compounds are used as melt additives for thermoplastic fibres to produce such fibres.
3. *Topical application of finishes*: Flame resistant fabrics could also be produced by applying a suitable finish by pad-dry-cure method. The chemicals of the finish adhere to the surface of the fibre through various physical or chemical bonding or penetrate into the fibre structure and prevent combustion from occurring. Some flame retardant chemicals/finishes have been listed below:
   - Mixtures of borax-boric acid and ammonium chloride-ammonium sulphate are non-durable but effective flame retardants for cotton. These finishes function by removing heat and enhancing the decomposition temperature. On heating, these chemicals coat the fibre with a glassy insulating polymer film and cut off the heat and oxygen supply.
   - Aluminium hydroxide and calcium carbonate also non-durable finishes that act by decomposing through strongly endothermic reactions. These endothermic reactions involve the absorption of heat, and thereby the pyrolysis temperature is not reached and no combustion takes place.
   - Phosphorus- and nitrogen-based chemicals can be used to create non-durable, semi-durable and durable finishes that are flame retardant. Although nitrogen alone is not an effective flame retardant, it acts synergistically with phosphorous to lend flame retardancy by

producing non-combustible decomposition products when exposed to heat. Diammonium phosphate, ammonium sulphamate and ammonium bromide are commercially important non-durable flame retardant finishes for cellulosics. Some durable phosphorous- and nitrogen-based flame retardant finishes are THPC (tetrakis hydroxymethyl phosphonium chloride), THPOH-NH_3 (tetrakis hydroxymethyl phosphonium hydroxide plus ammonia), Pyrovatex CP (N-methyloldimethylpropionoamide) and Proban (THPC-urea based system).

- Halides such as chlorine and bromine operate by forming free radicals that combine with the hydrogen and hydroxyl radicals which are the major reaction entities for combustion. The efficacy of halides as flame retardants can be greatly enhanced when used with antimony which has a synergistic effect.
- Hexafluoro zirconate and titanate salts based finishes are important finishes (Zirpro process) that are used for wool.

The requirements for flame retardant finishes are:

- Little or no adverse effect on the textile's physical properties.
- The textile should retain its aesthetic and functional properties.
- The product should be produced by a simple process with conventional equipment and inexpensive chemicals.
- The finish should be durable to repeated home launderings, tumble drying and dry cleaning.

Flame retardant fibres find use in the making of uniforms for fire fighters, astronauts, emergency personnel and the military. Floor coverings, upholstery and drapery also need to exhibit flame retaradant properties, especially when used in public buildings and aircrafts.

Antimicrobial Finish

Micro-organisms can be found almost everywhere in the environment. The basic nutritional requirements for their growth and multiplication are water, a source of carbon, nitrogen and some inorganic salts. Textiles provide suitable environment for the growth of micro-organisms which can lead to functional, hygienic and aesthetic difficulties. The most problematic organisms are bacteria and fungi. Textiles provide ambient conditions in terms of moisture, temperature and nutrients for the growth of bacteria and fungi. Fungi also produce multiple problems on textiles including discolouration, coloured stains and fibre damage. In textiles, bacteria cause slimy feel, unpleasant odour and loss of functional properties such as elasticity and tensile strength. Antimicrobial finishes form a very important part of textile processing because of two reasons: (*a*) protecting the textile from damage caused by mould, mildew and rot producing micro-organisms; and (*b*) protecting the textile user against pathogenic or odour causing micro-organisms.

Antimicrobial finishes are very important for industrial fabrics that are exposed to weather. Fabrics used for awnings, screens, tents, tarpaulins and ropes need protection from rotting and mildew. Home furnishings such as carpeting, shower curtains, mattress and upholstery also need antimicrobial finish. Antimicrobial clothing is also used in areas where the danger of infection from pathogens is large such as in schools, hospitals, nursing homes, hotels and crowded areas. Textiles in museums are also given this finish for the purpose of preservation. Microbial growth increases in the presence of size/starch. Textiles left wet between processing steps for an extended time also need an antimicrobial treatment.

The following are some properties of an effective antimicrobial finish:

- Fast acting since the bacterial population grows at an astounding rate.
- Kill or stop the growth of microbes.
- Durable even after multiple cleaning cycles or outdoor exposure.
- Safe for manufacturer to apply and consumer to wear.
- Meet strict government regulations and have a minimal environmental impact.
- Easily compatible with other finishing agents.
- Have no adverse effect on the fabric properties.

Antimicrobial finishes that control the growth and spread of microbes are called **bacteriostats** and products that actually kill microbes are **bacteriocides**. Antimicrobial products are divided into two types based on the mode of attack on microbes:

1. *Controlled release mechanism*: The antimicrobial agent is released slowly from a reservoir either on the fabric surface or in the interior of the fibre. Also known as **leaching type of antimicrobial**, these can be very effective against microbes on the fibre surface or in the surrounding environment. However, the reservoir will eventually deplete and the finish will no longer be effective. A widely-used biocide and preservation product is formaldehyde. Bound formaldehyde is released in small amounts from common easy-care and durable press finishes. Quarternary ammonium compounds also fall in this category. Triclosan (2, 4, 4'-trichloro-2'-hydroxydiphenyl ether), another most widely used antimicrobial product, is extensively used in mouthwash, toothpaste, liquid hand soap, deodorant, etc. Most recent approach for the controlled release of antimicrobial agent is **microencapsulation**. In this method, microcapsules are incorporated either in the fibre during primary spinning or in coatings on the fabric surface.
2. *Bound mechanism*: The antimicrobial agent is bound to the fibre surface and controls only those microbes that are present on the fibre surface, not in the surrounding environment. Because of their attachment to the fibre, bound antimicrobials can potentially be abraded away or deactivated and loose long-term durability. Examples of antimicrobial agents depicting bound mechanism are octadecylamino-dimethyl-trimethoxy-silylpropyl-ammonium chloride, polyhexamethylene biguanide and chitosan.

Chitosan is a modified biopolymer that is increasingly becoming popular as antimicrobial finish because it is non-toxic and biodegrdable.

Agents Used for Antimicrobial Finishing

Popular antimicrobial agents used are metals/metal salts, cationising agents, N- halamines, organic chemicals and natural polymers. Silver-based compounds are very promising candidates for antimicrobial finishes. Some cationic softeners also show varying antimicrobial action. A vast variety of natural products such as neem, aloe vera, tulsi, eucalyptus, clove (*Curcuma longa L.*), Gall nut (*Quercus infectoria*), Indian madder (*Rubia cordifolia*) and *Rumex maritimus*, exhibit good antimicrobial activity.

Insect- and Mite-resistant Finish

Insect-resistant finishes are chemical treatments that protect keratin-containing wool and other animal fibres from attack by the larvae of certain moths and beetles. Mites are not insects, they belong to the arachnid (spider) family. Mite-resistant finishes protect from dust mites that proliferate in bedding, mattresses and quilts. The carpet industry is the most important market for insect-resistant finishes. Other applications include upholstery fabrics, blankets, uniforms, apparel and furs. Practices such as repeated airing of products, maintaining lower room temperature, regular washing and cleaning of products, especially before storing them can go a long way in protecting textile products from damage by insects and micro-organisms. Insect resistant finishes are of two kinds:

1. *Digestive poisons*: These species-specific poisons interfere with the keratin-digesting process of the larvae and then kill them by blocking enzymes needed for digestion; e.g. chlorinated triphenylmethane, chlorphenylids, sulcofenuron. These poisons present low environmental hazard.
2. *Nerve poisons*: These chlorinated hydrocarbons attack the nervous system of the insects; e.g. dieldrin, permethrin, hexahydropyrimidine. These poisons are more effective than digestive poisons, but were discontinued in the 1970s because of their persistence in the environment and danger to aquatic life.

Ultraviolet Protection

Solar rays reaching the earth's surface is composed of light waves with wavelengths ranging from infrared to ultraviolet (UV). Though, the intensity of UV radiation is much less than the visible or infrared radiation, the energy per photon is significantly higher. The long-term exposure to UV light has a number of negative impacts such as acceleration of skin ageing, photodermatosis, phototoxic reaction to drugs, erythema, sunburn, increased risk of melanoma, and eye and DNA damage. UV rays are divided into UV-A (320–400 nm wavelength), UV-B (290–320 nm) and UV-C (200–290 nm). The damage to human skin from UV radiation is a function of the wavelength and the intensity of the incident radiation. Based on these two factors the wavelengths that are of maximum danger to skin are 305–310 nm. Therefore, UV-resistant textiles will be effective if they protect the wearer from solar UV radiation in the wavelength range of 300–320 nm.

The protective effect of textiles to UV rays is quantified as Ultraviolet Protection Factor (UPF) or Sun Protection Factor (SPF). UPF or SPF is determined as below:

$$\text{UPF} = \frac{\text{Potential erythemal effect by the radiation}}{\text{Actual erythemal effect transmitted through the fabric by the radiation}}$$

The higher the UPF, the more protective the fabric is to UV radiation. The term SPF is used for sun-blocking creams, where it tells for how long a person can be exposed to sunlight before skin damage occurs. A fabric with SPF > 40 is considered to provide excellent protection against UV radiation.

When light falls on a surface, either it can be reflected, absorbed or transmitted through the fabric. The amount of radiation reflected, absorbed or transmitted depends on many factors such as fibre type, fibre texture, fabric cover factor and the presence or absence of fibre delusterants, dyes and UV absorbers. Cotton and silk fibres offers little protection against UV radiation since the radiation can pass through without being markedly absorbed. Wool and polyester have high UPF since these fibres absorb UV radiation. Synthetic fibres such as nylon and polyester absorb UV radiation due to the presence of delusterant titanium dioxide. Compactly constructed fabrics with good cover factor provide better protection against UV rays as these fabrics do not allow the UV rays to reach the skin.

Tight microfibres provide a better protection than fabrics made from normal-sized fibres with the same specific weight and type of construction. Fabrics dyed in deep shades show higher UPF as compared to light shades. Some examples of chemicals providing UV protection are phenylsalicylades, benzo phenones, benzotriazole derivatives and oxalic acid dianilide derivatives.

Crease-resistant/Durable Press Finish

Cellulosic fibres, despite many advantages, have the biggest disadvantage of wrinkling after washing. This problem has been addressed by the development of durable press finish, also termed as crease-resistant, easy-care, wrinkle-resistant and wrinkle-free finish. The primary effects of durable press finish are reduction in swelling and shrinkage, improved wet and dry wrinkle recovery, smoothness of appearance after drying and retention of intentional creases and pleats. Apart from these benefits, these finishes maintain the dimensional stability of cellulose fabrics and its blends, the sheen of calendered fabrics (permanent chintz) and the stand and handle of pile fabrics.

Mechanism of Durable Press

Cellulosic fibres tend to wrinkle badly during use. The molecular chains are held together by weak H-bonds which break easily with stress of bending or creasing. Once they break, the polymer chains are free to move to relieve that stress. Hydrogen bonds can reform between the polymer chains in their shifted positions, in effect locking in the new configuration. Water molecules further facilitate the breakage of H-bonds in the fibre. This is the reason why cellulosic fibres exhibit excessive creasing after wet treatments.

Finishes used to impart wrinkle recovery work on the principle of forming crosslinks between molecular chains. This causes molecules to experience strain in the new/bent position but no breakage of bonds. Therefore, when the strain is removed the fabric returns to its original state. The finish is applied by padding the fabric with the chemical, drying on a frame, curing at high temperature, scouring to remove residual chemicals and drying. It can be applied by a pre-cured or post-cured process. The pre-cured process consists of padding followed by curing. The fully-finished fabric is thus available to be made into garments. The post-cured process involves padding the fabric followed by drying, cutting, sewing and then curing to impart the wrinkle resistance.

Initially, urea and formaldehyde-containing products such as N, N'-dimethylol ethylene urea (DMEU), melamine formaldehyde, N, N'- dimethylol-4, 5-dihydroxyethylene urea (DMDHEU) were used as durable press finish. However, these finishes had the problem of free formaldehyde that irritates mucous membranes, cause teary eyes and leads to difficulties in breathing and headaches. Formaldehyde is also a suspected carcinogen. Therefore, non-formaldehyde products such as N, N'- dimethyl-4, 5-dihydroxyethylene urea (DMeDHEU) and 1, 2, 3, 4 Butane tetracarboxylic acid (BTCA) are preferred. Finishing of textiles with citric acid or mono ester of citric acid also provides crease resistance. Crease-resistant finishes have undesirable side effects, such as considerable loss of strength and fabric deterioration.

Abrasion-resistant Finish

Natural fibres have a tendency to abrade whereas man-made fibres have inherent resistance to abrasion. There are two ways by which fabric damage can be reduced:

1. Fibres of high abrasion resistance can be blended with fibres of low abrasion resistance.
2. A thermoplastic resin binds the fibres more firmly into yarns and thus increases the time and amount of abrasion required to roughen the surface by fibre breakage; thus, application of thermoplastic resins can reduce abrasion damage.

Water-repellent Finish

The finish that repels water is termed as water-repellent finish. Going by the literal meaning of the term, a drop of water should not spread on the surface of textile and should not wet the fabric as well. The primary requirement of a repellent finish, whether it repels water, oil or soil is that it should not affect the air permeability of textiles—i.e. the fabric should prevent the penetration of water droplets but allow the transfer of air and moisture. This is achieved by a combination of finish and fabric construction. However, this is different from waterproofing. A waterproof fabric will not wet regardless of the amount of time it is exposed to water or the force with which the water strikes the fabric. When a uniform coating of suitable substances such as rubber is produced on the surface of a fabric, the interstices between the warp and weft yarns are blocked by the continuous film or substance and both water and air cannot pass through the treated fabrics. The waterproof coating makes the fabric stiff, non-permeable to air and moisture and consequently results in poor wear comfort. Water proofing is generally used in tarpaulin, umbrella cloth, rain coat fabrics, etc. The comparison of water-proof and water-repellent fabrics is given in Table 10.1.

Table 10.1: Comparison of water-proof and water-repellent fabrics

Water-proof fabrics	*Water-repellent fabrics*
No water can penetrate, non-permeable to air and moisture	Repels water, but heavy rain will penetrate, allows transfer of air or moisture
Films may stiffen in cold weather	Fabric is pliable, similar to untreated fabric
Cheaper to produce	More expensive to produce
Fabric does not breathe, uncomfortable to wear	Fabric breathes and is comfortable for rainwear
Permanent	Durable or renewable finish

Source: Compiled by the authors.

Water repellency depends on surface tension and fabric permeability. It is achieved by a combination of fabric structure and finish. Silicones and fluorochemicals give durable water repellency while zirconium compounds impart semi-durable effects. Wax emulsions and aluminium compounds are generally used as renewable water repellents. An important area in water-repellent textiles is microfibre fabrics finished with fluorocarbon polymers. A group of fluorocarbon-finished articles with special importance are ballistic fabrics, that provide protection against bullets, splinters and cutting.

Water-repellent finishes also enhance other fabric properties such as durable press, rapid drying and ironing and better resistance to acids, bases and other chemicals. However, there are also some undesirable properties associated with repellent finish, such as static electricity, poor soil removal in aqueous laundering, stiff fabric and increased flammability.

Lotus-leaf Effect

The lotus-leaf effect has been named after the unusual properties of the leaf surfaces of the lotus plant, which are remarkably water-repellent and soil-repellent. The surface of the lotus leaf is covered by a thin extracellular membrane termed as cuticle, which is covered by waxes forming characteristic microstructures due to self-organisation. The waxy layers generate super hydrophobicity and therefore adhesion of water is markedly diminished and water just rolls away from the surface. This remarkable self-cleaning effect is being harnessed to transfer the lotus effect into products.

Stain- and Soil-resistant Finish

Stain is a mark or discolouration on the fabric. In daily wear, the clothes get stained, and removal of stains completely has always been a challenge. To overcome this problem of staining, stain-resistant finish is applied to reduce the rate of soil deposition on fabric and help prevent spot staining. However, the stains should be removed quickly as they have a tendency to embed in the fabric and then these are difficult to be removed. Oil- and water-repellent finishes also work as stain- and soil-resistant finishes. Fluorochemicals impart repellency against both water and oil while silicone and other finishes are effective against water-based stains.

Antistatic Finish

Static, which literally means 'stagnant', is the net charge on an object in terms of electricity. When two different materials are pressed or rubbed together, the surface of one material will generally gain some electrons from the surface of the other material. The material that gains electrons will be negatively charged after the materials are separated and the other material will have an equal amount of positive charge. Some textile materials, especially hydrophobic synthetic fibres, have the tendency to build-up charge through this process; and textile materials are listed based on the polarity of charge separation when they are touched with another object in a series known as the **triboelectric series.**

Static build-up is a big problem and leads to other problems such as increased soiling, clinging of garment to the body, small electrical shocks caused by walking on carpets in low humidity conditions. In most dry textile processes, fibres and fabrics move at high speeds over various surfaces which can generate electrostatic charging from frictional forces. This electrical charge can cause fibres and yarns to repel each other, leading to ballooning. Therefore, antistatic agents are needed in order to reduce or eliminate build-up of electricity. The electrical charging of fibres by friction is affected by the nature of the mechanical contact, the ranking of the fibres in the triboelectric series, the humidity of the environment and the presence and absence of moisture on the fibres. The synthetic fibres are imparted spin finish which contains lubricating oils and other components, including antistatic agents. Antistatic finishes work by following mechanism:

1. By improving the surface conductivity.
2. By making the surface hydrophilic. The surface will then attract more water molecules, and lead to an increase in conductivity and carry away the static charge.
3. By developing an electric charge opposite to that on the fibre, thereby neutralising the electrostatic charges.

Common antistatic agents are based on long-chain aliphatic amines (optionally ethoxylated) and amides, quaternary ammonium salts, esters of phosphoric acid, polyethylene glycol esters and polyols. Conducting polymer nanofibres, particularly polyaniline nanofibres, are also being used to provide antistatic property. Electrically-conductive fibres can be produced by incorporating conducting materials, such as, carbon particles or other antistatic agents in the spinning dope. Carbon and metallic coatings can also be applied on the fibre surface. Antistatic yarns and fabrics can be made by incorporating fibres made from stainless steel, aluminium or other metals during spinning, weaving or knitting.

Antistatic finishes are generally applied to carpets, upholstery fabrics and airbags for automobiles, conveyor belts, filtration fabrics, airmail bags, parachutes, fabrics for hospital operating rooms and protective clothing. Fabrics made of cotton, rayon and wool may also be given antistatic finish depending on the intended use. However, most antistatic finishes are not durable and should be replaced after each laundering.

Soil-release Finish

Textile materials are prone to soil deposits containing both oily and particulate matter. Oily matter contains fatty materials. Particulate soil may be clay, soot or metal oxides. The use of durable press finishes and synthetic fabrics has further aggravated the problem of soiling and difficulty in removing it. So, the purpose of soil-release finishing is to facilitate the removal of soiling matter during laundering of the textile material.

Mechanism of Soil-release

There are several mechanisms responsible for removal of soil from fabrics. Two of these are listed below:

1. Adsorption of detergent and absorption of water leading to:
   (a) Penetration of detergent in the soil-fibre interface
   (b) Solubilisation and emulsification of soils
   (c) Removal of oily soil
2. Mechanical work leading to:
   (a) Surface abrasion to remove soil physically
   (b) Flexing between the fibres to remove soil
   (c) Swelling of finish to reduce inter-fibre spacing
   (d) Water carrying away the removed soil

The soil particles are removed from fibres by a two-step process. When the textile material is soaked in the wash liquid, a thin layer penetrates between the soil particle and the fibre surface, enabling the swelling of the particle surface. The soil particles are then transported away from the fibre and into the bulk of the wash liquid by mechanical action (Figure 10.4). The removal of soils is determined by the composition of the detergent and the mechanical action employed.

Soil-release finishes primarily work by making the fabric surface hydrophilic. This reduces the attraction between the fibre and oily soil and improves the action of water and detergents

Figure 10.4: Mechanism of release of soil particles

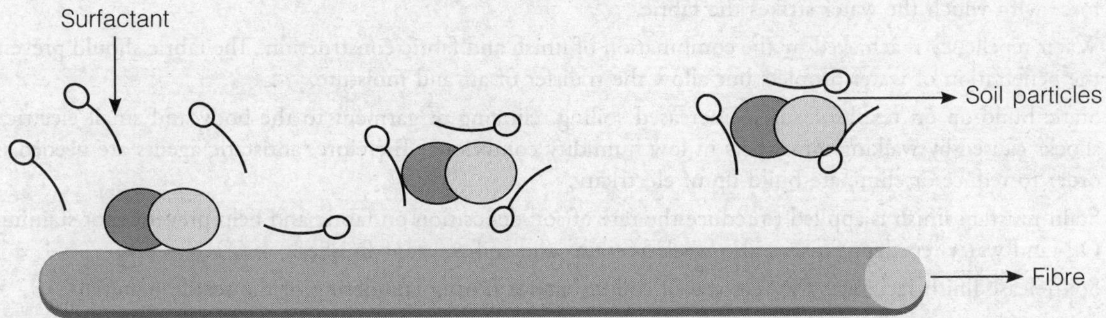

Source: Drawn by the authors.

in removing the soil. Soil-release finishes also prevent redeposition of soil, introduce antistatic properties and improve softness of the fabric. Soil-release finishes comprise chemicals such as, polymers of acrylic monomers and carboxylic groups, fluorochemicals, etc. Dual-acting fluorochemicals have been developed that impart both soil-repellent and soil-release properties. These are generally block copolymers of fluorocarbons and polar segments such as esters. When the fabric is dry, the fluorocarbon sections are on the surface providing soil repellency. In water, the hydrophilic segments come to the surface facilitating removal of soil.

SUMMARY

- Special finishes include those finishes that are not applied in routine to the fabrics but are given to obtain a specific effect or property depending on the end use.
- Designs can be imparted onto the fabric with the help of certain special calendering techniques such as schreinering, embossing, friction calendering or moiréing.
- The moiré finish is characterised by a soft lustre and a watermark effect created by differences in light reflection.
- Parchmentised or transparent effect is produced by treating the cotton fabric with strong sulphuric acid.
- A three-dimensional effect called plissé is produced by imprinting the cotton fabric with a chemical especially alkali.
- Flame-retardant finish is essential for uniforms of fire fighters, military and emergency personnel.
- Textiles provide suitable environment for the growth of micro-organisms which can lead to functional, hygienic and aesthetic difficulties.
- Antimicrobial finishes are very important because these impart protection to the textile user against pathogenic or odour causing micro-organisms as well as protection to the textile itself from damage caused by micro-organisms.
- The protective effect of textiles to UV rays is quantified as Ultraviolet Protection Factor (UPF) or Sun Protection Factor (SPF).
- Crease-resistant/durable press finish is applied to cotton and other fabrics to impart anticreasing properties to the fabric.

- A waterproof fabric is one that will not wet regardless of the amount of time it is exposed to water or the force with which the water strikes the fabric.
- Water repellency is achieved by the combination of finish and fabric construction. The fabric should prevent the penetration of water droplets but allow the transfer of air and moisture.
- Static build-up on textiles leads to increased soiling, clinging of garment to the body and small electrical shocks caused by walking on carpets in low humidity conditions. Therefore, antistatic agents are needed in order to reduce or eliminate build-up of electricity.
- Stain-resistant finish is applied to reduce the rate of soil deposition on fabric and help prevent spot staining. Oil- and water-repellent finishes also work as stain- and soil-resistant finishes.
- Soil-release finish facilitates the removal of soiling matter during laundering of the textile material.

KEY WORDS

Napping: This is a finish in which raised fibre ends are produced on the fabric surface.

Emerising/sueding/sanding: Mechanical finish in which open width fabric is rotated over one or more rollers covered with abrasive material such as sand paper or emery paper to produce suede-like finish.

Flocking: Finishing process in which short monofilament fibres, usually nylon or rayon are applied onto the substrate which has been previously coated with an adhesive to a give a pile effect.

Shearing: Mechanical finish to cut or shear the undesirable staple fibres from the fabric surface.

Beetling: Mechanical finish to increase the lustre of the cotton and linen fabrics.

Embossing: Special calendaring finish that produces three-dimensional designs on fabrics.

Limiting Oxygen Index (LOI): Mnimum content of oxygen in an oxygen/nitrogen atmosphere that is required to keep the sample burning.

Lotus leaf effect: is the name of the self-cleaning property that is imparted to textile surfaces. It has been named after the unusual properties of the lotus leaf surfaces, which are remarkably water-repellent and soil-repellent.

EXERCISES

1. What are the various special calendering techniques? Explain embossing.
2. Differentiate between napping and flocking.
3. Explain the different methods used to produce a flocked surface.
4. Explain the different methods used to produce a flame-retardant fabric.
5. Describe the modes of action of antimicrobial finish.
6. Enumerate the factors responsible for UV protection.
7. Explain lotus leaf effect.

REFERENCES

Joseph, M. L. 2003 [1976]. *Essentials of Textiles*. New York: Holt, Rinehart and Winston.

Kadolph, S. J. 2009. *Textiles*. New Delhi: Pearson Education.

Lewin, M. and S. B. Sello. 1983. *Handbook of Fiber Science and Technology. Volume II: Chemical Processing of Fibers and Fabrics: Functional Finishes*. New York: Marcel Dekker Inc.

McCrum, N. G., C. P. Buckley and C. B. Bucknall. 1998. *Principles of Polymer Engineering*. Second edition. Oxford: Oxford University Press.

Rouette, H. K. 2000. *Encyclopedia of Textile Finishing*. London: Springer.

Schindler, W. D. and P. J. Hauser. 2004. *Chemical Finishing of Textiles*. London: Woodhead Publishing.

Scott, R. A. 2005. *Textiles for Protection*. London: Woodhead Publishing.

ONLINE SOURCES (all accessed 18 June 2016)

www.textilelearner.blogspot.com/raising

www.swicofil.com/flock.html

11

DYEING

HIGHLIGHTS

- Basics of colouration
- Stages of dyeing
  - Dope dyeing
  - Fibre dyeing
  - Yarn dyeing
  - Piece dyeing
  - Product dyeing
- Colour fastness of dyed fabrics and their evaluation

Colour is the most visually recognisable finish with unparalleled aesthetic appeal. The appeal is universal although a particular hue may be in or out of fashion at a particular time. Its importance can be gauged from the fact that it can render a completely functional textile redundant if it is of a colour which is not in vogue. Moreover, all ancient cultures have ascribed meaning to various colours and they hold symbolic and ritualistic significance. In India too, this is not any different. The sacred thread *mouli*, which is tied before initiating any religious ceremony, for instance, is tinted with two auspicious colours—red and yellow.

According to Webster's dictionary, dyeing is 'the process of colouring fibers, yarns or fabrics by using a liquid containing colouring matter for imparting a particular hue to a substance.' Dyeing as a process is very old, and indeed predates written records. Historically, the primary source of dye has generally been nature, with the dyes being extracted from animals or plants. Since the mid-18th century, however, artificial dyes have been produced to achieve a broader range of colours and to render the dyes more stable to washing and general use.

How We See Colour

The human eye and brain together translate light into colour. Light receptors within the eye transmit messages to the brain, which produces the familiar sensations of colour. Colour is not inherent in objects. Rather, the surface of an object reflects light of some wavelength and absorbs all the others. We perceive the reflected light as different colours. Thus, red is not *in* an apple. The surface of the apple reflects the wavelengths that we see as red, while absorbing the rest. An object appears white when it reflects light of all wavelengths and black when it absorbs them all. Red, green and blue are the additive **primary colours** of the colour spectrum. By varying the amount of red, green and blue, all of the colours in the visible spectrum can be produced.

Colour is a visual sensation produced by the reflection of visible light rays that strike the cornea of the eye and are focused by the lens onto the retina. The retina is covered by millions of light-sensitive cells, some shaped like rods and some like cones. These receptors process the light into nerve impulses and pass them along to the cortex of the brain via the optic nerve. There are about 6 to 7 million cones, and over 120 million rods in each eye. Cones are concentrated in the middle of the retina, with fewer on the periphery while rods are concentrated around the edge of the retina. Rods transmit mostly black and white information to the brain. That is why our peripheral vision is less sharp and colourful than front-on vision. Also, rods are more sensitive to dim light than cones, so we lose most of our colour vision in dusky light. It is the rods that help our eyes adjust when we enter a darkened room. There are three types of cone-shaped cells, each sensitive to the long, medium or short wavelengths of light. These cells, work together with the connecting nerve cells to give the brain enough information to interpret and distinguish colours.

Dyes and Colour

Dyes are organic compounds which are used for giving colour to the textile materials. The purpose of a dye is to absorb light on selective basis so that the textile would then reflect back the unabsorbed wavelengths.

Dyes are able to absorb/reflect colours due to the presence of certain chemical groups in the dye molecule. These are called **chromophores**. Different chromophores and their combinations produce different colours. Some of the major chromophore groups are Azo group (–N=N–), thio group (=C=S), nitroso group (–N=O), nitro group (–NO_2), etc. The depth of a particular hue can be increased by increasing the number of chromophore groups present and/or by replacing the weaker chromophore groups with stronger ones. Also present in dye molecule are **auxochromes.** These are chemical groups that enhance the brightness of a colour or enhance the water solubility of a dye or provide the chemical group that can bind with the fibre. Their function depends on the number and type present. Examples of some auxochrome groups are: –OH, –COOH, –SO_3H, –$N(CH_2)_2$, –$NHCH_4$, –NH_2, –O_4SNa, etc. Dyes are generally soluble in water and have the affinity for the substrate on which they are being applied.

Pigments are also coloured molecules that are used for textile colouration. Pigments are insoluble in water and have no affinity for fibres. They can therefore be applied by adding them to the spinning solution or by using an adhesive, resin or bonding agent to attach them to the fibre. These colours are economical and easy to apply. However, they are not colourfast to laundering, dry-cleaning or rubbing; the colours produced are relatively permanent except when the bonding agent wears off due to abrasion. Also, the fabric becomes stiff or harsh after application of pigments. Pigments are mainly used for printing and these can be applied on all substrates.

STAGES OF DYEING

Dyeing can be carried out at different stages of the textile production process, viz. dyeing of fibres, yarn, fabric or finished product. In the case of man-made fibres, colour can also be added to the polymer solution prior to the extrusion of fibres. Each of these, however, requires a specific dye bath and specialised equipment for dyeing. If dyeing is done at an earlier stage, dye penetration is better and thereby the resultant fastness. However, the disadvantage is the inflexibility to change as per the fashion trend. Dyeing done at a later stage provides flexibility to meet colour demands as fashion changes.

Table 11.1: Various stages of dyeing textiles

Textile form	*Dyeing method*
Spinning solution	Dope dyeing or mass pigmentation
Fibre	Stock dyeing
Yarn	Skein dyeing, package dyeing and warp beam dyeing
Fabric	Piece dyeing: beam dyeing, beck/winch dyeing, jig dyeing, jet dyeing, pad dyeing
Finished product/garment	Product dyeing (paddle dyeing)

Source: Compiled by the authors.

Dope/Solution Dyeing

It is a good method and sometimes the only option available for dyeing of certain difficult-to-dye fibres, such as polyester and polypropylene. It is used only for dyeing of man-made fibres (MMF). Here the dye is added to the spinning solution before the filament is extruded through the spinneret. This gives evenly dyed textiles with good to excellent colourfastness to washing and light. Dope-dyed textiles, however, cannot be dyed again.

Fibre/Stock Dyeing

Dyeing at the fibre stage is also known as stock dyeing. Here dyeing is done in big enclosed containers called **kiers**. Loose fibres are placed in perforated containers which are placed inside the kiers. The dye liquor is circulated through the fibres till the desired colour is achieved. Man-made filaments can be coloured by this method after cutting them into small staple lengths. After dyeing, the excess dye liquor is removed and fibres are washed and dried. The advantages of this method are large quantities of fibres can be dyed at one go and also as dye penetrates individual fibres it gives uniform colour with good to excellent colour fastness. Stock dyeing can also be used to produce interesting colour effects by blending of different coloured fibres. Tweed fabrics are made by using dyed fibres. Fibre stage dyeing, however, is a time consuming and expensive process. Another disadvantage is that this method cannot accommodate rapid changes in fashion. The stock once dyed can lead to inventory pileup and wastage if not consumed immediately. Stock dyeing is usually suitable when colour effects (multi-coloured) are desired; for example, a black-dyed wool fibre might be blended and spun with undyed (white) wool fibre to produce soft shade of grey yarn. Tweed fabrics with colour effects such as Harris Tweed are examples of stock-dyed material. Dyeing is sometimes also done just before a yarn is formed. Wool slivers are frequently dyed at this stage. Wool slivers are wound on the package (tops) and dyed by the process called **top dyeing**. These slivers are then drafted (drawn) and spun into a yarn. If slivers of a different colours are combined into a yarn it can give a distinguished mottled or heather effect.

> When a yarn is spun using pre-dyed fibres of different colours, it gives a multi-coloured effect known as **heather effect**.

Yarn Dyeing

It is one of the oldest systems of colouring textiles. In general, yarn dyeing is less labour-intensive than fibre dyeing. Yarn dyeing can be done in various forms as follows:

- *Skein dyeing*: In skein dyeing, yarn is loosely coiled on a reel and then dyed. The coils, or skeins are hung over a rod which is immersed in a dye bath. The dye liquor is pumped in and out of the yarn to get a uniform shade. Skein dyeing is the most expensive yarn dyeing method. Skein-dyeing is popularly used for bulky acrylic and wool yarns.
- *Package dyeing*: The yarns are wound on spools, cones or similar units and these packages of yarn are stacked on perforated rods in a rack and then immersed in a tank. The dye is then forced outward from the rods under pressure and then forced back through the packages toward the centre to fully penetrate the entire yarn. Package-dyed yarns do not retain the softness and loftiness that skein-dyed yarns do. However, they are widely used for knitted and woven fabrics. Carded and combed cotton yarn used for knitted outerwear is mostly package dyed. Typical capacity for package dyeing equipment is 1210 pounds (550 kg) and for skein dyeing equipment is 220 pounds (100 kg).
- *Warp beam dyeing*: It is similar to package dyeing but more economical. Here, the yarn is wound on to a perforated warp beam which is then immersed in a tank for dyeing. Pressure is applied for better and deeper penetration of the dye. It is a popular method for the manufacture of denims where warp beams are traditionally dyed with indigo and interlaced with white weft using warp-faced twill weave. That is why the fabric appears white on one side and blue on the other side. Beam dyeing is more economical than skein or package dyeing, but it is only used in the manufacture of woven fabrics where an entire warp beam is dyed. Knitted fabrics, which are mostly produced from the cones of the yarn, are not adaptable to beam dyeing.

Designing Fabrics through Yarn Dyeing

Coloured yarns can be used to create interesting designs in the fabrics, such as stripes, checks, plaids, tone-on-tone effect and iridescent effects (colours that shimmer and change as the observer's position changes). A unique resist-dyeing technique can be employed on yarns to introduce elaborate, multi-coloured geometrical patterns into the fabric. Examples of such fabrics are the indigenous techniques in India such as Ikats of Orissa, Patan Patolas of Gujarat and Pochampally of Telangana. During these processes, a tie-dye is done on either warp or weft yarns or both. In traditional ikat, only the warp *or* the weft is dyed. In warp ikat the patterns are clearly visible in the warp threads on the loom even before the plain-coloured weft is introduced to produce the fabric. In weft ikat, the weft thread carries the dyed patterns, and therefore the pattern appears only as the weaving proceeds. Double ikat is a technique in which both warp and the weft are resist-dyed. The design in such fabrics is identical on both sides and can be worn on either side. Double ikat is only produced in three countries, India, Japan and Indonesia. In India it is done in Patan, Gujarat and in Pochampally in Telangana. The procedure is very time consuming and therefore the final product is expensive.

Piece Dyeing

Dyeing at the fabric stage is known as piece dyeing. The fabric so dyed is generally solid coloured, i.e. uniformly dyed in a single colour. It is relatively inexpensive compared to dyeing done at the fibre or yarn stage. It is also a more versatile and flexible method where the response to changing fashion trends can be made faster, is easier to do and costs less. The only problem is when dyeing is done in the rope form there may be irregularity of colour, and in continuous piece dyeing the ends of the fabric may not pick the same colour as the rest of the fabric. Piece dyeing is done on various kinds of machines. Some of the commonly used ones are discussed here.

- *Beam dyeing*: Woven or knitted fabrics are wound on beams and dyed in a process similar to warp beam dyeing of yarns. Dye is circulated through stationary beam of fabric in open width (Figure 11.1).
- *Beck/Winch dyeing*: It is a versatile, continuous process used to dye long yards of fabric. It is the oldest form of piece dyeing. This is called a **winch** as the winch mechanism is

Figure 11.1: Warp beam dyeing

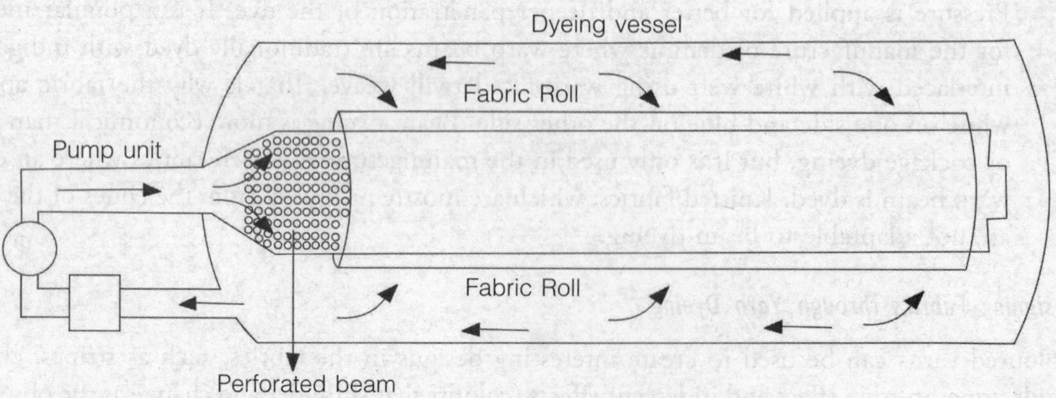

Source: Drawn by the authors.

used to move the fabric. The fabric is passed in rope form through the dye bath. The two ends of fabric are sewn together to form a loop. This fabric loop moves over a rail onto a reel which immerses it into the dye and then draws the fabric up and forward to the front of the machine. The reel pulls the fabric out of the dye liquor and over an idler roll. The idler roll presses some of the excess liquor from the fabric improving exchange of liquor from fabric to dye liquor. This process is repeated till the material is dyed uniformly to the desired colour intensity. About 900 kg or 1000 m of fabric can be dyed on the beck equipment at a time. The machine may be an open or an enclosed HTHP (high temperature high pressure) beck. Several loops of fabric of about the same length can be dyed simultaneously. The individual loops are however separated from one another by a dividing device (Figure 11.2). Chemicals and dyes are added in a compartment situated at the back of the beck. This is separated from the dye trough by a perforated baffle plate (about 6–9 inches thick). This compartment also has heated pipes, the steam vigorously agitates the compartment and aids in mixing the dyes and chemicals into the dye liquor.

- *Jig dyeing*: Jiggers are very popular machines because of their multipurpose usage not only in dyeing but also in fabric processing, viz. desizing, scouring, bleaching, etc. It uses the same procedure as of beck dyeing, however, the fabric is held on rollers at full width rather than in rope form as it is passed through the dye bath. This reduces the fabric's tendency to crack or crease. The fabric is dyed by letting off the fabric from one roller into the dye bath and winding it onto the other roller. After one cycle the process is reversed and the fabric is passed back to the first roller. This is repeated till the desired shade is achieved. It is a preferred method for heat-sensitive thermoplastic fabrics which cannot be dyed in rope

Figure 11.2: Beck/Winch dyeing

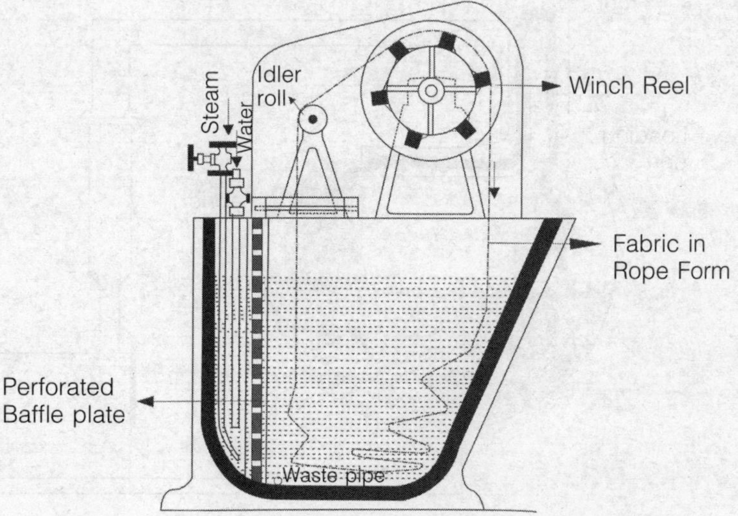

Source: Adapted by the authors from *Encyclopædia Britannica*, 1911 (CC0; public domain).

form as they will form permanent creases or colour streaks in the heated dye bath. Jig dyeing equipment can handle 550 pounds (250 kg) of fabric (Figure 11.3).

- *Jet dyeing*: The fabric is dyed in a completely tensionless rope form. The fabric is placed in a heated tube or column where jets of dye solution are forced through it at high pressure. The fabric too moves along the tube. The solution moves faster than the cloth while colouring it thoroughly. The dye solution is continually recirculated as the fabric is moved along the tube. Up to 1100 pounds or 500 kg of fabric can be jet-dyed. This machine is suitable for polyester and its blends, viscose rayon, wool and cottons, and woven, knitted and textured fabrics (Figure 11.4).

Figure 11.3: Jig dyeing

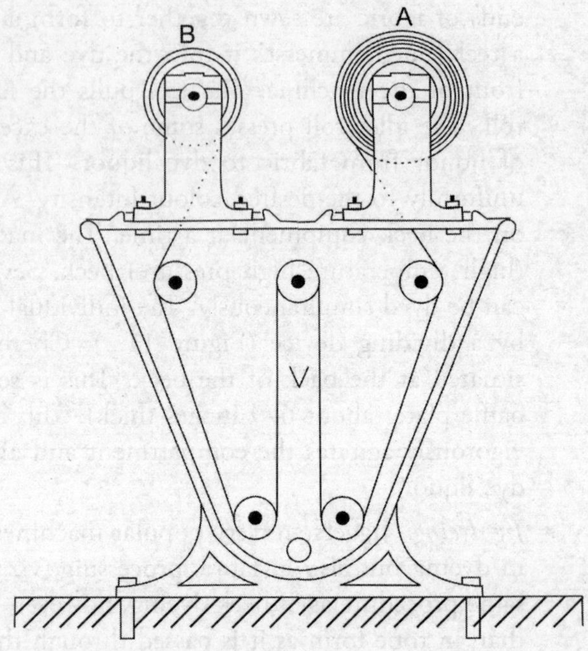

Source: Adapted by the authors from *Encyclopædia Britannica*, 1911 (CC0; public domain).

Figure 11.4: Jet dyeing

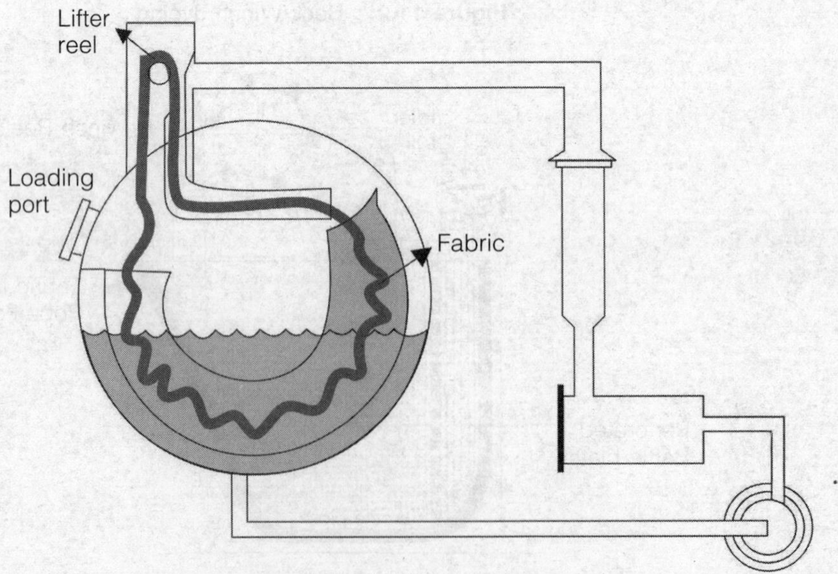

Source: *Drawn by the authors.*

- *Pad dyeing*: Padding is also done while holding the fabric at full width, like jig dyeing. The fabric is passed through a trough containing the dye. It is then passed between two heavy rollers which force the dye into the cloth and squeeze out the excess dye and then through a heat chamber for the dye to set. It is finally passed through a washer, a rinser and a dryer.

Most of the coloured fabrics are piece-dyed since these methods give the manufacturer maximum inventory flexibility to meet colour demands as fashion trends change. In terms of overall volume, the largest amount of dyeing is performed using beck and jig equipment.

Dyeing of Blends

When a fabric contains more than one type of fibre, as in a blend, its dyeing is more complex. There are primarily two options available: union dyeing and cross dyeing.

1. *Union dyeing*: This is a method of dyeing a fabric containing two or more types of fibres or yarns to the same shade so as to achieve the appearance of a solid-coloured fabric. The dye has to be selected carefully and applied properly to achieve uniformity in colour. Fabrics can be dyed using a one bath or two bath process. In one bath process, dyes which can be applied simultaneously are used. This however is not easy or always possible. In two bath process, dyes are applied individually to the different fibre components subsequent to one another, as per the prescribed procedure.
2. *Cross dyeing*: This is a method of dyeing blend or combination fabrics such that different fibre components are dyed in different colours or different shades of the same colour. It is also possible to leave one component fibre white and dye the other one, as is commonly seen in spun fabrics. Cross dyeing can produce multicoloured effect, heather effects, plaid, checks or striped fabrics. Cross-dyed fabrics may therefore be mistaken for fibre- or yarn-dyed materials as the fabric is not a solid colour, a characteristic considered typical of piece-dyed fabrics. It is not possible to visually differentiate between cross-dyed fabrics and those dyed at the fibre or yarn stage.

Tone-on-tone effect can be created on blended fabric after dyeing in single bath. They are light and dark shades of the same colour in a fabric generally containing one generic fibre such as by combining two different types of polyesters with varying dye affinity. The component with higher affinity will pick up the darker shade while the one with less affinity will pick up the lighter shade of same hue. Tone-on-tone effect can be achieved on specific varieties of nylon, polyester and acrylics. These varying dye properties are imparted during manufacturing. Nylon for instance can be made in regular, deep dye and ultra deep dye variety. This provides the possibility for three tone effects of the same colour. This effect is widely used in the carpet industry to produce tweed effect designs on them.

Product Dyeing

Also known as **garment dyeing**, this is the process of dyeing products such as hosiery, sweaters and carpets after they have been produced. This stage of dyeing is suitable when all components (including threads) are dyed to the same shade. A textile which will be dyed at product stage requires fabric which has been prepared accordingly. Typically, the fabric is bleached, mercerised and finished by applying wetting agents which will assist in dye penetration. Paddle dyeing is the most common method for product dyeing. Here garments are placed in net bags and 10 to 50 of these bags are placed in large tubs containing the dye bath. The dye is kept agitated by a motor driven paddle in the dye tub, and this improves circulation and ensures better dye penetration. This method is used to dye sheer hosiery since it is knitted using tubular knitting machines and then stitched prior to dyeing. Tufted carpets, with the exception of carpets produced using solution-dyed fibres, are often dyed after they have been tufted. This method is not suitable for apparel with many components such as lining, zippers and sewing thread, because of two reasons: each component may dye differently, and the components shrink variably and thereby distorting and misshaping the article. Dyeing at this stage is ideal for quick response. Many T-shirts, sweaters and other types of casual clothing are product-dyed for speedy response to demand for certain fashionable colours. Thousands of garments are constructed from prepared-for-dye (PFD) fabric, and then dyed to colours that are in vogue. However, it has to be noted that garment-dyed products may be more prone to colour bleeding and fading.

COLOUR FASTNESS OF DYED FABRICS

Fastness of coloured fabrics is of utmost importance for their efficient end use. The ability of a fabric to retain its original colour is one of the most important parameters which can affect consumer's satisfaction. Colour fastness of textiles is evaluated to a variety of agents such as washing, light, rubbing, perspiration, bleaches, dry cleaning, heat, etc. A coloured textile is generally assessed for specific fastness parameters which depend on its intended use. For instance, fastness to washing is important for clothes which are washed frequently, while fastness to sunlight is important for curtains. Similarly, fastness to bleach is important for swimwear, as water in swimming pools contain chlorine used as disinfectant, while fastness to perspiration may be more relevant to test for socks, shirts and other such apparel.

The various parameters that effect fastness are fibre(s) used, dyes and chemicals used and the method of application. Poor wash fastness will cause colour bleeding (discharge of colour into water) and the migration of this colour will result in fading (loss in depth of colour) of the article under wash and staining of adjacent articles. A material is deemed colourfast if it does not bleed, fade or show colour migration. Generally, a fabric may be colourfast to some conditions and not all.

Standard tests that are regularly updated and published are available from various organisations such as AATCC (American Association of Textile Chemists and Colourists), BIS (Bureau of Indian Standards), ISO (International Organization for Standardization). Any one of these standard tests can be used to determine the colour fastness of the dyed materials.

Wash Fastness of Dyed Fabric

The loss of colour during laundering is referred to as bleeding or the lack of wash fastness. Wash fastness of any material is the resistance of a material to the change in colour, change in surface or staining of the adjacent surface during laundering.

Factors Affecting Colour Loss

- Dye molecules may be held by weak Van der Waals forces or hydrogen bonds. Dye molecules may not have penetrated completely into the fibre polymer and remain on the surface.
- The presence of a large number of auxochromes. The auxochromes contribute to the water solubility of the dyes and thus poor colour fastness.
- Washing temperature also has an effect on the colour fastness. Often the colour gets removed from the fabric by the action of hot water.
- Concentration of the detergent solution also effects colour loss. Very high concentration of soap solution often adversely affects colour. Also a highly alkaline soap solution may attack the dye on the fabric.
- Abrasive action may also result in loss of colour.

A **launderometer** is used to test wash fastness of dyed fabric. It is a machine consisting of stainless steel canisters which rotate in a thermostatically controlled water bath. Conditions of temperature, detergent solution and abrasive action similar to those at home and commercial laundering are created by selecting suitable standard test methods so as to obtain appropriate results. The change in colour or staining in the launderometer can be due to the abrasive action against canister walls and steel balls.

The material to be tested is placed between undyed pieces, one of which is the same fabric as the sample while the other is a different fabric as mentioned in standard test methods. Assessment of fastness involves a visual determination of either change in shade or staining of an adjacent undyed material washed along with it. Grading for fastness is done by the use of Grey Scales through which it is possible to compare loss of colour or staining of any colour irrespective of its depth. It is necessary to use two scales, one for assessing change in colour and the other for staining (Figures 11.5a and b). The original dyed material and the tested sample are placed alongside each other. The material and scale are viewed simultaneously and the fastness rating is determined by that pair on the scale which shows a contrast equal to that between the original dyed sample and the tested portion. The Grey Scales for staining is used in exactly the same way as done for Grey Scales for fading but with the original white sample and the newly tinted sample. The rating/grade is given on a rating scale of 1 to 5.

Crock Fastness of Dyed Fabrics

Crocking is the transfer of colourant from the surface of a coloured yarn or fabric to another surface or to an adjacent area of the same fabric principally by rubbing. **Crock fastness** or **rub fastness** refers to the ability of a dyed specimen to resist any change in colour that may be brought about by the action of rubbing. Crock fastness is of great concern to the consumer as rubbing action is unavoidable in the normal wear of fabrics and this should not cause staining. Certain dyes tend to rub off when in frictional contact with another surface as there may be some loose, unfixed dye on the surface which gets further loosened by the rubbing action and then stains the adjacent fabric. This staining is generally enhanced when the friction is encountered in a wet atmosphere.

Figure 11.5a: Grey Scale for change in colour

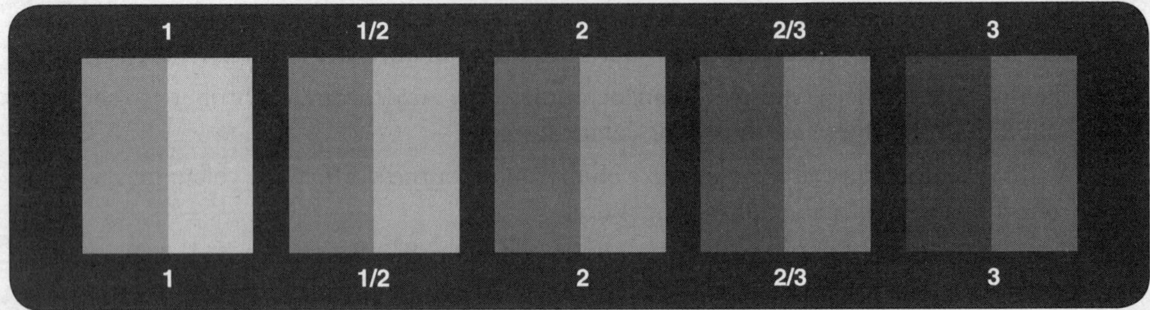

Figure 11.5b: Grey Scale for staining on white

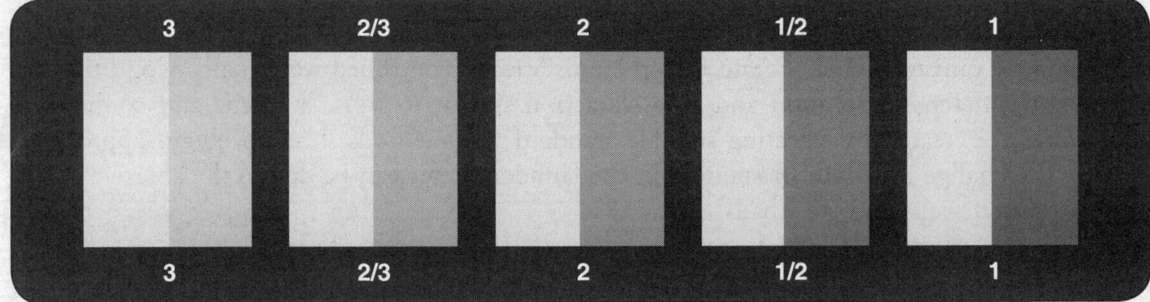

Source: Authors.

Assessing the change in the colour of the dyed fabric and the staining of the white surface can be used as a measure of its rub fastness. This is also done by using the Grey Scales mentioned earlier. The instrument used for the determination of crock fastness is referred to as a crock meter. The **crock meter** is a simple device which is used to stroke the dyed fabric with a uniform pressure and a uniform stroke length each time the rub fastness is to be determined. The test method is designed to determine the amount of colour transferred from the surface of coloured textile material to other surfaces by rubbing. It is applicable to textiles made from all fibres in the form of yarn or fabrics whether dyed, printed or otherwise coloured. The rubbing is carried out with a white cotton cloth in both dry and wet conditions to determine both the dry and wet rub fastness.

Light Fastness

The purpose of colour fastness test to light is to determine how much the colour will fade when exposed to a known light source. The test measures the resistance to fading of dyed textile when exposed to Xenon arc lamp. The spectral energy distribution of this lamp is quite close to natural light. In this test, a prepared specimen of fabric to be tested is half covered and exposed to artificial light along with eight light-sensitive Blue Wool Standards. These are eight swatches of wool dyed using acid blue or vat blue dyes. The dyes have been chosen so that each blue wool sample takes about two to three times longer to begin fading as compared to the previous one, i.e. Sample #8 will take the longest time to fade while Sample #1 will fade the fastest. The test specimens along with eight blue wool standards are mounted on a frame for exposure to light such that only half of each sample and standard is exposed to light while the other half is covered with an opaque sheet. The test sample is exposed to light for a certain period of time. When the first blue wool standard fades to a rating of 3 on the Grey Scale all the test specimens that have also faded to the same extent are given the light fastness rating of 1. The samples and standards are further exposed to light. The blue wool standard #2 fades next and test specimens fading along with it are given the light fastness rating of 2 and so on. Light fastness rating 1 indicates very poor light fastness and 8 indicates excellent light fastness.

A textile **swatch** refers to a small sample piece of cloth or fabric that is representative of a larger whole.

Perspiration Fastness

The areas of a garment that come in contact with body parts where there is heavy sweating such as underarms, etc., many a time, fade more than the rest of the garment. Two artificial perspiration solutions are made, one acidic and the other alkaline. The material to be tested is placed between undyed pieces one of which is the same fabric as sample while the other is a different fabric as mentioned in standard test methods. These pieces are sewn together and dipped in one of the perspiration solutions for 30 minutes at room temperature. This specimen is then placed between two glass plates and pressed together with force and kept in an oven at 37°C ± 2°C for 4 hours. The dyed and undyed samples are separated and dried. The procedure is repeated with the second perspiration solution. The change in colour and staining are the assessed using the Grey Scales.

SUMMARY

- The surface of an object reflects light of some wavelengths and absorbs the others. We perceive the reflected light as different colours.
- Dyeing is the process of colouring fibres, yarns or fabrics by using a liquid containing colouring matter for imparting a particular hue to a substance.
- Dyeing can be carried out at different stages of the textile production process, that is at the fibre, yarn, fabric or finished product stages.
- Dope dyeing is done for man-made fibres by adding the dye to the spinning solution before the filament is extruded through the spinneret.
- Colourfastness of a textile product can be evaluated against a variety of agents such as washing, light, rubbing, perspiration, bleaches, dry cleaning, heat, etc.

KEY WORDS

Chromophores: Chromophores are groups in a dye molecule that are responsible for the colour produced by that dye. Different chromophores and their combinations produce different colours. Some of these groups are: azo group, thio group, nitroso group, nitro group, etc.

Auxochromes: These are chemical groups that may enhance the brightness of a colour, improve the water solubility of a dye or provide the chemical group that can bind with the fibre. Examples of some auxochrome groups are: $-OH$, $-COOH$, $-SO_3H$, $-N(CH_2)_2$, $-NHCH_4$, $-NH_2$, $-O_4SNa$, etc.

Chromophores: Chromophores are groups in a dye molecule that are responsible for the colour produced by that dye. Different chromophores and their combinations produce different colours. Some of these groups are: azo group, thio group, nitroso group, nitro group, etc.

Pigments: Pigments are coloured molecules that are insoluble in water and have no affinity for the substrate.

Dyeing: The process of imparting a particular hue to fibres, yarns or fabrics by using a liquid containing the colouring matter.

Colour fastness: It is the resistance of a material to change in colour and staining of the adjacent surface during use and care.

Stock dyeing: Dyeing at the fibre stage is known as stock dyeing.

Union dyeing: The method of dyeing a blend fabric containing two or more types of fibres or yarns to the same shade so as to achieve the appearance of a solid coloured fabric is known as union dyeing.

Cross dyeing: The method of dyeing different components of a blend or a combination fabrics to two or more shades or hues by the use of dyes with different affinities for the different fibres is known as cross dyeing.

EXERCISES

1. What are chromophores and auxochromes?
2. What are the methods of fibre dyeing? What kinds of designs are possible in fibre dyeing?
3. What are the various methods of yarn dyeing?
4. What is the difference between union dyeing and cross dyeing?
5. What are the design possibilities in yarn dyeing?
6. What do you understand by crock fastness?
7. What are the two types of Grey Scales? Why are they used?
8. What are the factors that affect colour loss during washing?

REFERENCES

Chakraborty, J. N. 2010. *Fundamentals and Practices in Colouration of Textiles.* New Delhi: Woodhead Publishing India Pvt. Ltd.

Corbman, B. P. 1983. *Textiles: Fibre to Fabric.* Sixth edition. New York: McGraw Hill Book Co.

Hollen, N. and J. Sadler. 1973. *Textiles.* Fourth edition. Singapore: The Macmillan Company.

Joseph, M. L. 1988. *Essentials of Textiles.* Fourth edition. New York: Holt, Rinehart and Winston, Inc.

Kadolph, S. J. 2009. *Textiles.* Tenth edition. New Delhi: Dorling Kindersley (India) Pvt. Ltd.

12

DYESTUFFS AND THEIR APPLICATION

HIGHLIGHTS

- Theory of dyeing
- Classification of dyes
- Properties, application and classification of
  - Direct dyes
  - Reactive dyes
  - Azoic colours
  - Vat dyes
  - Sulphur dyes
  - Basic dyes
  - Acid dyes
  - Chrome mordant dyes
  - Metal complex dyes
  - Disperse dyes
  - Natural dyes

In the last chapter we have understood the basics of colour, dye components that are responsible for its hue and solubility, and the various stages at which dyeing can be carried out in the making of textiles from fibres. As we have already seen, a **dye** is a chemical substance that is used to impart colour to a fabric. Until the 1850s, dyes were derived from nature. In 1856, Sir William Perkin accidently produced the first synthetic dye called 'Mauve', which produced a purple colour. Since then continuous developments in this field have added hundreds of dyes that meet the demands of the manufacturers, designers and customers on the one hand and the textile industry which keeps developing new fibres and blends on the other. The application of dyestuff to textile is a skilled job that requires knowledge about the properties of chemicals used. This chapter discusses the chemistry, mechanism and procedure of applying each class of dyestuff that can be used for textile dyeing.

Theory of Dyeing

Theory of dyeing

1. Migration
2. Adsorption
3. Diffusion
4. Substantive attachment

A dye is taken up by a fibre as a result of the chemical attraction between them. The various stages in dyeing are: (*a*) *Migration* of the dye molecules/ions from the dye solution to the interface of the fibre and dye solution. All dyes reduce the surface tension at the water interface, leading to increase in the number of dye molecules/ions (i.e. the solute) near the fibre than elsewhere in the solution. (*b*) *Adsorption* of the dye molecules/ions on the surface of the fibre. The adsorption of dye molecules onto the fibre surface from the interfacial layer is extremely rapid and generally exceeds the speed at which they can be replaced from the bulk of the solution. Thus, agitation or turbulence of the dye solution is very important. (*c*) *Diffusion* of the dye from the surface of the fibre to its centre. The rate of this diffusion process within the fibre might be quite slow and vary widely from dye to dye and fibre to fibre. Temperature plays an important role in diffusion as it increases the kinetic energy of the dye molecules and opens up the fibre structure. (*d*) *Substantive attachment* of dye molecules/ions by various forces of attraction such as hydrogen bonds, ionic bonds, Van der Waals forces of attraction, etc.

Forces Responsible for Fixation

Various types of bond formations can occur between the dye molecules/ions and the fibre. These are:

- **Covalent bonds**, which are strong bonding forces that are formed due to sharing of electron pairs between atoms. This is observed when reactive dyes react with cotton. These are the strongest bonds of the five bonds.
- **Ionic bonds** are formed due to the electrostatic force of attraction between oppositely-charged ions. These are weaker than covalent bonds but stronger than the other three. Ionic bonds play an important role in dyeing of protein fibres.
- **Hydrogen bonds** are the most common bonds between dyes and fibres. Both dyestuff and fibres have various groups capable of entering hydrogen bonds.
- **Van der Waals forces** are related to substantivity of dyes wherein they have no connection with other bonds such as ionic bonds or hydrogen bonds.
- **Hydrophobic interaction** is a phenomenon caused by hydrocarbon groups in an aqueous phase being surrounded by more highly-oriented clusters of water molecules. The dye absorbed inside the fibre is prevented from migrating back to solution by the hydrophobic interaction.

> The affinity of a dye for its substrate and its retention within the substrate through various dye-fibre interactions is termed as **substantitvity**.

For successful dye uptake it is important for optimum conditions to be maintained in the dye bath. It is therefore important, to know the following: the nature of the fibre structure, the nature of the dye as part of the bath solution and in the fibre, the rate of dyeing process (and how this is influenced by the transfer of dye from the solution to the dye-fibre interface first and then by the diffusion of the dye from the interface to the centre of the fibre). The reactions/processes prior to transfer and after diffusion vary depending on the dye class. To facilitate the process though, the coloured chemical component should be in the form of molecules or ions.

Classification of Dyes

Dyes can be classified in various ways—on the basis of the chromophores, on the basis of their chemical constitution or on the basis of their application conditions. The most popular system of classifying dyes is on the basis of their application conditions. Dyes that have similar application conditions and that show similar affinity towards certain fibre(s) are grouped under a particular **dye class**; e.g. direct dyes, reactive dyes, acid dyes, etc. These dye classes can be further grouped according to the fibre types for which they are most widely used. The dye classes used for different fibres are listed in Table 12.1.

Table 12.1: Dye classes for different textile substrates

Cellulosics	*Protein*	*Nylon*	*Acrylic*	*Polyester, Actetates*
Direct dyes	Acid dyes	Acid dyes	Basic dyes	Disperse dyes
Reactive dyes	Metal complex dyes	Metal complex dyes	Disperse dyes	
Azoic dyes	Reactive dyes	Reactive dyes		
Vat dyes	Basic dyes	Basic dyes		
Sulphur dyes	Direct dyes	Disperse dyes		
Natural dyes	Chrome mordant dyes	Chrome mordant dyes		
	Natural dyes			

Source: Compiled by the authors.

DIRECT DYES

Three decades after the advent of synthetic dyes, cotton still had to be dyed by a lengthy and complicated method as the existing dyes had no affinity for cellulosics. The Industrial Revolution resulted in the production of huge quantities of cotton, and there was an urgent need for an easier and quicker method of dyeing mass-produced cotton. In 1884, Böttiger prepared the dye Congo Red, which could colour cotton by merely boiling the material in a solution of the dye. Dyes such as Congo Red that are marked by simplicity of application belong to the class of dyes called **direct dyes**. According to AATCC, direct dyes are 'anionic dyes that are substantive to cellulose when applied from an aqueous bath containing electrolyte.'

Chemical Structure of Direct Dyes

Dyes owe much of their character to the presence of aromatic rings. Direct dyes are mainly derivatives of benzene (C_6H_6; 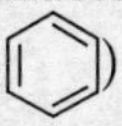) and naphthalene ($C_{10}H_8$; 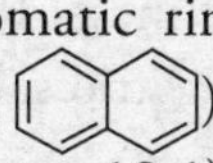). Sodium sulphonate groups ($–SO_3Na$) are attached to benzene or naphthalene rings (Figure 12.1). It is this sulphonate group which gives water solubility to the dye. In water the dye ionises into sodium cations (Na^+) and coloured sulphonate anions ($Dye–SO_3^-$). The higher the number of these groups present, higher the solubility of the dye in water. In fact, an electrolyte is needed to force the dye out of water and into the fibre. Other substitute groups found in direct dyes are hydroxyl group ($–OH$), amino group ($–NH_2$) and amido group ($–CONH_2$).

Figure 12.1: Structure of Congo Red, a direct dye

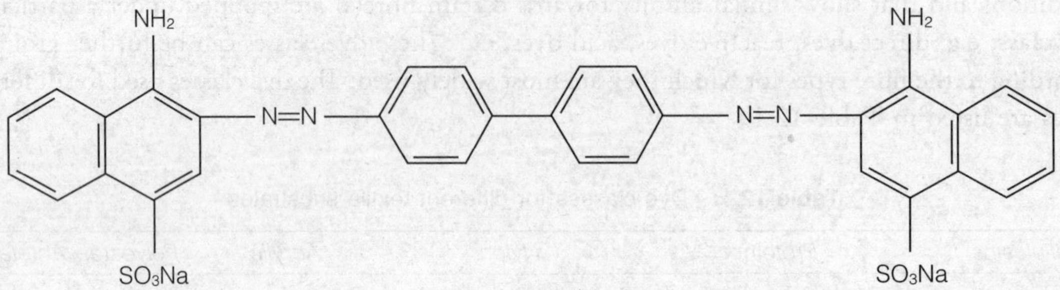

Classification of Direct Dyes

The Society of Dyers and Colourists (SDC) classifies direct dyes on the basis of their dyeing characteristics as follows:

- *Class A – Self Levelling Direct Dyes*: Dyes in this group have poor affinity towards the cellulosic fibres. These dyes therefore require relatively large amounts of salt to exhaust well. These dyes show good levelling characteristics and are capable of dyeing uniformly even when the electrolyte is added at the beginning of the dyeing operation.
- *Class B – Salt Controllable Direct Dyes*: These dyes have relatively poor levelling or migration characteristics. Therefore, these dyes require controlled addition of salt for uniform results. Salt is usually added after the dye bath has reached the dyeing temperature.
- *Class C – Salt and Temperature Controllable Direct Dyes*: These dyes have very high affinity for the substrate, and hence show poor levelling and migration behaviour. Their substantivity increases very rapidly with increasing temperature. Their rate of dyeing is controlled by controlling the rate of rise of the temperature of the dye bath as well as controlling the addition of salt. Since these dyes have very high affinity for the fibre, light shades of these dyes may require very little salt for exhaustion.

Levelling is the ability to dye all parts of a material uniformly.

The absorption of dye from the dye bath by the substrate being dyed is termed as **exhaustion**.

Migration is the ability of a dye to move from one part of the fibre to the other once inside the fibre.

The above three classes of direct dyes require different dyeing parameters in terms of quantities of salt, time of addition of salt and rate of increase of temperature. It is therefore, preferable to use direct dyes of the same class when using a mixture of dyes to get a particular colour/shade. If need be then dyes of Class A and Class B or dyes of Class B and Class C can be used together, but not dyes of Class A and Class C as their performance in the dye bath will be very different.

Application of Direct Dyes

The exact method of application of direct dye would depend on the class of dye used.

Class A Dyes

The dyestuff is mixed with a small amount of cold water and an anionic or non-ionic wetting agent. To this, sufficient boiling water is added and continuously stirred to get the dye paste. This solution is then added to the dye bath passing through a strainer. In general, soft water is preferable but is not essential for most direct dyes. If soft water is not available, a sequestering agent or soda ash (sodium carbonate), 1–3 per cent on weight of cotton, is added to the solution for assistance. Sodium chloride or Glauber's salt ($Na_2SO_4.10H_2O$) is also added for higher exhaustion of dye. The cotton substrate is then introduced into the dyebath at 40–50°C and the dye bath temperature is raised to boil over a period of 30–40 minutes and continued at boil for about 45–60 minutes thereafter.

Class B Dyes

For dyeing of dyes of Class B, the dye bath is prepared in a manner similar to that for Class A dyes, but without the presence of salt. Salt is added during the time the dye bath is raised to boil or while approaching the boiling temperature.

Class C Dyes

In the case of the application of dyes of Class C, it is important that the dyeing process should commence at a low temperature and without the addition of any electrolyte. The temperature of the dye bath is raised very slowly and dyeing continued for 45–60 minutes. The electrolyte should be added in portions after the liquor has reached boil. There is often a narrow range over which the exhaustion is most rapid and thus the control of temperature is often very important.

Effect of Dyeing Parameters

Effect of Temperature

The amount of dye adsorbed by the fibre at equilibrium decreases with the increase in dyeing temperature. However, as the temperature increases, equilibrium is achieved faster due to increased kinetic energy of the molecules. Every dye has different temperature of maximum exhaustion. Exhaustion rises to a maximum as temperature increases and then falls with a further increase in temperature. The optimum temperature may vary for different dyes. This causes a change in shade when a boiling liquid cools and can often turn a dyeing matched at boil to become off shade because dyes start exhausting with decrease in temperature. For a particular dye, maximum exhaustion may be achieved at 20°C, but it may take 24 hours to reach equilibrium. This is not practical for commercial dyeing and, therefore, dyeing is usually not carried out at optimum temperature till equilibrium but carried out at a temperature where fairly good exhaustion can take place in 1–2 hours. If more than one dye is used then care has to be taken that all dyes give maximum exhaustion in similar temperature range, otherwise shade variation may occur as all the dyes will not exhaust fully.

When an undyed piece of material is placed in a suitable dye bath, the uptake of dye is relatively rapid at first. This initial rate decreases, until eventually no further dye is absorbed. At this point, the rate of absorption of dye is equal to the rate of desorption of the dye by the material. This state is known as the **state of equilibrium**.

Effect of Electrolyte

Addition of electrolyte to the dye bath promotes exhaustion of direct dyes. Cellulosic fibres assume a negative charge when immersed in water. This repels the similarly-charged ions of the direct dye. Electrolytes reduce the charge on the fibre and thereby facilitate the approach of dye ions to such a close range of fibres that hydrogen bonds or Van der Waals forces can become effective. As mentioned earlier, the higher the number of sulphonic acid groups in the dye, higher the repulsion and therefore the requirement of more amount of salt to nullify the repulsive effect. The electrolyte also reduces the solubility of the dye. Generally, the amount of sodium chloride used is 10 times the amount of dye (% on weight of fabric [o.w.f.]). Although economical, sodium chloride causes corrosion of stainless steel under high temperature conditions. Sodium sulphate (Na_2SO_4) or Glauber's salt ($Na_2SO_4.10H_2O$) is therefore preferred despite higher costs. For Class A direct dyes, salt is added at the beginning whereas for Class B and Class C, it is better to add salt in 3–5 instalments after the dyeing temperature has been reached.

Effect of pH Value

Direct dyes are invariably applied from neutral solutions. The addition of alkali has a retarding effect on the rate of adsorption. At alkaline pH solubility of the dye is increased, also –OH groups on cellulose may convert to hydroxylate ions which are negatively charged and may repel the dye anions. Alkali, therefore, acts as a levelling agent, or is added to remove hardness in water or to improve solubility of some direct dyes such as Benzopurpurine 4B and Chlorazol Orange PO.

Levelling agents promote uniform distribution of dye throughout fibre structure by reducing the strike rate and enhance the solubility of dye.

Effect of Liquor Ratio

In dyeing, the amount of water to be taken is calculated as a ratio of the weight of the material known as material to liquor ratio. For example, if the material weight is 1 g and the amount of water to be taken is 30 g or 30 mL, the liquor ratio is 1:30. Thus for 500 g of the material, the amount of water for the dyebath would be 500 x 30, which is 1500 mL. Material to liquor ratio may be different for different machines and is abbreviated to MLR or M:L ratio. Since percentage exhaustion—i.e. the percentage of the total amount of dye originally in the dye bath that is taken up by the substrate by the end of the dyeing cycle—increases as the concentration of dye in dye bath increases, it is preferable to use low liquor ratios.

Mechanism of Substantivity of Direct Dyes

The substantivity of direct dyes for cellulosic fibres is facilitated by the alignment of the linear, coplanar aromatic structure of these dyes parallel to the cellulose chains. The more linear the dye structure, the more easily it can align itself on the surface of the fibre, thus increasing substantivity. These are effected by Van der Waals forces and hydrophobic interactions. The role of hydrogen bonding or the possibility of weak acid-base reaction between the –OH group of cellulose and amino group in dye molecule has also been suggested.

Evidence has also been presented to show that the dye molecules exist in a higher state of aggregation in the fibre than in the solution. This means that the dye enters the fibre in a state of low aggregation and once inside, the molecules become more aggregated and permanently locked. The existence of aggregates in the fibre has been confirmed by electron microscopy.

Fastness Properties and Usage

Direct dyes do not possess good fastness to washing or other wet processes such as scouring, etc. Its wash fastness varies for different dyes but it generally ranges from 1–4 and with 3 being the average. Its light fastness varies from very good to moderate. Its average light fastness is 6. Direct dyes can be applied on a wide variety of textiles intended for end use such as draperies, upholstery fabrics, linings, automotive fabrics and apparels.

After-treatments for Direct Dyes

Because of their relatively poor wet fastness, direct dyes are frequently after-treated. More than one of the following after-treatments can also be used for any one dyeing:

- *After rinsing*: Rinsing of dyed goods with sodium chloride or sodium sulphate (10 g/L) is carried out. Magnesium salts can also be used at lower concentration of 1–2 g/L. This treatment does not improve the wash fastness of the dye. It only prevents migration of the dye especially during the period between dyeing and drying.
- *Cationic fixing agents*: These are chemicals with large molecules which dissolve in water and disassociate into a large positive ion/cation and a small negative ion; e.g. quartenary ammonium compounds. The positive ion combines with the negatively-charged direct dye and forms a large complex molecule which has low solubility and therefore improved fastness. The amount of chemical required would depend on the depth of the shade. These compounds tend to reduce the light fastness of direct dyes.
- *Treatment with Potassium Bichromate*: Wash fastness can be improved by treating with potassium bichromate ($K_2Cr_2O_7$). The dyed material is immersed in 1–3 per cent of $K_2Cr_2O_7$ along with 1–2% of 30% acetic acid at 60–80°C for 20–30 minutes.

- *Diazotisation and coupling*: This requires a free aromatic amine ($-NH_2$) in the dye which can be treated with sodium nitrite ($NaNO_2$) and a strong acid followed by reaction with a compound containing an amino group or hydroxyl group attached to a benzene structure. A new azo group (–N=N–) is produced. This improves wash fastness as the size of the dye increases, but the shade changes.
- *Treatment with formaldehyde*: Some direct dyes when treated with formaldehyde form methylene bridges between dye molecules. However this method is not popular as formaldehyde use is restricted in the industry.

REACTIVE DYES

Dyes which contain ions and molecules that react with other groups in the fibre to form covalent dye fibre bond, and thereby become an integral part of the fibre, are known as **reactive dyes**. They are water soluble dyes that can be applied using a variety of methods, and are used for dyeing cellulose, protein and polyamide fibres. They are ecofriendly and have a full range of bright shades across the spectrum. Additionally, they have good to excellent wash fastness (owing to strong dye–fibre interaction) and moderate to good light fastness. The general structure of reactive dyes can be represented as S–C–B–R (Figure 12.2) where 'S' stands for the solubilising group, 'C' stands for the colouring part (chromophore), 'B' stands for the bridging part and 'R' stands for the reactive part.

Figure 12.2: General chemical structure of reactive dyes

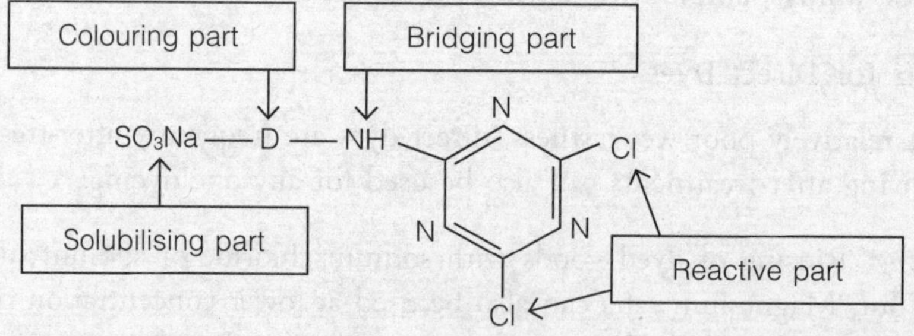

Source: Drawn by the authors.

Classification of Reactive Dyes

Reactive dyes can be classified into three categories: (*a*) Nucleophilic substitution-based dyes; (*b*) Nucleophilic addition-based dyes; and (*c*) dyes based on several addition and elimination steps.

Nucleophilic substitution-based dyes include mono and dichlorotriazines, trihalogenopyrimidines, 2-methylsulphonyl-4-methyl-5-chloropyrimidine dyes, etc. Out of these, mono and dichlorotriazine dyes are the most popular. Dichlorotriazine dyes possess two reactive chlorine atoms and are called 'cold brand' or 'M brand' as they are applied at room temperature. Monochlorotriazines possesses only one chlorine atom and are called 'hot brand' or 'H dyes' as they are applied under application of heat. Figure 12.3 shows the structure of dichlorotriazine and monochlorotriazine dye.

Figure 12.3: Structures of dichlorotriazine and monochlorotriazine dyes

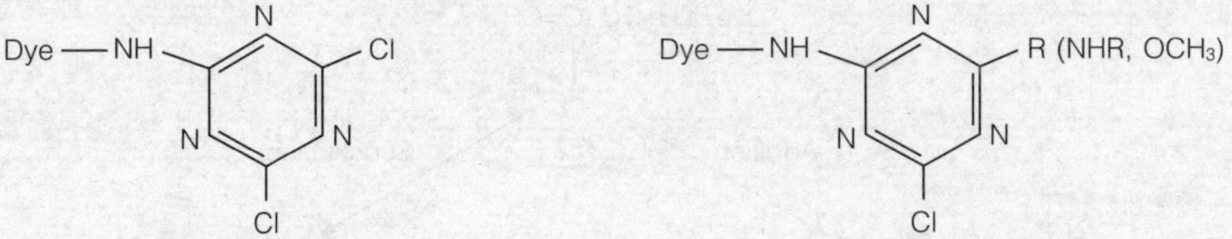

(a) Dichlorotriazine dye

(b) Monochlorotriazine dye

The reaction that takes place between dichlorotriazine dye and cotton can be represented as follows (Figure 12.4):

Figure 12.4: Reaction of dichlorotriazine dye on cotton

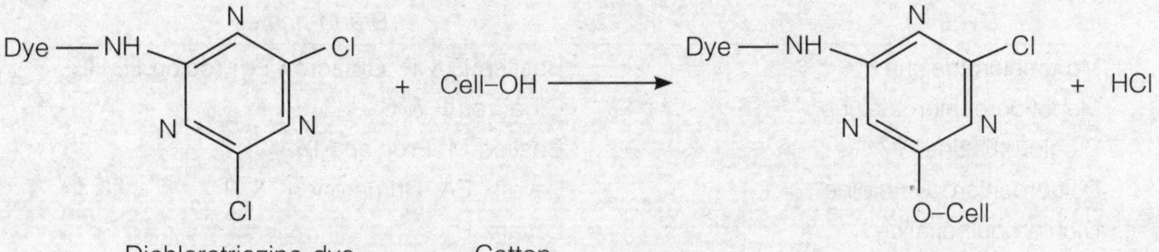

Dichlorotriazine dye + Cell–OH (Cotton) → Dye–NH–(triazine)(Cl)(O–Cell) + HCl

Nucleophilic addition-based dyes include sulphuric acid esters of β-amino ethylsulphones (Remazol, Navictive), sulphuric acid esters of β-hydroxypropionamides, etc. Vinyl sulphone dyes (Remazol dyes), with the general formula $DSO_2CH{=}CH_2$, are the most important dyes in this class. These vinyl sulphone dyes do not give any byproduct while reacting with cotton or water. They follow nucleophilic addition mechanism and are applied through application of heat and alkaline medium. However, they are rarely found in their reactive form (i.e. $DSO_2CH{=}CH_2$) and are generally present in the form of their precursor, β-sulphatoethylesulphone derivatives which release the reactive vinyl sulphone groups in the presence of an alkali.

$$D\text{–}SO_2\text{–}CH_2\text{–}CH_2\text{–}OSO_3Na + NaOH \longrightarrow D\text{–}SO_2\text{–}CH{=}CH_2 + Na_2SO_4 + H_2O \quad \text{(Eqn 12.1)}$$

The addition reaction of vinyl sulphone dye that takes place with the fibres is shown below:

$$D\text{–}SO_2\text{–}CH{=}CH_2 + HO\text{–}Cell \longrightarrow D\text{–}SO_2\text{–}CH_2\text{–}CH_2\text{–}O\text{–}Cell \quad \text{(Eqn 12.2)}$$

$$D\text{–}SO_2\text{–}CH{=}CH_2 + HO\text{–}Wool \longrightarrow D\text{–}SO_2\text{–}CH_2\text{–}CH_2\text{–}O\text{–}Wool \quad \text{(Eqn 12.3)}$$

Dyes based on several addition and elimination steps include bromo-acrylamido dyes (Figure 12.5) which are especially used for wool. They undergo a nucleophilic addition reaction as well as a nucleophilic substitution reaction.

Figure 12.5: Bromo-Acrylamido dyes

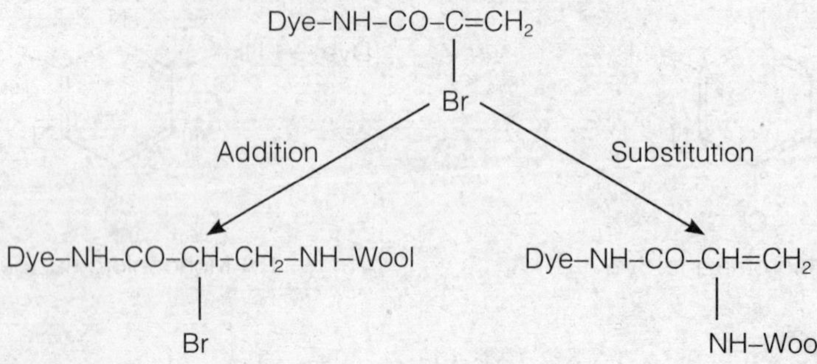

Table 12.2 summarises various kinds of reactive dyes and their commercial brand names.

Table 12.2: Various reactive dyes and their commercial brand names

Dyes	*Brand names*
Monochlorotriazine	Basilen E & P, Cibacron E, Procion H, HE
Monofluorochlorotriazine	Cibacron F & C
Dichlorotriazine	Basilen M, Procion MX
Difluorochloropyrimidine	Levafix EA, Drimarene K & R
Dichloroquinoxaline	Levafix E
Trichloropyrimidine	Drimarene X & Z, Cibacron T
Vinyl sulfone	Remazol
Vinyl amide	Remazol
Acrylamide	Lanasol, Lanasyn

Source: Compiled by the authors.

Bifunctional Reactive Dyes

Reactive dyes have several noteworthy deficiencies as a group. These include generally low fixation with its resulting highly-coloured waste and the large amount of salt needed for their fixation. One of the promising approaches to overcome these issues is the development of bi and poly-functional reactive dyes.

Advantages of Bifunctional Reactive Dyes

Bifunctional reactive dyes are those which contain two reactive groups. The presence of two reactive groups of equal reactivity increases the probability of fixation of these dyes because at least one will surely react. For instance, if three reactive groups with a fixation efficiency of 55 per cent in the monofunctional dye are all present in the same trifunctional dye, the dye fixation efficiency could be 91 per cent. This is not so simple, however. One cannot add more reactive groups to a dye and still have the same dye. Each time the overall structure is changed, the physical and chemical properties of the product are also changed. There are two types of bifunctional reactive dyes: (*a*) **homobifunctional dyes** which contain two same reactive groups; and (*b*) **heterobifunctional dyes** which contain two different reactive groups.

Dyeing Mechanism of Reactive Dye

Application of reactive dyes is generally carried out in three stages—exhaustion, fixation and washing off.

> **Dyeing mechanism of reactive dye**
> - Exhaustion of dye in the presence of an electrolyte or dye absorption
> - Fixation under the influence of an alkali
> - Washing-off the unfixed dye from the material surface

Dye Absorption / Exhaustion

The dye is added to the dye bath and circulated through the textile substrate at ambient temperature. The temperature of the dye bath is then raised to the dyeing temperature recommended for that dye class. An electrolyte is added to assist the exhaustion of dye. The higher the affinity, lesser the liquor ratio; the lesser the shade depth, lesser the salt dose. Generally, NaCl or Na_2SO_4 are used as electrolytes. Cold dyes which can be applied at room temperature require salt concentration of 50 g/L and hot brand dyes that are applied at a medium temperature of around 70°C require salt concentration of 75 g/L. Rapid exhaustion is essential to resist colour loss through hydrolysis.

Fixation

Fixation of dye and the rate of hydrolysis depends on the presence of reactive groups in the dye molecule. The reactive group of the dye reacts with the terminal –OH or $–NH_2$ group of the fibre and forms a strong covalent bond. This is an important phase, which is controlled by maintaining proper pH. In the case of cellulosics, the reaction with –OH groups takes place at alkaline pH. Rise in pH up to 11 increases exhaustion and reactivity, but beyond this, exhaustion decreases; and, excess alkali in bath promotes hydrolysis of dye. Therefore, the dye bath should be free of alkali till exhaustion is complete to resist premature fixation of dye and the corresponding uneven dyeing and excessive dye hydrolysis. Ideally, pH of around 10.5–11 with only Na_2CO_3 and 11–12.5 with a combination of NaOH and Na_2CO_3 are suitable for cold and hot brand dyes, respectively. Fixation of reactive dyes with the $–NH_2$ groups of protein fibres can take place in a neutral medium. Mild alkaline medium can be used to involve the –OH also in reaction with the dye.

An example reaction of vinyl sulphone dye during fixation is shown below:

$$D{-}SO_2{-}CH_2{-}CH_2{-}OSO_3Na + OH{-}Cell \rightleftharpoons D{-}SO_2{-}CH_2{-}CH_2{-}O{-}Cell + NaHSO_3 \quad \text{(Eqn 12.4)}$$

$$D{-}SO_2{-}CH_2{-}CH_2{-}OSO_3Na + H_2N{-}Wool \rightleftharpoons D{-}SO_2{-}CH_2{-}CH_2{-}HN{-}Wool + NaHSO_3 \quad \text{(Eqn 12.5)}$$

Wash-off

With dyeing completed now, a good wash must be applied to remove any traces of unfixed dye from the material surface. This is necessary for level dyeing and good wash fastness. This may be done by a series of hot wash, cold wash and soap solution wash. All reactive dyed goods reveal true hue and tone in the absence of alkali and so neutralisation with dilute acetic acid solution (1 mL/L) at 40–50°C is essential after fixation and prior to soaping.

Let us look at the mechanism of application of reactive dyes on cellulose and wool.

Reactive Dyes on Cellulose

Reactive dyes can be represented as Dye–X, where 'Dye' includes all the physical and structural components of a water-soluble cellulose reactive dye, except the leaving group X. Cellulose contains a considerable number of –OH groups which are less reactive. All reactive dyes react with nucleophilic reagents, and regardless of their mechanisms the overall reactions can be represented as follows:

$$\text{Cell–OH} + \text{OH}^- \rightleftharpoons \text{Cell–O}^- + H_2O \qquad \text{(Eqn 12.6)}$$

$$\text{Dye–X} + \text{Cell–O}^- \longrightarrow \text{Cell–O–Dye} + \text{X}^- \qquad \text{(Eqn 12.7)}$$

$$\text{Dye–X} + \text{OH}^- \longrightarrow \text{Dye–OH} + \text{X}^- \text{ (Hydrolysis)} \qquad \text{(Eqn 12.8)}$$

The reaction represented by Eqn 12.6 is a reversible equilibrium reaction between cellulose and alkali, to generate the fibre nucleophile (Cell–O$^-$) with which the dye can react. This equilibrium is established immediately after the alkali is added to the system. The reaction represented by Eqn 12.7 is the reaction of the dye with the fibre nucleophile, Cell–O$^-$ within the fibre and is key to the formation of covalent dye–fibre bond. The reaction represented by Eqn 12.8 is the hydrolysis reaction by which the reactive dyes are wasted by reacting with water to form byproducts which will no longer react with the fibre. This reaction takes place both in the fibre and in the solution in the dye bath. The general conditions of dyeing of cotton by different reactive dyes are given in Table 12.3.

Table 12.3: General conditions of dyeing of cotton by different reactive dyes

Dye	*Dyeing temperature (°C)*	*Salt (NaCl) (g/L)*	*Alkali (Na_2CO_3) (g/L)*
Dichlorotriazine	30	25–55	2–15
Monochlorotriazine	80–85	30–90	10–20
Trichloropyramidine	80–90	30–100	30–50
Vinyl sulphone	60	50	5

Source: Compiled by the authors.

Reactive Dyes on Wool and Silk

Reactive dyes form covalent bonds with the amino and hydroxyl groups of protein fibres in neutral and mild alkaline medium, respectively. In addition, in the acidic medium, these dyes can also form ionic bonds of relatively higher bond energy with the amino groups. Reactive dyes produce very bright shades on wool with light fastness ratings of 5–7 and wash fastness ratings of 4–5. The affinity of reactive dye for wool is very high, and poses the problem of level shades during production. Non-uniform initial absorption of dye on fibre can be controlled with the usage of non-ionic surfactants.

AZOIC COLOURS

An **azoic colour** (ingrain dye) is an insoluble pigment formed *in situ* by the means of a chemical reaction between colourless, soluble precursors already inside the particular fibre, mainly cellulosics. The intermediates/precursors forming an azoic colour are coupling components (also known as naphthols) and diazo components such as primary amines or stabilised diazonium compounds (Figure 12.6). Thirty-four naphthols and 50 bases/salts are available and their combinations can produce about 2000 shades; but only a few of them are used because they are either very expensive and cheaper substitutes exist, or they have poor reproducibility and brightness. The base is first diazotised and then coupled with the naphthol to form the colour. Since the colour is formed inside the fibre and the formation is termed *in situ* (Figure 12.7).

Figure 12.6: Structures of β-naphthol and p-nitro aniline

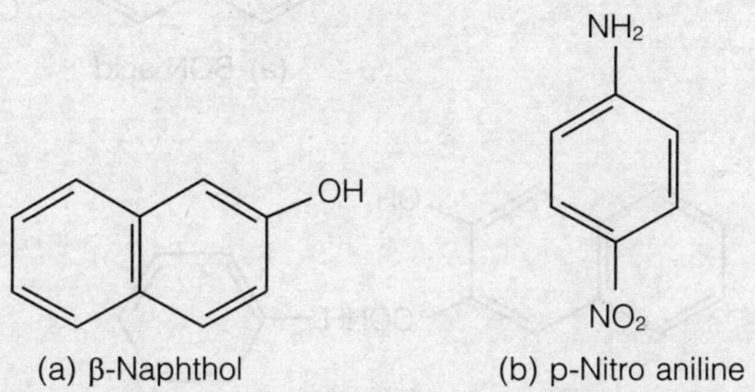

OH

NH_2

NO_2

(a) β-Naphthol (b) p-Nitro aniline

Figure 12.7: Example of azoic colour formation

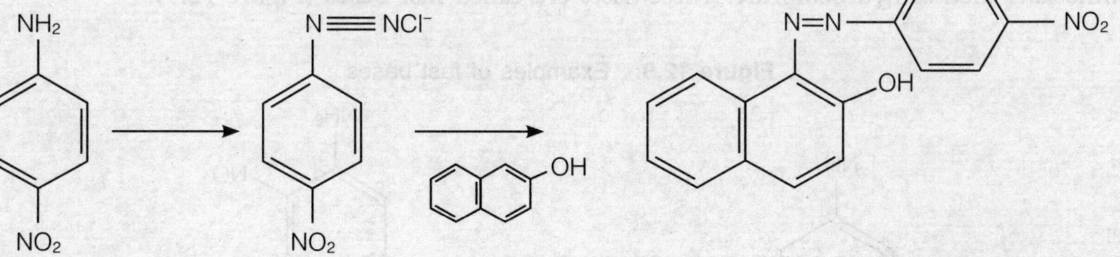

NH_2 $N{\equiv}NCl^-$ N=N NO_2

OH OH

NO_2 NO_2

Coupling Components

As mentioned above, the azoic coupling components are called naphthols. The first-used naphthol, β-naphthol (or 2 naphthol), had poor substantivity for cotton and tended to bleed out the fabric into the wet diazonium ion bath causing pigment formation on the fibre surface. This problem has now been overcome by the usage of coupling components with much higher substantivity. In 1912, Napththol AS—the commercial name of the anilide of 2-hydroxy-3-naphthoic acid (beta-oxynaphthoic acid [BON acid])—was introduced. This was soon followed by the introduction of other anilides of BON acid which are soluble in an alkaline solution and have substantivity for cotton, and thereby doing away with the requirement of intermediate drying of the naphtholated fabric before development with the diazonium ion solution (Figure 12.8).

Figure 12.8: BON acid and anilides of BON acid

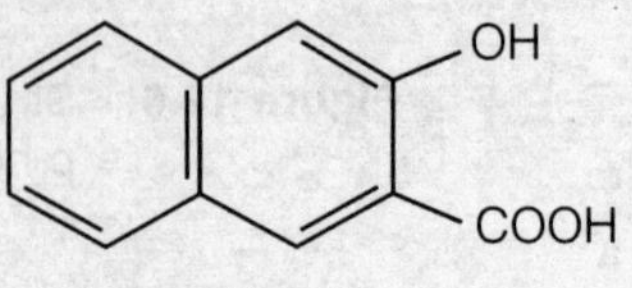

OH
COOH
(a) BON acid

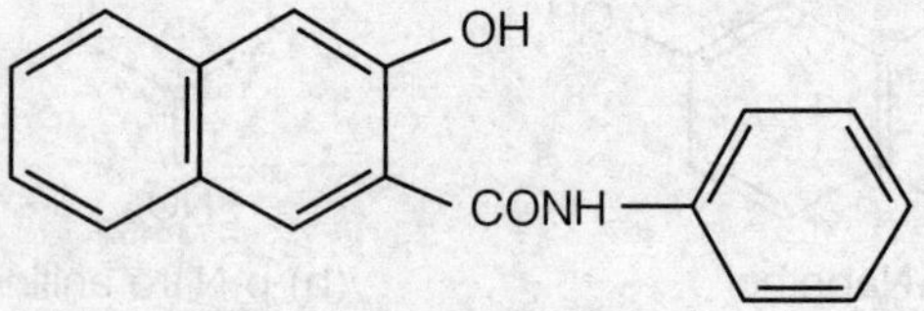

OH
CONH
Naphthol AS

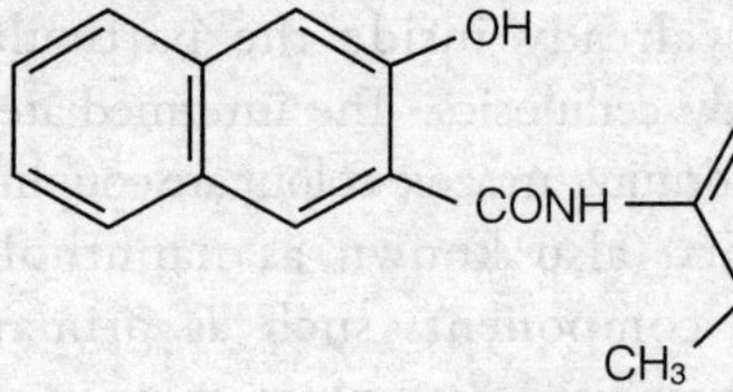

OH
CONH
CH_3
Naphthol ASD

(b) Anilides of BON acid

Diazo Components

As mentioned above, the diazo component couples with the naphthol to produce colour. The azo group in the diazonium solution is the colour-forming component (or chromophore). A solution of a diazonium ion can be obtained by the diazotisation of a primary aromatic amine. Many of these amines are simple substituted aniline derivatives and are available as free amine base or as amine salts such as hydrochloride. These bases are called **fast bases** (Figure 12.9).

Figure 12.9: Examples of fast bases

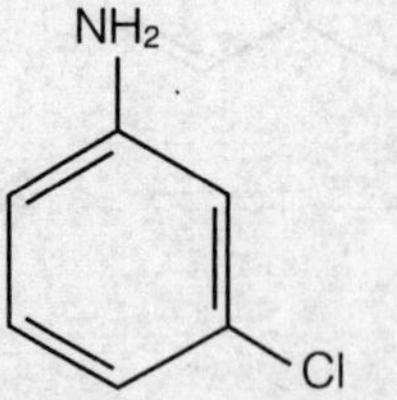

NH_2
Cl
(a) Orange GC base

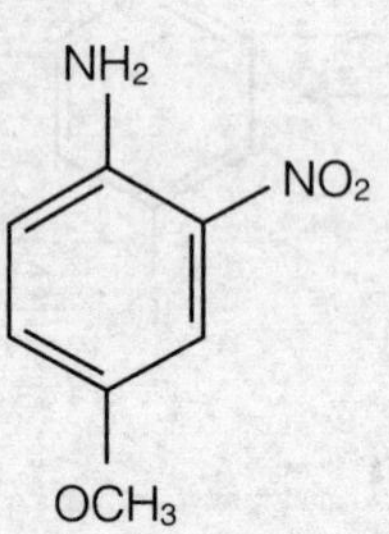

NH_2
NO_2
OCH_3
(b) Bordeaux GP base

Diazotisation of a primary aromatic amine is often difficult and solutions of diazonium ions are inherently unstable—they undergo decomposition even at low temperature and particularly on exposure to light. Thus, a variety of stabilised diazonium components which are soluble in water and ready for coupling reaction are now used; these are called **fast salts** (Figure 12.10).

Figure 12.10: Examples of fast salts

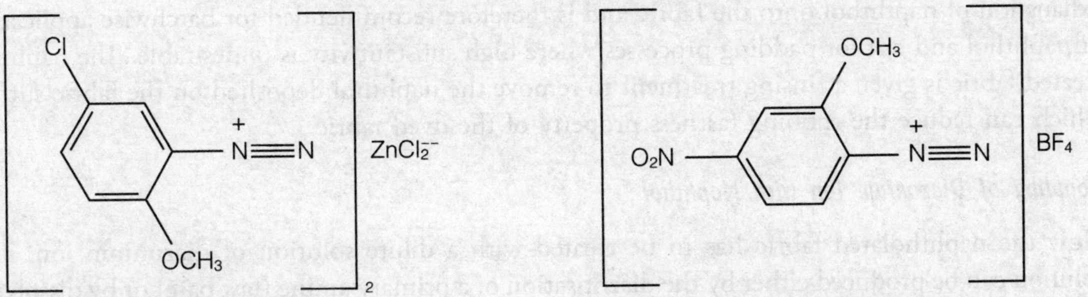

(a) Fast Red Salt RC (b) Fast Red RL

Application of Azo Dyes

Application of azo dyes

1. Dissolution of the coupling component (naphthol)
2. Application of naphtholate ion onto fabric
3. Coupling of diazonium ion solution with naphthol
4. Soaping of fabric to remove superficial pigment

Batch dyeing method of azoic colour application involves four steps:

Dissolution of Naphthol

Naphthols are insoluble in water but in their sodium salt form they are soluble in water (Figure 12.11). The method of solubilising a naphthol is called **dissolution**. There are two methods of carrying out this dissolution: hot dissolution method and cold dissolution method.

Figure 12.11: Dissolution of naphthol

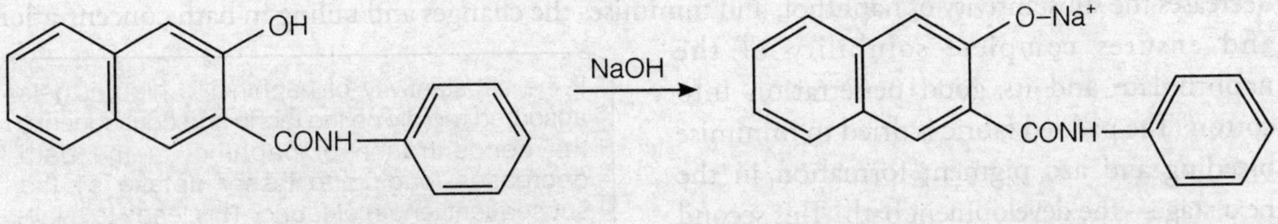

In **hot dissolution method**, the naphthol is treated with a solution containing a dispersant and heated to boil. Then, the required amount of alkali (NaOH) is added and the mixture is continuously stirred. The solution is then cooled and diluted to the required amount. In **cold dissolution method**, the naphthol is dissolved with industrial alcohol and treated with alkali to convert to its naphtholate ion. Then, cold water is added to get the required concentration of naphthol. Naphthol has the tendency to react with carbon dioxide and get precipitated. To protect it from this reaction, formaldehyde is added as a stabiliser.

Application of Naphtholate Ion onto Fabric

When cotton is immersed in the naphtholate solution, exhaustion begins according to the substantivity of the naphtholate anion. Generally, treatment for 30 minutes is adequate for reasonable levelling of the naphthol. Naphthols of high substantivity are used for batch processes while that of low substantivity are used for continuous method. Addition of salt increases exhaustion of napththol onto the fabric and is therefore recommended for batchwise application of naphthol and not for padding processes where high substantivity is undesirable. The naphthol treated fabric is given a rinsing treatment to remove the naphthol deposited on the fabric surface which can reduce the rubbing fastness property of the dyed fabric.

Coupling of Diazonium Ion with Naphthol

Now the naphtholated fabric has to be treated with a dilute solution of diazonium ion. This solution can be produced either by the diazotisation of a primary amine (fast base) or by dissolving a fast salt directly in water. Diazotisation involves the reaction of the fast base with sodium nitrite in an acidic solution at or below room temperature. The fast base is pasted with aqueous sodium nitrite followed by the addition of dilute hydrochloric acid. An excess of free hydrochloric acid and sodium nitrite must be maintained. Spotting onto starch-iodide paper is carried out to test for excess of nitrous acid. This diazotisation reaction is carried out at 5–10°C which is attained by adding ice. Due to the involvement of ice, these colours are also known as **ice colours**. A solution of the fast salt is prepared by pouring lukewarm water (25–30°C) containing a non-ionic dispersant over the fast salt and stirring the mixture until it completely dissolves. The naphtholated fabric is then passed through the diazotised base solution or dissolved fast salt solution at room temperature for coupling to take place to produce colour.

For continuous dyeing of cotton with azoic colours, a fairly common method employed is pad–dry–pad process. Since the naphtholate ions used are of low to medium substantivity, a small pad trough is used and the fibre is padded with a solution as hot as 80°C. This further decreases the substantivity of naphthol, and minimises the changes and tailing in bath concentration and ensures complete solubility of the naphtholate and its good penetration into cotton. The padded fabric is dried to minimise bleeding and azo pigment formation in the next stage—the development bath. This second pad bath contains a non-ionic surfactant along with the diazonium ion solution to ensure rapid fabric re-wetting and dispersion of any azo pigment formed in the previous bath. After padding with the diazonium component, the fabric is aired for a minute or so to allow time for the coupling reaction, and this is followed by rinsing and soaping.

> If the substantivity of naphthol is high, it gets absorbed rapidly on the fabric and consequently the concentration of naphthol in the bath decreases leading to lesser uptake by the subsequent length of fabric. This leads to a tail-like effect; i.e. the concentration on the fabric decreases or changes from one end to the other. This effect is known as **tailing**.

Soaping of the Fabric

After coupling, the goods are rinsed and soaped under alkaline conditions with good mechanical action to remove any azoic pigment that has formed or deposited on the cotton fibre surfaces. This ensures optimum fastness to washing and rubbing. Rinsing and soaping may involve both crystallisation and aggregation of the dye particles inside the fibre. Thus, this stage also helps develop the intended dye shade since the shade of dyeing is associated with changes in the physical form of the dye particles within the fibres. Soaping, therefore, helps achieve maximum light and chlorine fastness.

Fastness of Azoic Colours

The fastness to washing of azoic dyeings on cotton is usually very good to excellent; however, careful elimination of azo pigment particles loosely held onto fibre surfaces must be ensured. Intermediate drying or rinsing of fabric containing the naphthol, and the soaping of the final dyeing are key processes that ensure optimum fastness. Rubbing fastness is affected if deep dyeings have not been well soaped; they can easily transfer colour onto adjacent white fabrics, even under conditions of gentle rubbing.

Colour Range

The actual hue of azoic colours depends on the choice of the diazonium and coupling components. Azoic combinations are still the only class of dyes that can produce very deep orange, red, scarlet and bordeaux shades of excellent light and washing fastness. The pigments produced have bright colours and include navies and blacks. However, there are no greens or bright blues in this class of dyes.

VAT DYES

Vat dyes are among the oldest natural colouring materials that have been used on textiles. Indigo, an example of vat dyes, has been used in India for many millennia. The chemical structure of vat dyes retain the $>C=O$ chromophore in their structure, and they are insoluble in water. Therefore, the dyes are treated with a reducing agent that converts them into leuco compounds ($>C=OH$ group) which, in the presence of alkalis, show solubility in water. Traditionally, in the absence of chemical reducing agents, naturally available vat dyes (organic matter) were fermented in a wooden vessel known as the **vat**. Hence, the name 'vat dyes'; and the process of converting these dyes into their soluble form is known as **vatting**.

The leuco compounds are either colourless or show a colour quite different from the colour of the final product which is achieved after oxidation. Though vat dyes are non-ionic in nature, they convert into anions on solubilisation. In this form, they have excellent affinity for the cellulosics. Once the dye anion is attached to the fibre, the dye is oxidised to convert it into an insoluble coloured derivative which gets trapped inside the fibre forming linkage through H-bonds and Van der Waals forces. Dyeing with vat dyes thus involves three basic steps:

- **Vatting**: Converting the water insoluble vat dye molecule into a soluble form—the leuco compound.
- **Dyeing** of goods with the leuco compound.
- **Oxidation** of the leuco dye to form the insoluble colour pigment inside the fibre.

Classification of Vat Dyes

Based on their chemical structure, vat dyes can be classified as ingoid and anthraquinoid dyes. The classification and examples are shown in Figure 12.12.

Figure 12.12: Classification of vat dyes based on their chemical structure

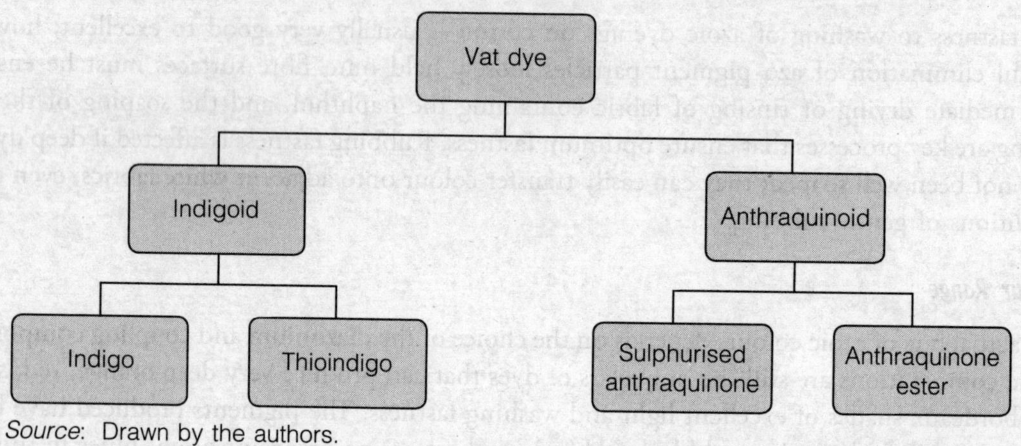

Source: Drawn by the authors.

The oldest known classes of organic vat dyes, **Indigoid dyes** are made up of $>C{=}O$ group as the chromophore and –NH or –S– groups as the auxochromes (Figure 12.13). The reduced and solubilised form of indignoid dyes have limited affinity for cellulosic fibres and get oxidised very quickly into insoluble blue indigotin which provides colour to the fibre. **Anthraquinone dyes** have anthraquinone groups in their strucuture with made up of $>C{=}O$ groups as the chromophores and $-NH_2$, –OH, alkylamino (–NHR, $-NR_2$), benzamide (–NH–CO–RH) or alkoxy (–OR) groups as auxochromes (Figure 12.14). These dyes exhibit excellent fastness properties, but are more expensive.

Figure 12.13: Chemical structures of indigoid and anthraquinone dyes

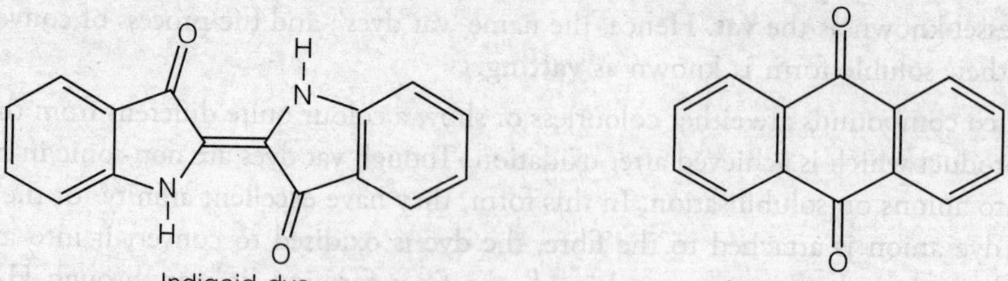

Indigoid dye

Anthraquinone dye

Vat dyes can also be classified based on the method of application with respect to vatting and dyeing conditions and the chemicals required for completion of the dyeing process. Under this system of classification, there are four sub-groups:

Classification of vat dyes

1. Indanthrene cold (IK) dyes
2. Indanthrene warm (IW) dyes
3. Indanthrene normal (IN) dyes
4. IN special dyes

- *Indanthrene cold (IK) dyes*: Dyes in this category are easily reduced but have poor substantivity. These dyes require relatively lower concentration of caustic soda (NaOH) and Sodium Dithionite ($Na_2S_2O_4$) and low vatting temperature (35–50°C) and dyeing temperature (30°C). However, they require a higher concentration of common salt or anhydrous sodium sulphate for improving dye exhaustion. C.I. Vat Blue 1(indigo) belongs to this category.
- *Indanthrene warm (IW) dyes*: Dyes in this category possess moderate affinity and hence require comparatively less salt for complete exhaustion. These dyes are slightly difficult to reduce, and therefore require little more alkali and $Na_2S_2O_4$ and slightly higher vatting and dyeing temperature (45–50°C). Examples of dyes in this category are C.I. Vat Blue 6 (Blue BC), Brown 3 (Brown R), Orange 15 (Orange 3 G) and Red 10 (Red 6B).
- *Indanthrene normal (IN) dyes*: Dyes under this category possess very good affinity for cellulosics but are not easily reduced. These require relatively higher concentration of alkali and $Na_2S_2O_4$ and higher vatting temperature (55–60°C) for proper reduction and higher dyeing temperature (55–60°C). No salt is required for these dyes. Examples include C.I. Vat Violet1 (Purple 2R), Green 1 (Green XBN), Yellow 2 (Yellow GCN) and Direct Black 8 (Grey 2B).
- *IN special dyes*: Dyes under this category show excellent affinity for fibre and require exceptionally higher concentrations of alkali and $Na_2S_2O_4$ and high vatting and dyeing temperatures (above 60°C). No salt is required for these dyes as well. Examples include C.I. Vat Brown 5 (Brown G) and Green 9 (Black 2B).

The dyeing parameters of the four classes of vat dyes are detailed in Table 12.4.

Table 12.4: Dyeing parameters for different classes of vat dyes

Parameters	*IK*	*IW*	*IN*	*IN special*
Vatting (x°C)	35–50°C	45–50°C	55–60°C	≥ 60°C
Dyeing (x–5°C)	35°C	45–50°C	50–55°C	≥ 60°C
$Na_2S_2O_4$ (g/L)	0.4–4.5	0.4–6.25	1.5–10.0	15
NaOH (g/L)	0.4–3.0	0.4–4.5	1.5–10.0	15
NaCl (g/L)	6–50	3–25	–	–

Source: Compiled by the authors.

Vat dyes are available under various brand names such as Navinon, Algol, Cibanone, Chemithrene, Helanthrene, Intravat, Paradone, Sandothrene, etc.

Application of Vat Dyes on Cellulosics

The various stages involved in application of vat dyes on natural and man-made cellulosic fibres are as follows.

Steps in the application of vat dyes

1. Reduction and solubilisation
2. Dyeing
3. Oxidation
4. Soaping

Figure 12.14: Examples of vat dyes

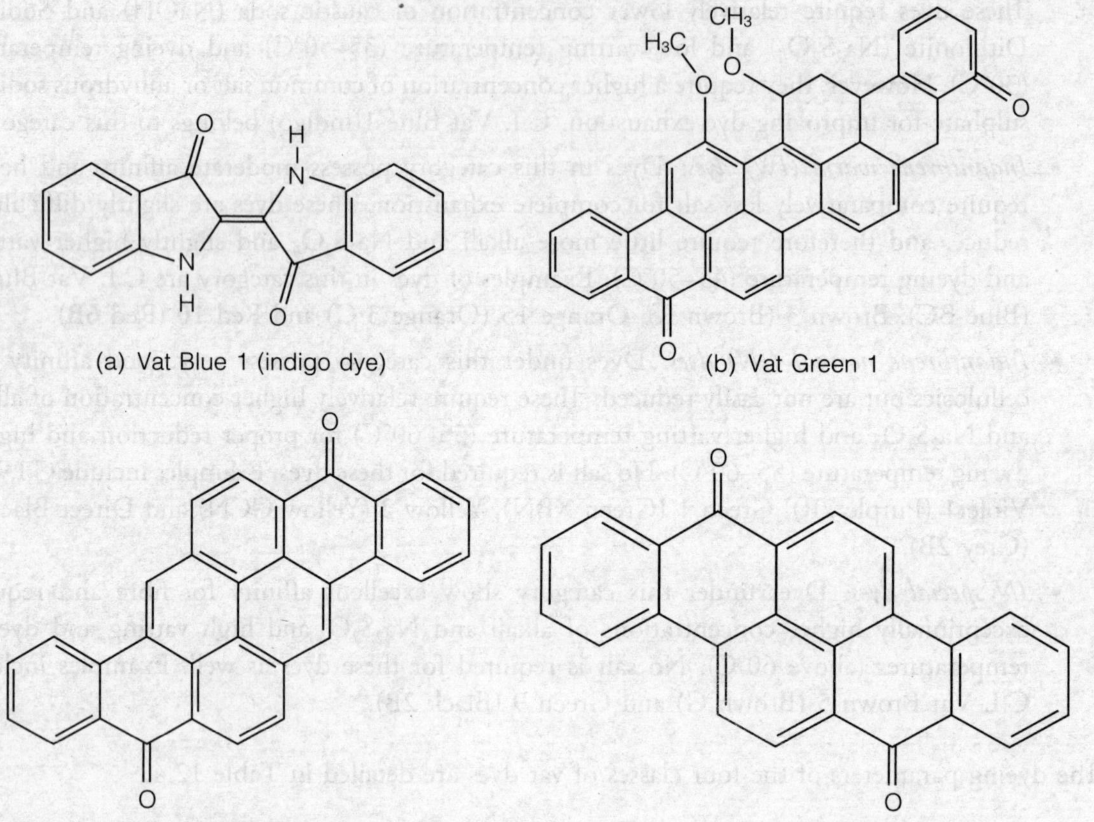

(a) Vat Blue 1 (indigo dye)

(b) Vat Green 1

(c) Vat Orange 9

(d) Vat Yellow 4

Reduction and Solubilisation of the Dye Molecule

A typical dyeing recipe for vat dyes would include 1% of the dye pasted with a little bit of Turkey Red Oil. To this water is added, and the mixture is heated to maintain the required vatting temperature as given in Table 12.4. To this one-third amount of NaOH and sodium hydrosulphite ($Na_2S_2O_4$) are added and the solution is continuously stirred for about 10 minutes. $Na_2S_2O_4$, also known as sodium dithionite, is the commercially used reducing agent for large-scale dyeing because it produces a clear reduction bath without any sediments; and adding a little more of dithionite than required helps maintains stability by acting on the dissolved oxygen in the water so that the reduced dye does not precipitate due to oxidation. When $Na_2S_2O_4$ is mixed with water or alkali, nascent hydrogen is released (Eqns 12.9, 12.10).

$$Na_2S_2O_4 + 4H_2O \rightleftharpoons 2NaHSO_4 + 6[H] \quad \text{(Eqn 12.9)}$$

$$Na_2S_2O_4 + 2NaOH \rightleftharpoons 2Na_2SO_3 + 2[H] \quad \text{(Eqn 12.10)}$$

In its reduced form, vat dyes show complete change in colour and a layer of coloured foam forms above the solution. In the presence of NaOH, the reduced form of the dye is quite stable.

But, $Na_2S_2O_4$ is very unstable in the presence of water and gets decomposed into various byproducts through hydrolytic, thermal, oxidative or other reactions. To deal with these problems, two to three times more $Na_2S_2O_4$ is required which adds to the cost and moreover being unstable, its storage is further a problem.

To minimise the harmful impact of the byproducts formed during the reduction process, eco-friendly reducing systems have been recently developed for vat dyes. One of these is the hydroxyacetone method, where hydroxyacetone is used to reduce indigo dye. It shows 20 per cent better efficiency and also reduces the consumption of additional chemicals. Another eco-friendly method is electrochemical reduction, in which the dye is reduced when it comes in contact with an electrode.

The problem of over-reduction may occur if excess amount of $Na_2S_2O_4$ is used or the vatting temperature is not maintained as per the requirement. As a result, dye structure is permanently changed and hinders in the development of the exact shade of dye. Therefore, it is important to follow the prescribed parameters. Addition of glucose or formaldehyde to the bath may reduce the chances of over-reduction to some extent.

Once the dye is reduced, its solubility in the dye bath should be maintained to prevent the oxidation of leuco vat dye back to its original structure. Addition of prescribed amounts of NaOH helps maintain this solubility and also increases the dye affinity for the fibre. It also maintains the pH of the bath by neutralising acidic byproducts ($NaHSO_3$, Na_2SO_3) formed due to the oxidation of $Na_2S_2O_4$. NaOH is further required to react with atmospheric CO_2 and thus retain the dye in its reduced form. However, excess amounts of alkali lowers the colour yield as the repulsion between the dye and the cellulose molecule increases.

Dyeing

Since cellulosic fibres have great affinity for solubilised leuco vat dyes (sodium salts), controlled conditions should be maintained to promote level dyeing. This can be achieved by adding the dye paste in two instalments and by adding non-ionic dispersing agents which reduce the exhaustion rate. Depending upon the substantivity of the dyes for the fibre, an electrolyte may be added to improve the final dye uptake. In order to obtain composite shades where dyes are used in combination, these should belong to the same class and show compatibility.

Oxidation of Dyed Textiles

The oxidation of dyed textiles is carried out in the presence of open air, sodium perborate/hydrogen peroxide at 50–60°C or hypochlorite at normal room temperature for 15–20 minutes. This causes the dye molecule to get trapped inside fibre and also develop the true shade by restoring the dye structure to its original form. After the completion of dyeing, a thorough wash is imparted for the removal of the reducing agent and alkali ($Na_2S_2O_4$ and NaOH) completely. This is essential for getting patch-free quality shades, else free oxygen would not be released by the oxidising agents, leading to lower fastness properties. Preferably, the oxidation process should take the least time at the lowest possible temperature so that cellulosic fibre is not damaged due to its oxidation, especially in the presence of hypochlorite. Vat dyes from the IK and IW classes which have moderate affinity take longer to get fixed inside the fibre. In these cases, hydro extraction should be done instead of rinsing and then oxidation in the presence of open air.

Soaping

Soaping is the final treatment given to a dyed textile to develop brighter shades. Soaping involves boiling the fabric with 0.5 g/L each of soap and soda ash. During this process, the dye molecules align themselves parallel to the fibre axis. Besides developing brighter shades, soaping also enhances wash fastness by promoting the aggregation of dye molecules to form large crystals. It also helps remove superficial dye molecules that are loosely attached to the fibre surface.

Fastness Properties of Vat Dyes

Overall light fastness of vat dyes is above satisfactory and are rated at 5–6 on a scale of 8. However, few shades of yellow, brown and orange show photochemical degradation (fading of shades) when exposed to light for a long duration. Vat dyes show very good wash fastness rated at 4–5 on a scale of 5. These dyes are also very fast to industrial laundering with hypochlorite. Besides, vat dyes also possess outstanding fastness to peroxide and hypochlorite bleaches and perspiration.

Colour Range

The range of colours available in vat dyes include blues, browns, blacks, oranges, reds, greens, violets and yellows. Among these, blues, greens and earth tones have good colour palette. The colour gamut (range of colours dyed with any of the application class) for reds and turquoise hues are quite restricted due to the absence of bright colours.

Solubilised Vat Dyes

In order to overcome the cumbersome reduction and subsequent stabilisation of the reduced vat dye, solubilised vat dyes (a derivative of vat dye) were manufactured. These dyes are available in powder or paste form under the brand name of 'indigosol'. They are readily water soluble and do not require any reduction or solubilisation process during application. However, they possess very low affinity for cellulosics as compared to the parent vat dye. Therefore, Na_2CO_3 is added to increase the dye intake and higher temperatures are maintained for promoting levelled dyeing of shades.

These dyes are comparatively expensive and application range is limited to lighter shades of blue, purple, orange, yellow, pink, green and golden. These dyes are mainly used on cotton and polyester/cotton blend and show excellent fastness to washing and sunlight. After application and oxidation, the original vat dye structure is recovered and any further treatment that is required will be same as used for vat dyeing.

Dyeing with Indigo

Indigo, a member of indigoid vat dye class, is used to produce brilliant shades on denim with desirable fading effect. Unlike other vat dyes, reduced and solubilised indigo possesses no affinity for cottons and hence exhaust dyeing method cannot be used for dyeing denims. On the contrary, multi-dip/nip padding technique is used for dyeing followed by air oxidation in between for gradual development of required shade. Generally 6-dip 6-nip padding cycles are carried out for complete dyeing of denim. In each cycle, the fabric is dipped in dye bath for 30 seconds, padded with this liquor at 80% pick-up and then aired for one minute. After six consecutive cycles, the fabric is aired for three minutes so that all the reduced dye gets converted into its oxidised form. It is then given five cold rinses and five hot washes (50–60°C) and finally dried for two minutes at 100°C.

Exhaust dyeing is that method where dyeing is carried out from a dye bath for a period of time till maximum possible dye has exhausted/transferred onto the material.

Muti-dip/multi-nip technique is a process where the fabric is dipped in an indigo dye bath for 5–15 minutes, then passed through a padding mangle (nipped) to remove excess dye liquor, and finally exposed to air for the colour to develop by oxidation of the dye. The process is repeated multiple times till the required shade is obtained.

Various factors which affect indigo dye uptake in each cycle include the pH of the dye bath, the time of immersion of the fabric in the dye bath, airing time, dye concentration and temperature. All these factors further impact the degree of penetration of dye molecules and its fastness.

SULPHUR DYES

All the sulphur dyes contain sulphur linkages as a characteristic feature of their molecular configuration. Like vat dyes, these are insoluble in water but can be reduced to their soluble form in the presence of sodium sulphide or sodium hydrosulphite. Sodium carbonate is also added to maintain the required alkalinity of the solution. The reducing agent causes the dye molecule to break down into water-soluble simpler components which show affinity for cellulosics. After dyeing, the fabric is exposed to air or a mild oxidising agent which causes the dye to oxidise back to its original insoluble form.

$$\underset{\text{(Insoluble stage)}}{\text{Dye–S–S–Dye}} \xrightarrow[2[H^+]]{[Na_2S + Na_2CO_3]} \underset{\text{Leuco/Thiol (Soluble stage)}}{\text{Dye–S–Na + Na–S–Dye}} \xrightarrow[\text{Oxidation}]{[O^-]} \underset{\text{(Insoluble form)}}{\text{Dye–S–S–Dye} + H_2O}$$

Sulphur dye was first made by heating saw dust along with sulphur and caustic soda in 1873 to produce a brown shade. However, Vidal is considered the pioneer when they produced Vidal Blacks in 1893. Later many new dyes were added to this class of sulphur dyes. Various brand names under which sulphur dyes (available in powder or paste form) have been marketed include Sulphur, Asathio, Coranil, Sulphol, etc. Sulphur dyes are not suitable for application on woollens as the increased concentration of the reducing agent (sodium sulphide) and high temperature of dye bath would adversely affect the fibre leading to hydrolysis.

Chemical Nature

The chemical nature of sulphur is unknown as these dyes do not have any well-defined structure and consistency in composition. In general, they are synthesised by melting together (180–350°C) organic nitrogenous compounds, alkali and sulphur which is usually in the form of sodium sulphide or polysulphite.

Application of Sulphur Dyes on Cellulosics

Sulphur dyes can be applied on cellulosics (cotton and rayon) quite easily. They are considered as a cheaper alternative to direct dyes for colours like navy blue, black, khaki and olive green. Bright red, violet and orange cannot be synthesised in sulphur dyes; also their light shades are relatively dull in appearance and hence not commercially produced. Dyeing properties of sulphur dyes in reduced form are mostly similar to those of direct dyes and the presence of an electrolyte improves exhaustion rate. However, temperature of the dye bath may also vary the exhaustion rate of the dye. For dyeing with sulphur dyes, the exhaust method is extensively used due to the ease in application and ability to rectify problems related to uneven dyeing. The various stages which lead to final dyeing of textiles are as follows:

Application of sulphur dyes on cellulosics

1. Reduction and solubilisation
2. Dyeing
3. Oxidation
4. After treatment

Reduction and Solubilisation

Dye is pasted with little Turkey Red Oil, sodium sulphide and sodium carbonate along with some water. Boiling water is then added to the dye paste and, if need be, the dye solution is boiled again. This process results in complete reduction and solubilisation of sulphur dyes to form stable compounds called leucothiols (Eqn 12.11).

$$\text{D–S–S–D} \underset{[\text{NaHS} + \text{NaOH}]}{\overset{\text{Na}_2\text{S}}{\longleftrightarrow}} \text{D–S–Na} + \text{Na–S–D} \qquad \text{(Eqn 12.11)}$$

This solubilised dye is then added to the dye bath. It is preferable to strain the dye before adding to the dye bath. Na_2S is cheaply available and quite effective, but causes enormous water wastage. There are other chemicals which could be used as reducing agents. Sodium hydrosulphide (NaHS) which is considered powerful in comparison to Na_2S can also be used as the reducing agent, but it may lead to over-reduction of the dye. Thiosalicylic acid ($C_7H_6O_2S$) and thioglycolic acids may also be used but these may interfere with the development of the actual shade. If sodium bisulphite ($NaHSO_3$) is used, it retains the dye bath in a reduced form thus preventing bronziness and also prevents oxidation at the

If proper dyeing conditions are not maintained, H_2S liberated during dyeing reacts with the metal vessel containing the dye bath and produces corrosive metallic sulphides that are insoluble and dark coloured. These stain the dyed goods, and are referred to as **bronziness**. To avoid this only stainless steel vessels are preferred for dyeing.

surface of the dye bath. Besides, few sulphur-free reducing agents such as iron salts along with alkali, hydroxyacetone and molasses–NaOH combination, can also be used. Glucose can be used as a reducing agent for efficient and odourless dyeing in exhaust and pad-dry-steam method, but it adds to the cost of dyeing.

During the reduction process care should be taken to avoid over-reduction as it would lower the affinity of the dye towards cellulosic fibres. Moreover, in the case of hard water, EDTA (ethylene diamine tetra acetic acid, a chelating agent) must be added to overcome the harmful impact of metals on the fibre.

The **chelating agent** is a chemical compound that coordinates with a metal to form stable water-soluble metal complexes. It is often used to remove heavy metal ions which can interfere with the dyeing process. These are also known as **sequestering agents**.

Dyeing

After the reduction and solubilisation process is over, the soaked cotton fabrics are added to the dye bath. The temperature of the dye bath is maintained at 90–95°C for about 30–60 minutes so that dye molecules can penetrate adequately and dyeing takes place at a satisfactory rate. The fabric should be completely submerged in the dye bath during dyeing to avoid its oxidation when it is exposed to air. Common salt is then added for increased dye affinity, and the process is continued for the next 1.5–2 hours. A longer dyeing time is required as these dyes show moderate affinity for cellulosics. The dyed samples are taken out and washed under cold water thoroughly. The dye bath is kept standing so that it can be used for subsequent dyeing as a lot of dye is still left in the dye bath due to incomplete exhaustion.

Oxidation

The oxidising agents most commonly used include either $Na_2Cr_2O_7$ alone or $K_2Cr_2O_7$ along with an acid (H_2SO_4 or CH_3COOH). In order to achieve the final colour, the dyed fabrics are oxidised at 50–60°C for half an hour. H_2O_2 which is eco-friendly is preferred as an alternative oxidising agent. After oxidation, soaping of the dyed fabric is carried out using sodium perborate.

After-treatments

Cottons dyed black may get damaged on storage in moist conditions as the dye gets oxidised to form sulphuric acid and due to the formation of tiny holes due to local hydrolysis. In order to overcome this problem, sulphur-dyed materials are after-treated with bichromate solution followed by immediate rinsing of the byproducts. Light fastness of the dyed fabrics can be improved by either treating these with metal salts in acidic medium or by applying dye-fixing agents. Even the intensity of the hues can be enhanced to some extent further dyeing with basic dyes. Sulphur blacks may show a tendency to have **bronziness**. This is due to the presence of insoluble dye on the fibre surface. This can be due to incomplete reduction of the dye in the bath or exposure of the goods to air while being dyed, incomplete removal of dye after dyeing or excessively heavy dyeing. Bronziness can be removed by treating the dyed goods in a dilute solution of sodium sulphide at 30°C. Goods dyed in black are also given an oiling treatment to improve the intensilty and brightness of the colour. Dyed goods are treated in a solution of 2–5 g/L olive oil soap, 1–2 g/L olive oil and 1–2 g/L sodium carbonate at 60°C for 15–30 minutes. Treated goods are hydroextracted and dried without rinsing.

Fastness Properties

The wash fastness of sulphur dyes is rated as excellent; they are better than direct dyes because the dye molecules are comparatively larger, insoluble in water and held with Van der Waals forces. Sulphur dyes, except black, possess moderate fastness to light. Hence, they can be applied for effective use on products which do not require prolonged exposure to light. These dyes are not fast to chlorinating agents and are hence not used for swimwear.

BASIC DYES

Basic dyes possess cationic/basic groups in their structures which enable the dyes to react and form electrovalent bonds with acidic groups available in the fibres. In the dye paste, the basic dye molecule ionises and causes the formation of the coloured component which is a cation or positively-charged radical, and hence these dyes are also known as **cationic dyes**. The anion of basic dyes is usually a chloride ion. Once in the dye bath, the dye dissociates into the chromophore cation and the chloride anion. Basic dyes are derived from organic bases having the general formula R(OH)–C₆H₄–NH₂ which is capable of salt formation as shown in Eqn 12.12.

$$R(OH)\text{-}C_6H_4\text{-}NH_2 + HCl \longrightarrow R{=}C_6H_4{=}NH_2Cl + H_2O \qquad \text{(Eqn 12.12)}$$

(Colourless base) (Coloured salt)

Due to the absence of the chromophore, the base is colourless and the colour only appears on salt formation. These are usually chlorides but can also be oxalates, sulphates, nitrates and zinc chloride double salts sometimes.

Properties of Basic Dyes

- *Ionic nature*: Basic dyes are cationic in nature. Due to their cationic nature, basic dyes can be precipitated by direct or acid dyes which are anionic. The precipitation is due to the formation of basic–direct dye or basic–acid dye complex which is insoluble in water.
- *Solubility*: Basic dyes are readily soluble in alcohol and methylated spirit but not easily soluble in water. Unless care is taken when dissolving them, they may form a sticky mass which can be difficult to bring in to the solution. Basic dyes can be better solubilised in hot water along with a little acetic acid or in cold water along with methylated spirit.
- *Colour intensity*: Basic dyes are characterised by the brilliance and intensity of their colours, which do not usually occur with other dye classes.

- *Levelling properties*: These dyes have a very high strike rate, therefore levelling is poor. This may often lead to uneven dyeing if not suitably controlled.
- *Reaction with alkali*: Basic dyes are highly sensitive to alkalis and on reaction they decompose with the liberation of the colourless dye base. Thus, water containing alkali or temporary hardness should not be used with basic dyes without first neutralising it with acetic acid. The presence of calcium or magnesium bicarbonates, metal hydroxides or carbonates in hard water causes precipitation of the dye into a colourless mass.
- *Reaction with tannic acid*: All basic dyes can combine with tannic acid to form tannic acid–dye complex. This property has an important application in practical dyeing. Cellulose is dyed after mordanting with tannic acid while protein fibre is first dyed with the basic dye and then it is followed by back tanning with tannic acid to improve wash fastness.
- *Action of reducing agents*: When treated with reducing agents, most basic dyes are easily converted to colourless compounds which can be easily reoxidised to the dye, even on exposure to air.

Affinity of Basic Dyes for Textile Substrates

These dyes show good affinity towards wool, silk and acrylic, but have no affinity for cellulose. This affinity for protein fibres is due to the presence of –COOH groups in their polymer chain. The dye cation reacts with these –COOH sites to form electrovalent bonds with the fibre. When –COOH groups in the fibre are ionised, negative sites are developed in the fibre which react with the cations of the basic dye. HCl that is formed as a byproduct retards the reaction resulting in even dyeing (Eqn 12.13).

$$H_2N\text{–Protein–}COOH + DNH_3^+Cl^- \longrightarrow H_2N\text{–Protein–}COO^-DNH_3^+ + HCl \qquad \text{(Eqn 12.13)}$$

Basic dyes are invariably used in dyeing of acrylic fibres which have anionic groups, most commonly the sulphonate ($-SO_3$) group, attached to it. The positively-charged cations of the basic dyes get attracted towards the negatively-charged anions in the acrylic fibre. This reaction of the cation and anion results in salt linkages (Eqn 12.14).

$$AcSO^-_3 \quad + \quad D^+ \quad \longrightarrow \quad AcSO^-_3D^+ \qquad \text{(Eqn 12.14)}$$

$AcSO^-_3$	+	D^+	⟶	$AcSO^-_3D^+$
Acrylic fibre with dye site (negatively charged sulphonate group)		Dye Cation		Dye attached to dye site on acrylic fibre

Application of Basic Dyes

Dyeing of Wool and Silk

Due to their brightness and clarity of shades, basic dyes have been used on silk but these are very rarely used on wool. Their method of application is as follows. The dye is pasted with acetic acid (30%) that weighs about the weight of dye used and sufficient hot water. The basic dye possesses higher affinity for protein fibres, and therefore, the dye paste is added in three instalments. The dye bath is prepared with water and 1–2% acetic acid on the weight of the goods at room temperature. One-third of the dye paste is added, and goods are introduced when the bath is at normal room temperature. The temperature is then raised to 80°C for silk and to 100°C for wool. During this time, the other two instalments of dye are added and dyeing is continued for the desired time. After dyeing, the goods are rinsed, hydro-extracted and dried. The wash fastness of basic dyes can be improved by treating the dyed goods with a solution of 1 per cent tannic acid at 60°C for 20 minutes, squeezing and then treating with a lukewarm solution of 0.5% tartar emetic (potassium antimony tartrate; $C_8H_{10}K_2O_{15}Sb_2$) which is used as a mordant for certain dyes.

Dyeing of Acrylic

The dye bath is made using 1 g/L of non-ionic dispersing agent, 1 g/L of 80% acetic acid and the same amount of sodium acetate. The pH of the dye bath is maintained at 4–5. The dye paste is added to the bath followed by the fabric that has to be dyed. Since there is very little absorption of dye below 75°C, the temperature of the bath is rapidly raised to 75°C. Thereafter, the temperature is raised very gradually to boil, maintained at that temperature and dyeing continued for 1–1.5 hours. If possible, dyeing can be carried out at 105–107°C because exhaustion is more rapid at this range and there is better migration of dye leading to level dyeing.

Dyeing of Cellulose

Cellulose has no acidic groups, and therefore has no affinity for basic dyes. To dye cellulose with basic dyes, there are two options: the material must either be treated with suitable mordanting agents such as tannic acid, or may be predyed with direct dyes to introduce anionic groups in the fibre. Direct-dyed cotton lacks brilliance in shade and fastness, and therefore basic dyes can be used as an after-treatment to brighten the shades and improve fastness. Practical application of the basic dye onto cellulose includes three steps:

1. *Mordanting*: Solution is made containing tannic acid (equal to twice the weight of the dye to be applied). The fabric is introduced and the solution is heated up to boiling temperature and is followed by cooling down for two hours or preferably overnight, as the greatest absorption of tannic acid by cellulose takes place during cooling.
2. *Fixing*: The goods are then taken out and squeezed and tannic acid is fixed by treatment with tartar emetic (potassium antimony tartarate). The material is agitated in cold aqueous solution containing tartar emetic (equal to half the weight of tannic acid used) for 15–20 minutes followed by rinsing and hydro-extraction.
3. *Dyeing*: The dye bath is made up of 1–2% of 40% acetic acid. The basic dye possesses higher affinity for tannic acid mordanted cellulose so the dye paste is added in three instalments. The dye bath is prepared with water and acetic acid at room temperature. One-third of the dye paste is added, mordanted cotton is introduced at room temperature and dyed for 15 minutes; the second one-third dye paste is added, and the temperature is raised to 40°C and dyed for a further 20 minutes after which the last one-third of the dye paste is

added, the temperature is raised to 70°C and dyed for the desired time. After dyeing, the fabric is rinsed, hydro-extracted and dried.

Fastness Properties

Basic dyes have poor to moderate light and wash fastness. The fastness properties are however better on acrylic due to the compact structure and hydrophobic nature of the fibre.

Modified Basic Dyes

These dyes, based on the chemistry of basic dyes, have longer molecular structures than traditional basic dyes and are essentially the only basic dyes still in use in textile dyeing and printing. These dyes have significantly improved properties like excellent substantivity, better light fastness and better levelling. They also ensure eco-friendly dyeing as the process produces clear backwaters which not only facilitate easier cleaning but also reduces the negative environmental impact that may be caused due to loss of colour in effluent waste water.

ACID DYES

Acid dyes are generally sodium salts of organic acids. Since the anion in the dye is the active colouring component and is generally applied from an acidic bath, these dyes are called **acid dyes**. Most acid dyes are sulphonic acid salts, but a few also contain carboxyl groups. Acid dyes have direct affinity for protein fibres and are the main class of dyes used to dye wool. Polyamide fibres also have an affinity for acid dyes. Acid dyes resemble direct dyes in chemical structure, however most of them do not dye cellulose fibres. The first acid dye was developed by Nicholson in 1862 by sulphonating aniline blue which was called Bleu de Lyon. Other basic dyes were subsequently sulphonated to convert into acid dyes, making them more easily applicable to wool.

Classification of Acid Dyes

There are four classes of acid dyes based on the affinity of the dye for the fibre (Table 12.5): (*a*) **Levelling acid dyes** which have poor or low affinity for fibre and can thus migrate from one place to the other during the dyeing process. Due to the low affinity, they require highly-acidic condition for dyeing. The dyeing output is very even, however these dyes have poor wash fastness. (*b*) **Intermediate acid dyes** are a very small group of dyes. Most of them can be classified either as levelling or milling acid dyes. (*c*) **Milling acid dyes** have good affinity and show good wash fastness, good fastness to milling and sometimes good light fastness. They are called milling dyes because they have resistance for the milling treatment given to wool. **Milling** is a treatment where wool is treated with alkali or acid in presence of steam for shrinking. The dye has good affinity for fibre and thus it rushes to the fibre at low pH which can result in uneven dyeing. Hence a pH of 4–5 is maintained during the dyeing process. (*d*) **Super milling dyes** have very high affinity for fibre. Hence it is very important that initially the dye is absorbed uniformly as levelling is not possible at later stage. For this purpose, a pH of 6–7 is maintained where not much positively-charged amino groups are present on the fibre. These dyes have medium to good wet fastness properties. However, some of the dyes have poor light fastness in pale shades.

Table 12.5: Classification of acid dyes based on affinity

Dye class	*Dye medium*	*pH of application*
Levelling acid dyes	H_2SO_4	2.5–3
Intermediate group acid dyes	HCOOH	3.5–4
Milling acid dyes	CH_3COOH	4–5
Super milling acid dyes	$(NH_4)_2SO_4$	6–7

Source: Compiled by the authors.

Effect of Various Auxiliaries added during Dyeing

Effect of Acids

Acid is the most significant assistant in the application of acid dyes. Numerous dyes do not exhaust on wool at all unless the dye bath has been acidified. The amount of acid not only influences the total amount of dye absorbed, but also the rate of exhaustion.

Effect of Glauber's Salt

Glauber's salt ($Na_2SO_4.10H_2O$) is used as a levelling agent with many acid dyes. A levelling agent can perform any one of the following functions:

1. *Provide competition to dye*: The sulphate anions, being smaller and more mobile than the dye anions, move faster towards the fibre and get adsorped rapidly and neutralise the charge on the fibre. Later, as the dye has more affinity for the fibre it slowly replaces the sulphate ions. For this purpose an anionic surface active agent, such as soap or Turkey Red Oil, can also be used. But if a large quantity of levelling agent is added there will be too much competition for the dye and this would result in lower dye uptake.
2. *Suppress ionisation of dye*: The levelling agent and dye combine to form a complex, which gradually breaks down with rise in temperature, progressively releasing the dye for more gradual adsorption by the fibre.
3. *Promote dye uptake*: In case of super milling dyes, salt plays an opposite effect i.e. it promotes dye uptake on protein fibres by reducing zeta potential. Wool possibly attains a negative electrical potential in neutral conditions which is used for dyeing of super milling dyes. Electrolyte reduces this surface negative charge between the dye and the fibre by absorption on the fibre surface of positive ions released by the salt.

Zeta potential is a measure of the electrostatic charge on particles that are suspended/immersed in a liquid and the resultant repulsion.

Theory of Acid Dyeing

Acid dyes fix to fibres by hydrogen bonding, Van der Waals forces and ionic bonds. Polyamide fibres such as wool, silk and nylon have –CONH– linkages (amide linkages) in them. When this fibre is inserted in water it attains a positive charge on the end amino group and a negative charge on the carboxylic group which is called Zwitter ion. When an acid is added, the net charge on the fabric remains positive as the negatively charged carboxylic groups get neutralised.

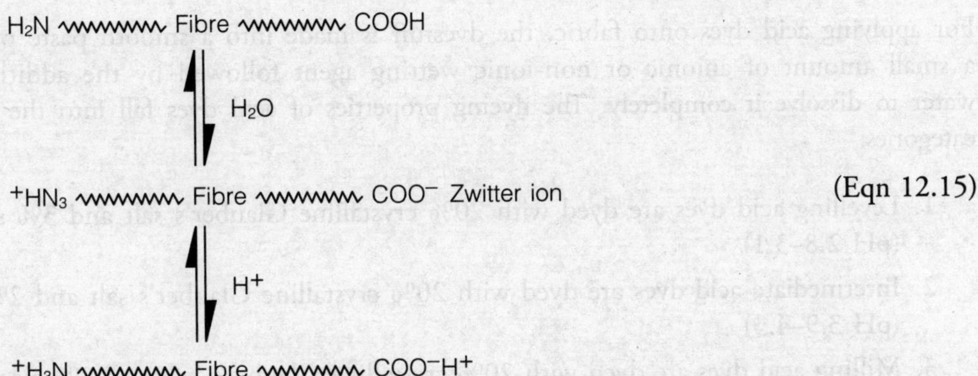

(Eqn 12.15)

The presence of acid in the dye bath has the effect of severing the salt linkages between the molecular chains and thereby increasing the available number of positively-charged sites. Such a system is capable of attracting dye anions. In a dye bath with dye molecules and hydrochloric acid (HCl), there is initially a rapid adsorption of both H^+ and Cl^- ions. Due to their higher mobility Cl^- ions reach the positively-charged sites in the fibre first. When the dye anions reach the fibre sites, they replace the chloride ion since their electrostatic attraction is further enhanced by other forces such as hydrogen bonds and Van der Waals forces. The reaction between an acid dye and a protein fibre such as wool, silk and nylon can be represented as follows:

$$\text{Fibre–NH}_2 + \text{HSO}_3\text{–Dye} \longrightarrow \text{Fibre–NH}_3^{+}\,{}^{-}\text{O}_3\text{S–Dye} \qquad \text{(Eqn 12.16)}$$

In the case of acid dyeing, dye absorption depends on the number of amino end groups present in the fibre. The total number of amino groups (in gram equivalents per kg) and crystallinity (which describes the accessibility of the fibre for dyeing) of a few fibres that are commonly acid dyed is given in Table 12.6.

Table 12.6: Number of NH_2 groups and crystallinity of polyamide fibres

Fibre	*Total No. of NH_2 Group*	*Crystallinity %*
Wool	0.8 g eqv/kg	20
Silk	0.2 g eqv/kg	60
Nylon	0.04 g eqv/kg	Variable

Source: Compiled by the authors.

Since the number of amino groups in wool are the highest, wool will be dyed deepest followed by silk and then nylon. Wool is less crystalline because it has a lot of open spaces, and so it is very easy to dye. In silk most of the crystalline regions are near the core of the fibre and the outer regions are amorphous, and so it can also be dyed easily. The crystallinity of nylon depends on the heat setting and other treatments given and thus its dyeing properties vary accordingly.

Application of Acid Dyes

For applying acid dyes onto fabric, the dyestuff is made into a smooth paste preferably with a small amount of anionic or non-ionic wetting agent followed by the addition of boiling water to dissolve it completely. The dyeing properties of acid dyes fall into the following five categories:

1. Levelling acid dyes are dyed with 20% crystalline Glauber's salt and 3% sulphuric acid (pH 2.8–3.1)
2. Intermediate acid dyes are dyed with 20% crystalline Glauber's salt and 2% formic acid (pH 3.9–4.3)
3. Milling acid dyes are dyed with 20% crystalline Glauber's salt and 2% glacial acetic acid (pH 4.7–5.1)
4. Super milling acid dyes are dyed only with 5% ammonium sulphate.
5. Cotton and wool union fabrics are dyed only with 10–40% crystalline Glauber's salt.

In general, dyeing is started at 40°C by introducing the fabric into a solution containing the acid and Glauber's salt. After 10 minutes, the acid dye is added. Then, the temperature is slowly raised to boil over a time of 40 minutes and dyeing is continued at boil (for wool and nylon) for about one hour. Silk is dyed at 80°C and not at boil so as to avoid damage to the fibre.

Fastness Properties

The fastness properties of acid dyes vary as per the dye class used. In general, acid dyes have good to very good wash fastness and good light fastness. The colours tend to be less fast on silk than on wool even though silk has affinity for acid dyes.

Colour Range

A large colour range is available with acid dyes starting from yellow to mustard, orange, pink, red, brown, green, blues, purple and black.

CHROME MORDANT DYES

Chrome mordant dyes have been used since ancient times. They combine with metallic oxides to form insoluble coloured complex which colour the fabric. Many examples of naturally-available

Figure 12.15: Examples of chrome mordant dyes

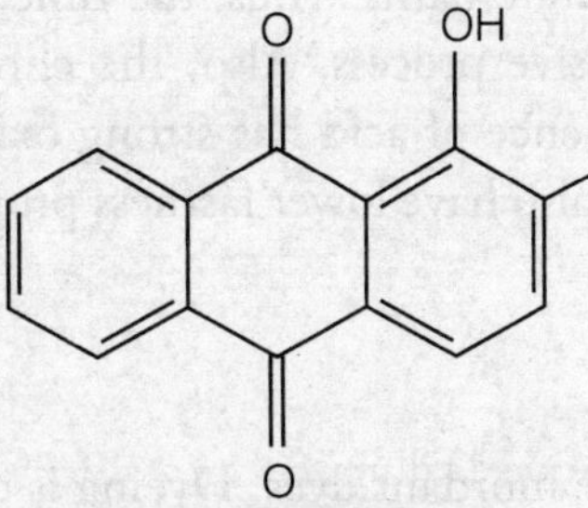

Alizarine (Natural Dye)

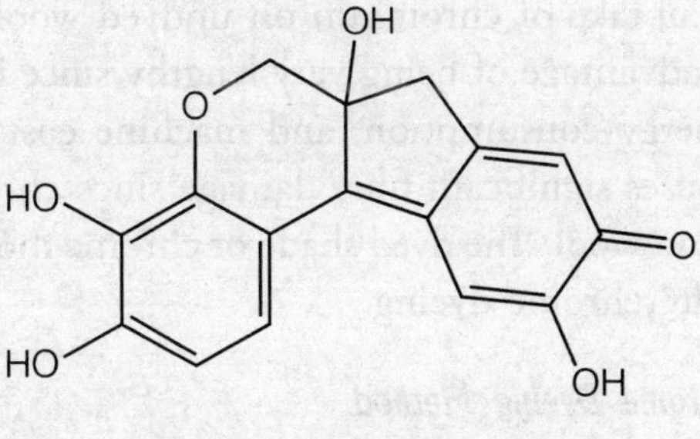

Heamatin

chrome mordant dyes are Alizarin, Logwood, Fustic and Weld (Figure 12.15) which are dyed on fibres that have already been mordanted with metallic hydroxides such as chromium, tin or aluminium. Alizarin, which is obtained from madder, is commonly known as polygenetic mordant dye since it develops a variety of colours with different mordants. For example, with aluminium mordant it gives red, with tin pink, with iron brown and with chromium puce-brown colour is obtained. However, with the development of synthetic dyes, chromium has become the most commonly used metal in mordant dyeing because of the increased light fastness and wash fastness property it lends. There are certain disadvantages such as long dyeing times, potentially high level of fibre damage and chromium residues in effluent. Chrome dyes hold a special position in the dyeing of wool fibres, since they give very level dyeing, good migration properties and excellent wash fastness properties. With these advantages, these dyes are widely used particularly for dyeing heavy shades, such as navy blue and black.

Application Methods of Chrome Mordant Dyes

Three methods have been followed in the application of these dyes.

Application methods for chrome mordant dyes

1. Chrome mordant dyeing
2. Metachrome dyeing
3. Afterchrome dyeing method

Chrome Mordant Dyeing

Chrome mordant dyeing or prechrome dyeing is the process in which the polyamide fibre is first treated with chromium metal followed by dyeing. During chroming, the fabric is treated with chromium salt, i.e. sodium or potassium dichromate, where hexavalent chromium is converted to trivalent. Following this, during dyeing chrome mordant dye is complexed with chromium. The process is started at 50°C with formic acid and dichromate salt. Then the temperature is raised to 98°C which is maintained for about 40 minutes and then the bath is emptied. The fibre is then rinsed to remove superficial chromium. Then dyeing is carried out as usual with acetic acid and sodium sulphate.

The advantage of this process is that it permits easier shade matching. Additionally, there is higher uptake of chromium on undyed wool rather than dyed wool. However, this process has the disadvantage of being very lengthy since it requires two separate baths. Thus, the time, water and energy consumption, and machine cost make it an expensive process. Also, the chroming step causes significant fibre damage since chromium in the presence of acid has strong oxidising action on wool. The dyed shade of chrome mordant dyeing may also have lower fastness properties than afterchrome dyeing.

Afterchrome Dyeing Method

This is the most commonly used technique for dyeing of chrome mordant dyes. Dyeing is carried out in two steps: dyeing in the presence of an acid, followed by chroming. Although two separate steps are involved, they are often carried out in the same bath and thus reduces dyeing time, water and energy requirements. This process yields better fastness properties than the other two dyeing methods. However, there are chances of shade going off as shade changes on complexing with chromium. Therefore, it is generally used for dyeing loose wool as it can later be blended with other coloured fibres before spinning yarn.

Metachrome Dyeing

Metachrome dyeing method relies on the co-application of dye and chromium from a dye bath at neutral pH. It is a single-bath single-stage process. In this method potassium dichromate and ammonium sulphate are added to maintain the pH at around 6–7. Dyeing is carried out at 60–70°C to avoid premature formation of dye–chromium complex in the dye bath. Therefore, only dyes having good affinity for wool can be used; there is a restriction on the number of dyes that are suitable for this method. Initially, a low temperature (60°C) is maintained and later the temperature is increased at which ammonium sulphate is broken to release ammonia and sulphuric acid, lowering the pH of the dye bath for exhaustion of dye.

This process has various disadvantages such as a limited number of suitable dyes, limited exhaustion at neutral pH leading to inability to achieve heavy shades, high residual levels of chromium and poor rubbing fastness properties. However, this technique is used for specific shades such as browns because of good levelling properties and wet fastness property.

Fastness Properties

Chrome mordant dyes have very good overall fastness properties such as wet fastness as well as light fastness. However, they produce dull shades. They are therefore majorly used for dyeing dark colours such as blue, browns and black and in dyeing of carpet wool.

METAL COMPLEX DYES

By the end of the 19th century, the use of chrome mordant dyes on wool became extensive due to their excellent all-round fastness properties. However, mordant dyeing of wool or polyamide fibres suffer from several disadvantages, such as difficulty in colour matching owing to the change of shade with mordanting and tendering of fibres due to the prolonged and relatively complicated two-stage, pre- and after-chrome dyeing processes. Efforts were therefore made in the early 20th century to overcome these disadvantages, and these resulted in the development of metal-complex or premetallised acid dyes.

As evident from the name, in metal complex dyes one metal atom, mostly chromium, is complexed with either one or two dye molecules of a typically monoazo dye that contains groups, such as hydroxyl, carboxyl or amino groups, that are capable of coordinating with the metal. If the metal is complexed with one dye molecule, it is called 1:1 metal complex dye; and if it is complexed with two dye molecules, it is called 1:2 metal complex dye. 1:1 and 1:2 metal complex dyes resemble levelling and milling acid dyes respectively in terms of general application conditions. In general terms, metal complex dyes yield shades on wool that are duller than those given by nonmetallised acid dyes, but slightly brighter than those obtained using mordant dyes.

1:1 Metal Complex Dyes

1:1 metal complex dyes are dyes where coordination of one dye molecule has taken place with a metal ion. The metal ion used in these dyes is most commonly trivalent and hexa-coordinate, and is usually either Cr^{3+} (predominantly) or Co^{3+}. The first 1:1 metal complex dyes produced were named Neolan and Palatine fast dyes. They are majorly based on the o,o'-dihydroxyl azo structure. Although the sulphonic acid group present in these dyes render dye solubility. they have to be applied from an acidic dye bath. An example of 1:1 metal complex dye is C.I. Acid Violet 58 which has one chromium atom attached to one molecule of the dye (Figure 12.16).

Figure 12.16: 1:1 metal complex dye

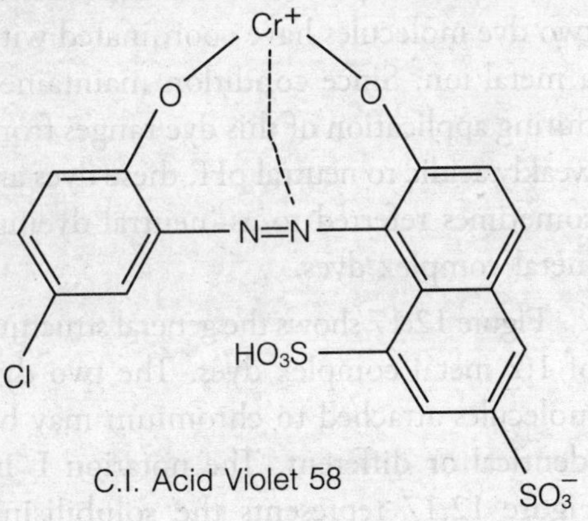

Cr^+, O, O, N=N, HO_3S, Cl, SO_3^-

C.I. Acid Violet 58

The advantages of this dye are excellent migration and penetration property, ease of application and excellent light fastness properties. However, they have wet fastness properties lower than that of mordant dyes. Moreover, a highly acidic pH is required for the application of these dyes.

Application of 1:1 Metal Complex Dyes

1:1 metal complex dyes are applied onto wool usually from a strong acidic dyebath of about pH 2. Under these conditions the dye exhibits excellent migrating and levelling properties. Given the high concentration of sulphuric acid, it is necessary to either neutralise or buffer the residual acid in the fibre at the end of the dyeing process. Also, since wool tends to get degraded at boil in the presence of sulphuric acid, either reduced amounts of sulphuric acid or other acids such as formic acid along with levelling agents can be used.

Dye-Fibre Interaction

In the case of 1:1 metal complex dyes, various dye–fibre bonds are possible, and they have been listed below:

1. Salt linkages or ionic linkages between sulphonate group of the dye and protonated amino groups of he fibre. Dye–$SO_3^{-}$$^{+}H_3N$–fibre
2. Salt linkages between chromium ion and the carboxylate ion of the fibre. Dye–Cr^{+-}OOC–fibre
3. Co-ordination bond between chromium ion and appropriate ligands such as carboxyl or imino groups (>C=NH) in the fibre.
4. Ion–dipole, dipole–dipole and other related forces.

Co-ordination bond is a type of a covalent bond between two atoms where the shared electrons are provided by only one atom.

Ligand is an ion or molecule attached to a metal atom by co-ordinate bonding. The bonding with the metal generally involves donation of one or more of the ligand's electron pairs.

Fastness Properties

1:1 metal complex dyes have excellent light fastness properties. However, their wet fastness is lower than that of mordant dyes.

1:2 Metal Complex Dyes

1:2 metal complex dyes are dyes where two dye molecules have coordinated with a metal ion. Since condition maintained during application of this dye ranges from weakly acidic to neutral pH, these dyes are sometimes referred to as 'neutral-dyeing' metal complex dyes.

Figure 12.17 shows the general structure of 1:2 metal complex dyes. The two dye molecules attached to chromium may be identical or different. The notation L in Figure 12.17 represents the solubilising group. If the solubilising group is either of sulpho methane group(SO_2CH_3), sulpho amide group (SO_2NH_2) or sulphonic acid (SO_3Na), the dye would be suitable for wool and silk. There can also be no solubilising group, i.e. hydrogen occupies the place of L; such dyes are used for dyeing nylon.

Figure 12.17: 1:2 metal complex dye

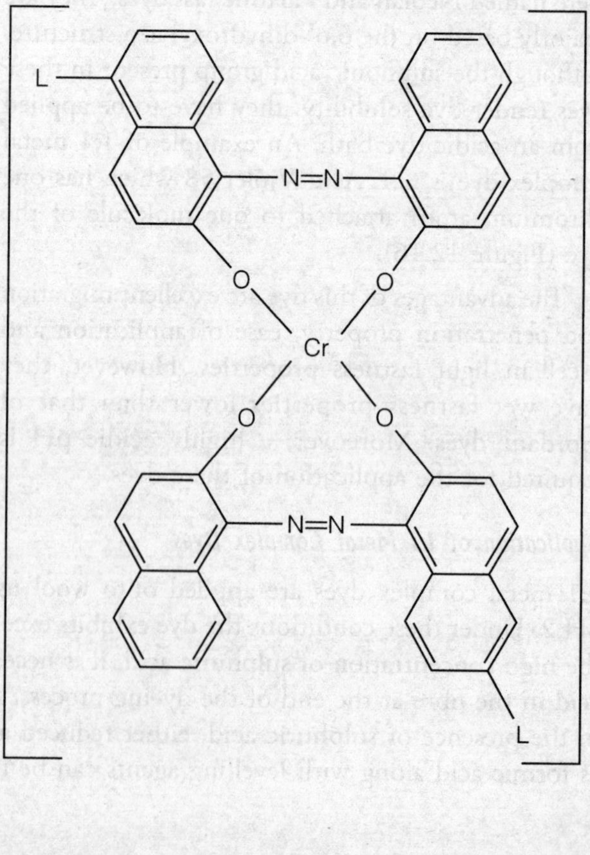

Application of 1:2 Metal Complex Dyes

1:2 metal complex dyes are usually applied onto polyamide fibres with the pH maintained at 6–7. Levelling of the dye is dependent on the pH and the temperature, and is generally enhanced by the use of a levelling agent and Glauber's salt. The dye bath is cooled to 80°C for shade matching. Ammonium sulphate or a combination of CH_3COOH and ammonium sulphate may be used to neutralise any residual acid due to previous treatments given to wool.

Dye–Fibre Interaction

All 1:2 metal–complex acid dyes carry one or two or three negative charges. The dyes are applied to wool under weakly acidic to near-neutral pH conditions. However, at such pH the number of protonated amino groups in the substrate is small; and consequently, ion–ion interaction will be limited. Thus the dyes behave as non-metallised (i.e. milling) acid dyes of large molecular size in their adsorption characteristics on wool and, as with milling nonmetallised acid dyes, forces other than electrostatic bonds will contribute to dye–fibre substantivity; i.e. hydrophobic interactions that operate between the dye and hydrophobic regions within the fibre make an important contribution to substantivity.

Fastness Properties

1:2 metal complex dyes have very good wash fastness and light fastness and can also withstand the alkali wash given to carpets. They have a very large dye molecule and high affinity towards the fibre, giving good fastness properties.

DISPERSE DYES

Dyeing of fibres such as cotton, wool, silk, etc. which are hydrophilic in nature, is done by direct, acid, vat, sulphur dyes, etc. When hydrophobic fibres made their appearance soon after the First World War, they could not be dyed with these conventional water soluble dyes. Hence scientists created a new class of dyes called disperse dyes.

Disperse dyes are water insoluble, non-ionic dyes suitable for dyeing hydrophobic fibres such as cellulose acetate, nylon, polyester, acrylic and other synthetic fibres, and are applied along with a dispersing agent to retain the dye in a fine dispersion. These dyes derive their name from their insoluble aqueous properties and the need to apply them from an aqueous dispersion. The pigments used in the preparation of disperse dyes are either azo compounds or anthraquinone derivatives.

Properties of Disperse Dyes

- *Ionic nature*: Disperse dyes are electrically neutral (i.e. non-ionic).
- *Solubility*: They are insoluble in water or have very low water solubility. They dissolve in organic solvents like benzene, toluene, etc.

- *Affinity*: They are suitable for dyeing hydrophobic fibres. Disperse dyes are used for dyeing man-made cellulose esters and synthetic fibres, specially acetate, polyester, nylon and acrylic fibres.
- *Molecular size*: Of all dyestuffs, disperse dyes are of the smallest molecular size.
- *Sublimation fastness*: Due to their stable electronic arrangement, disperse dyes have good sublimation fastness.
- *Gas fading*: In the presence of nitrous oxide, textile materials dyed with certain blue and violet disperse dyes with an anthraquinone structure tend to fade. This is called **gas fading** of disperse dyes which is a defect of this dye.

Requisites of Dyeing

1. *Dispersing agents*: Since disperse dyes are sparingly soluble in water and are often crystalline with varying particle size, dispersing agents, that ensure proper dispersion of the dyestuff and avoid settling of dye during dyeing, are applied in the dye bath. An example of a commercially-available dispersing agent is Setamoland Lyocol (Archroma).
2. *Levelling agents*: Levelling agents promote uniform distribution of dye throughout fibre structure by reducing the strike rate and enhancing the solubility of dye. Some commercially-available levelling agents are Lyogen and Eganal (Archroma) and Albatex (Huntsman).
3. *Carriers*: A **carrier** is an organic compound, which when dissolved or emulsified in the dye bath, interferes with the dispersion pattern of the dye and the physical characteristics of the fibre, and accelerates the rate of dyeing to cause better dye uptake at lower temperature. Carriers allow dyeing of even deep shades at boil within a reasonable dyeing time. Carriers are of two types: (*a*) *water soluble carriers* such as benzoic acid, salicylic acid and phenol which are not preferred because a high dose is required; and (*b*) *water insoluble carriers* such as mono/di and trichlorobenzene, diphenol, *o*-phyenylphenol which are mostly preferred.
4. *Temperature of dyeing*: Dyeing temperature varies from dye to dye; the larger the structure of the dye, more opening is needed, necessitating a higher temperature of dyeing. However, all disperse dyes exhaust well at or below 117°C and a little higher temperature is maintained to compensate for heat loss and ensure better migration of dye molecules for levelled shades. Nylon 66 and acrylic are dyed with disperse dyes at boil as adequate opening of fibre structure occurs at this temperature. Cellulose acetate is dyed at 85°C while polyester is dyed at 120–130°C.
5. *pH of dyebath*: Generally, in the commercial dyeing of polyester fibres with disperse dyes, dyeing is carried out within the pH range of 5.5–6.5. Strongly alkaline or acidic conditions, such as higher than pH 9 and lower than 4 induce hydrolysis of the fibre as well as decomposition of azo disperse dyes.

Mechanism of Dyeing

As disperse dyes have very low water solubility at room temperature, the solubility is increased either by heating the bath beyond 90°C or addition of a dispersing agent or both. Disperse dyes are specifically applied at high temperature, and therefore after the dye is added it remains in the bath in two basic forms: (*a*) very little part remains in the soluble form; and (*b*) the rest remains in a finely dispersed insoluble form. When the fibre is introduced into the dye bath, the dissolved dye molecules slide past the narrow pores in the fibre and get attached to it due to physical forces. This causes a reduction in the amount of dissolved dye molecules in the bath, and forces the dispersed insoluble dye particles to break up and dissolve in the solution thus maintaining the amount of dissolved dye in the bath. The overall rate of dyeing is controlled by the combined effect of temperature and the type of dispersing agent used. The various stages of the dyeing process are as follows:

1. Dispersion of the dye in the bath
2. Dissolution of dye molecules in water
3. Deposition of dye on fibre surface
4. Absorption of dye into the fibre and
5. Diffusion of dye inside

At the end of the dyeing process, equilibrium is achieved when the rate of dissolution, surface deposition, absorption and diffusion of dye become equal to each other with no further dye uptake.

Methods of Application of Disperse Dyes

There are three common methods of dyeing with disperse dyes and they are as follows:

Methods of dyeing with disperse dyes

1. Carrier method
2. High temperature high pressure method
3. Thermosol method

Carrier Method

In this method, the dye bath is set at a pH of 5.5–6.5 with acetic acid; a carrier is added and the textile is first treated with this carrier at 60°C. This is followed by the addition of the dye. The bath is then heated up to boil (100°C) and dyeing is continued for one hour. This is followed by soaping and washing to ensure complete removal of carrier. The advantages of this method are:

- Rate of dyeing can be increased by using carriers.
- Dyeing can be carried out at boil at atmospheric pressure, so no special equipment is required.

The limitations are:

- Carriers add to the production cost of dyeing
- Carriers are toxic
- Carriers affect the fastness properties of the dyed material

- This method is restricted to the production of only light shades as fibre opening as well as solubilisation of dye is not enough for diffusion of dye inside, causing poor wash fastness.

High Temperature High Pressure Method

This method is most commonly used for dyeing of polyester as high temperature is required for adequate opening up the fibre structure. Dyeing at high temperatures (120–130°C) requires pressurised equipment, and hence this process is known as High Temperature High Pressure method. It consists of three main stages:

1. *Exhaustion stage*: This is the most critical stage that determines the levelness of the dyed fibre, therefore the rate of heating should be appropriate to allow controlled adsorption of the dye. The bath is set at 50–60°C with the dispersing agent, CH_3COOH (pH 5.5–6.5) and water. The fabric is placed in the bath and the solution is agitated and temperature raised to 90°C, succeeded by control over rate of heating at 1°C per minute up to 120°C beyond which the temperature is raised to 130°C at one go and dyeing is continued for 1–2 hours.
2. *Diffusion stage*: At this stage the dye diffuses into the fibre and the time required is directly related to the diffusion characteristics of the dye being used and the depth of shade required. Generally, the time required is 10–20 minutes for pale shades, 20–30 minutes for medium shades and 30–35 minutes for deep shades. After dyeing is complete, the bath is cooled down to 80°C at the rate of 5°C per minute.
3. *Clearing stage*: Any surface-deposited dye as well as auxiliaries from the dyed polyester must be removed for better fastness. In the case of light shades, the dyed textile is soaped and washed. In the case of deep shades, reduction clearing is done in which the dyed fibre is treated with a reducing bath which contains $Na_2S_2O_4$ and NaOH. The bath is maintained at 60–70°C for 15–20 minutes.

Table 12.7: Dyeing temperatures for fibres that can be dyed using disperse dyes

Polyester	120–130°C
Nylon	85–120°C
Acrylic	70–100°C
Cellulose acetate	40–80°C

Source: Compiled by the authors.

This is the most widespread method of batch colouration because of the following advantages:

- Efficient levelling and covering of yarn irregularities with superior fastness through effective diffusion of dye can be achieved through this method.

- Process efficiency is higher due to better exhaustion of bath.
- There is no perceptible loss of elasticity or tensile strength when polyester fibres are dyed under neutral or slightly acidic conditions at 130°C.
- Auxiliary chemicals, except a little acid and the dispersing agent, are not required.

Thermosol Method

This is a continuous method of dyeing based on the sublimation property of disperse dyes. The textile is padded with the padding liquor (dye + dispersing agent + water), dried and exposed to hot air at 190–210°C for one minute. As the temperature of the fabric approaches the optimum temperature, the fibre structure opens up, and the disperse dyes begin to sublime. At this moment, the polyester fibres absorb their vapours, i.e. the dye molecules diffuse into the fibre. The efficiency of this method is based on the size of the dye molecule, extent of dispersion, sublimation property and solubility of dye. This method has restricted use due to poor migration of dye resulting in uneven dyeing.

Dyeing of Cellulose Acetate with Disperse Dyes

Disperse dyes can be readily applied to secondary cellulose acetate at 80°C for about an hour. Higher temperatures are avoided as the acetate groups on the cellulosic fibre will then be hydrolysed to hydroxyl groups and spoil the surface of the fibres and reduce their substantivity towards the disperse dyes. However, cellulose triacetate is more difficult to dye with disperse dyes as it has a more compact molecular structure, but they can be dyed at boil.

Dyeing of Nylon with Disperse Dyes

Nylon fibres can be dyed under conditions similar to those used for cellulose acetate fibres. However, the exhaustion rates vary from dye to dye, so the selection of dyes according to their compatibility is necessary. Disperse dyes on nylon are also more sensitive to fading by ozone and nitrogen dioxide. They are, however, economical and easy to apply.

Dyeing of Acrylic with Disperse Dyes

In the case of acrylic fibres, the presence of anionic groups such as $–SO_3H$ and –COOH permit only pale shades to be obtained under normal conditions with disperse dyes.

Fastness Properties

Disperse dyes have moderate to good wash fastness with rating of 3–4. However, deep shades are rarely dyed with disperse dyes because of their inferior wash fastness, and the dyeing of nylon with disperse dyes is limited mainly to pale shades. As polyester has a crystalline structure and is hydrophobic, it tends to have better wash fastness than nylon and acetate with a given dye. The light fastness of disperse dyes is fair to good with a rating of 4–5. Light fastness is highest on polyester, followed by di and triacetate, and then nylon.

NATURAL DYES

The art of dyeing is considered to be as old as the human civilisation and dates back to prehistoric times. Madder-dyed cloth have been found in the Harappan excavations and historical records attest to the use of natural dyes in China since at least 2600 BC. Description of natural dyes in *Atharvaveda* and the colourful wall paintings of Ajanta during the first century AD are also evidence to the use of dyed garments in ancient times. Colour is developed on the fibre or fabric as a result of a chemical reaction between the fibre structures and dye structures. Initially, these naturally-occurring dyes were used directly in their crude form without any chemical processing. However, gradually several sophisticated procedures were invented which led to the creation of a wide range of colours and improved of fastness properties.

Mordants

Mordants form an insoluble compound by reacting with the dyestuff as well as the fibre and hence hold the colour within the fibre. The effect of mordants was accidently discovered while washing undyed fabric in stream of water rich in some metallic compounds. The natural dyes which need mordanting for development of colour are known as **adjective** dyes, whereas dyes which do not require such treatment are known as **substantive** dyes. Mordants can be broadly classified into three categories—metal salts or metallic mordants, tannins or tannic acid and oils or oil mordants. Commonly-used metallic salts include alum, potassium dichromate, ferrous sulphate, stannic chloride, copper sulphate and stannous chloride. Tannins like myrabolan and sumach, which are secretions from the bark or other parts of a plant (leaves or fruits), are also used as mordanting agents. Oil mordants react with the main mordant (alum in the case of dyeing with madder dye) to form an insoluble compound, hence improving the wash fastness of the dyed fabric.

Based on their source, natural dyes can be classified into three categories.

Types of natural dyes
- Vegetable dyes
- Animal dyes
- Mineral dyes

Vegetable Dyes

Vegetable dyes are extracted from different parts of tree, i.e. bark, leaves, pods, flowers or fruits. A wide variety of vegetable dyes have been used which can be applied directly or in combination with various mordants. Colour can be developed by applying the mordant on the textile material before or after dyeing. Some of the commonly-used vegetable dyes have been briefly summarised below:

1. *Indigo (Neel)*: Indigo is a naturally occurring vat dye that imparts a blue colour that is fast to washing, perspiration and light. It is known as the **king of all natural dyes**. Other colours of Indigo include yellow green to blue green on mordanting with turmeric (poor fastness) and pomegranate rind (moderate fastness) after dyeing. Indigo is suitable for being used on cotton and wool and the major producers in India are Bihar, Assam and Tamil Nadu.

2. *Indian Madder (Majeeth/Manjith/Manjeeth)*: Madder, which contains the chemical pigment Alizarine, has traditionally been used to produce shades of red on textiles. It produces wide range of shades like red when topped with alum, pinkish red when mordanted with alum, crimson to dull red with sodium bichromate, maroon to reddish brown with iron sulphate, reddish orange with tin chloride and brick brown with 8% solution of oxalic acid. Madder is suitable for cotton and woollen fibres, and the major producers in India are Sikkim and the Assam region.
3. *Cutch (Khair/Katha)*: This dye gives a range of browns when the colour is developed on mordanting with various agents. These shades possess good fastness properties to washing and perspiration and show moderate fastness to light. This dye is suitable to be used on cotton, silk and woollen fibres. Nagpur and Mumbai are major centres of production of this dye.
4. *Tesu (Dhak Pas/Tesu-ka-phul, Palash)*: Yellow clay brown is developed on treatment of the dye with 2% potassium dichromate and light grey is developed when treated with 2% aluminium sulphate. Bright orange colour is obtained when the dye is applied after mordanting with alum. However, the shades by this dye produced are not very fast. It can be suitably applied on cotton, silk and woollen fibres and is produced abundantly in Uttar Pradesh and Bihar.
5. *Pomegranate rind (Anar-ka-per/Dhalum)*: Colours like golden brown, straw yellow, greyish to deep black, dark brown and khaki are developed when the fabrics dyed in this dye are treated with different mordants. All these colours—except black, which is moderately fast—show good fastness to washing, dilute acids and alkalis, perspiration and light. This dye can be applied on cotton, silk and woollen fibres and is produced in Punjab, Kashmir and Almora regions.
6. *Turmeric (Haldi)*: A range of colours are obtained from turmeric dye: yellow, green (when re-dyed with indigo), lemon yellow (when topped with bleaching powder), bright yellow (when pre-treated with alum), bright orange-yellow (when pre-treated with tin chloride) and olive (when pre-treated with iron sulphate). These colours show moderate fastness to neutral soaps and poor fastness to alkalis. Turmeric can also be applied on cotton, silk and woollen fibres. Important states which grow turmeric are Andhra Pradesh, Odisha, Tamil Nadu, West Bengal, Assam, Maharashtra and Karnataka.
7. *Walnut rind (Akhrot)*: It is known to produce dark olive directly, and warm brown, dark brown, olive brown, deep brown and bright brown when the dyed fibres are mordanted. These dyes are very fast to washing and moderately fast to light. This dye can be applied on cotton, silk and woollen fibres, and is found in temperate regions of India like Himachal Pradesh, Kashmir and Almora in Uttarakhand.
8. *French Marigold (Genda)*: This dye is extracted from the bright lemon or orange-coloured marigold flower commonly found all over India. It can be suitably applied using different mordants on both silk and wool fibres to get a wide range of fast colours.

9. *Henna (Mehndi)*: Derived from the leaves of the henna shrub, these dyes produce a variety of colours that can be developed on wool and silk fibres when combined with metallic salts. Combining with aluminium gives camel brown, with copper gives yellow ochre and with iron it produces blackish brown. Henna is mainly cultivated in Punjab, Gujarat, Himachal Pradesh and Rajasthan.

Animal Dyes

Dyes extracted from the bodies of some insects and invertebrates are categorised as animal dyes. The most common types include Cochineal and Tyrian purple. Besides, lac insect has been used to produce red and violet colour, murex snail to produce purple and octopus/cuttlefish to produce sepia brown.

1. *Cochineal*: This dye is obtained from the dried carcasses of the female red bug (*Dactylopius coccus*). These insects are native to Mexico and were brought to South India and Bengal in the late 1700s. It gives very fine crimson when mordanted with aluminium and scarlet when mordanted with tin oxide. The dyes possess excellent fastness properties.
2. *Tyrian Purple*: This dye is extracted from the white liquid which is secreted by a species of sea snails of the family Muricidae, and found in the Mediterranean Sea. This dye which produces a wide range of red-purple colours has been in use since at least 1000 BC. In ancient times, the process of dye-making was very time consuming and complicated, and the amount of dye produced was very limited. The dye was therefore very expensive and available only to the rich and the ruling class, and came to be known as **Royal Purple**.
3. *Lac*: This dye is extracted from the thick fluid secreted and deposited as hard continuous coverings over twigs of plants by the lac insect (*Laccifer lacca*). These twigs are removed, the encrustations scrapped off and processed to produce the dye. The colours include crimson and scarlet which show good fastness to light and washing.

Mineral Dyes

Dyes that are extracted from mineral sources are known as mineral dyes. The most common example in this category is Iron Buff. Scraps of iron are covered with vinegar and water and are allowed to stand for a specific duration. The material to be dyed is then soaked in this mixture and later treated with a solution of wood ash. A beautiful yellowish brown shade (iron buff) is developed on exposure of the material to air. With variations in this method of production, a variety of shades such as light yellowish brown, deep reddish brown and rust can be developed. Another example is Mineral Khaki, which is produced by impregnating the material to be dyed with iron sulphate and chromium sulphate mixture and then reacting with an alkali. The colour produced shows good fastness to light and washing but creates a rough texture on the fabric. Other important mineral colourants include chrome yellow, Prussian blue and manganese brown. However, mineral sources of colourants may be poisonous and are therefore not being commercially used.

Revival of Natural Dyes

Until the middle of the nineteenth century, natural dyes were being used for all types of dyed and printed textiles in India. With the introduction of synthetic dyes, which were cheaper and possessed excellent fastness properties, natural dyes lost their popularity. However, the art of natural dyeing is still being practiced at a smaller scale in remote areas and cultures of the country. Some of these places include Sanganer, Bagru, Jodhpur (Chhiponka), Barmer and Akola in Rajasthan; Ajrakh printing in Gujarat; Machilipatnam (Kalamkari) in Krishna district of Andhra Pradesh, etc. These constitute only 1 per cent of the total dyed textiles produced in the country. The growing awareness of the harmful impact of synthetic dyes on the environment over the past few decades has revived the interest in natural dyes. Consumers are realising the benefits of natural dyes, some of which are as follows:

- These are sourced from naturally-occurring renewable materials and hence eco-friendly and biodegradable.
- Very soft and soothing colours are produced which are close to nature.
- These dyes and the tannins used for mordanting provide good protection from ultraviolet light.
- Some of the natural dyes possess anti-microbial properties which lend an additional functional benefit to the dyed material.
- Other properties which these dyes might possess include the property of mosquito repulsion and flame resistance.
- Natural dyes like manjishth, turmeric, myrobolon, etc. possess medicinal properties.

In line with the above benefits, lot of research is being carried out in the field of natural dyes. Efforts are being made towards developing efficient procedures for extraction, standardisation and fixation of natural dyes. Besides being applied on natural fibres, these dyes are now being applied on synthetic fibres too. Leading apparel outlets such as Fabindia, Anokhi, Kilol, Suvasa, Mother Earth, etc. and designers and design houses such as Bina Rao (Creative Bee), Anita Dongre (Grassroot), Raw mango, Bhusattva, Tvach, etc. are further taking initiative in promoting the use of natural dyes by sourcing naturally-dyed fabrics. Conscious consumers are attracted towards such practices which in turn lead to popularity of the natural dyes.

SUMMARY

- Dyeing is a complex phenomenon which involves basic processes such as transfer of coloured chemical component from the dye bath to the fibre, adsorption at the fibre surface and then diffusion from the surface into the fibre.
- Direct dyes are anionic dyes that are substantive to cellulose when applied from an aqueous bath containing an electrolyte.

- Reactive dyes contain groups that react with the groups in the fibre to form covalent dye–fibre bonds.
- Acid dyes are generally sodium salts of organic acids that have a direct affinity to protein fibres and are the main class for dyeing wool.
- Dyeing with vat dyes involves three basic steps: preparation of the water-soluble sodium derivative of leuco compound, dyeing of goods with the leuco compound and oxidation of the trapped leuco compound to the insoluble pigment.
- Sulphur dyes contain sulphur linkages as a characteristic feature of their molecular configuration.
- Basic dyes possess cationic/basic groups in their structures. These dyes can be applied on acrylic and protein fibres.
- Disperse dyes are water insoluble, non-ionic (electrically neutral) dyes suitable for dyeing hydrophobic fibres such as cellulose acetate, nylon, polyester, acrylic and other synthetic fibres.

KEY WORDS

Diffusion: It refers to the transfer of dye from the surface of the fibre to the core of the fibre.

Exhaustion: The absorption of dye in the dye bath by the substrate being dyed is termed as exhaustion. Percentage exhaustion is the percentage of the total amount of dye originally in the dye bath that is taken up by the substrate by the end of the dyeing cycle.

Fixation: Attachment of dye molecules that are absorbed by the fibre to dye sites in the fibre through some kind of dye–fibre interactions is known as fixation.

Solubilising groups: These are ionic groups that are attached to dye molecules and impart solubility to the dye. The solubility of a dye is greatly influenced by the nature and number of these solubilising groups.

Substantivity: The affinity of a dye for its substrate and its retention within the substrate through various dye–fibre interactions is termed as substantitvity.

Leuco compounds: Vat dyes form leuco compounds when they are subjected to a reduction reaction. These compounds are either colourless or show a colour quite different from the original colour of the vat dye.

Carrier: A carrier is an organic compound which accelerates the rate of dyeing to effect better dye uptake at a lower temperature. These are used during the application of disperse dyes on polyester.

Levelling agents: Levelling agents promote uniform distribution of dye throughout the fibre structure by reducing the strike rate of the dye on the fibre.

Mordants; Mordants form a bridge link between the dye and the fibre by reacting with both, the dyestuff and the fibre, and thereby hold the dye within the fibre.

EXERCISES

1. Describe the various stages involved in dyeing.
2. What are the various types of dye–fibre bonds that can take place?
3. List the various dye classes that can be used for dyeing cellulosic fibres.
4. Give the classification of direct dyes.
5. Why is electrolyte added during dyeing with direct dyes?
6. Briefly describe the substantivity of direct dyes for cotton fibre.
7. Give the general structure of reactive dyes.

8. What are advantages of using a bifunctional reactive dye?
9. Describe the various steps involved in applying vat dyes on cotton fabrics.
10. Explain the most commonly used method for the dissolution of vat dyes.
11. Describe the role of soaping in the case of dyeing of cotton with vat dyes.
12. Elaborate the properties of sulphur dyes.
13. Briefly describe the steps involved in dyeing with sulphur dyes.
14. Why do basic dyes have affinity for protein fibre?
15. How can basic dyes be applied on cellulosic fibres?
16. Classify acid dyes based on affinity.
17. Explain the effect of use of Glauber's salt on dyeing with acid dyes.
18. Describe the various dyeing methods used for chrome mordant dyes.
19. List the various dye–fibre interactions between 1:1 metal complex dyes and protein fibres.
20. List the various properties of disperse dyes.
21. Explain the mechanism of dyeing with disperse dyes.
22. Elaborate the most commonly used method of dyeing polyester with disperse dyes.

REFERENCES

Aspland, J. R. 1997. *Textile Dyeing and Coloration.* North Carolina: American Association of Textile Chemists and Colorists.

Broadbent, A. D. 2001. *Basic Principles of Textile Colouration.* West Yorkshire: Society of Dyers and Colourists.

Chakraborty, J. N. 2010. *Fundamentals and Practices in Colouration of Textiles.* New Delhi: Woodhead Publishing India.

Gohl, E. P. G. and L. D. Vilensky. 1987. *Textile Science.* New Delhi: CBS Publishers and Distributors.

Lewis, D. M. 1992. *Wool Dyeing,* West Yorkshire: Society of Dyers and Colourists.

Shore, J. 1995. *Cellulosics Dyeing.* West Yorkshire: Society of Dyers and Colourists.

Trotman, E. R. 1964. *Dyeing and Chemical Technology of Textile Fibres.* Third edition, London: Charles Griffin and Company Limited.

ONLINE SOURCES (all accessed May/June 2016)

http://textilelearner.blogspot.in/2012/01/disperse-dye-history-of-disperse-dye.html

http://textilelibrary.blogspot.in/2009/03/disperse-dye.html

http://cdn.intechopen.com/pdfs-wm/25012.pdf

http://dyes4dyeing.blogspot.in/2012/07/dyeing-of-acrylic-fibres.html

http://dyes4dyeing.blogspot.in/search/label/Dyeing%20Nylon

13

TEXTILE PRINTING

HIGHLIGHTS

- Introduction to printing on textiles
- Methods and styles of printing
- New developments in printing
- Dye fixation

A variety of fabrics are available in the market. One may choose from fabrics in solid colours and fabrics with designs that are applied through printing. Patterns on fabrics can be created using dyes, pigments or other coloured substances by hand or machine processes. One of the earliest methods of decorating textiles is hand painting, but it is a time-consuming procedure and does not always result in uniform repeat of a motif. A variety of techniques have now been developed for printing designs. There are many pieces of evidence to show that the art of printing was known to ancient Indians and the Chinese. They used wooden blocks for hand printing. In fact ancient Indians knew the method of mordant application as is evident from the fragment of madder-dyed cloth excavated at Harappa.

Textile printing is essentially a design or decorative pattern that is applied to a constructed fabric. It is considered as part of the finishing industry and has often been described as localised dyeing, i.e. dyes or pigments are applied locally or discontinuously, to produce attractive designs with well-defined boundaries through the artistic arrangement of motif or motifs in one or more colours. It utilises the same dyes or pigments applied to produce a dyed fabric. When the process is properly carried out, the printed fabric is well protected from friction and washing because a strong bonding is formed between the fibre and the dye.

Steps in printing

1. Preparation of the fabric
2. Preparation of print paste
3. Printing of fabric
4. Drying after printing
5. Dye fixation
6. Washing off and soaping

Preparation of Fabric

Better quality of print and clarity of colour can be achieved if fabrics are pre-treated by desizing, scouring and bleaching. Desizing removes the size/starch that may be present on the fabric. Scouring refers to the removal of impurities such as oils, wax, gums, dirt and other natural and non-fibrous impurities present in the fabric. This is often followed by bleaching and the reagents used for bleaching depend on the origin of their component fibres. For cellulosic fibres, bleaching is carried out with a suitable oxidising agent such as sodium hypochlorite or hydrogen peroxide. These chemicals oxidise and destroy the coloured impurities, and as a result whiten the material. The preparatory processes for different fibres have been discussed in detail in Chapter Nine.

Print Paste

Dyes or pigments used in dyeing are usually applied from a water bath or solution. When the same dyes or pigments are used for printing, they must be thickened with gums or starches to prevent the wicking or flowing of the print design during printing and subsequent drying. This thickened solution, which is about the consistency of heavy buttermilk, is called the **print paste**. The print paste is prepared using printing ingredients that are compatible with each other. A well prepared paste gives the best results. It is made up of the following components:

The general composition of print paste:
- Dyestuff
- Water
- Thickener
- Hydrotropic/hygroscopic agents
- Auxiliary chemicals

1. *Dyestuff*: Many dyes can be used in printing pastes. Selection of the dye depends on the fibre composition of the substrate on which the design is to be printed.
2. *Thickener*: These may be natural, modified natural or synthetic polymers which make the print paste viscous or gel-like. Table 13.1 lists some important thickeners used in textile printing.

Table 13.1: Various thickeners used for textile printing

Type	*Source of extraction*	*Name of the gum*
Natural	Cereal	maize starch, wheat starch
	Trees and shrubs	gum arabic, gum senegal, gum tragacanth
	Plant or tree seeds	locust bean gum, guar gum and starches
	Seaweed	sodium alginate
Modified natural	Starch derivatives	British gum (dextrin), carboxy-methyl starch
	Cellulose derivatives	carboxy methyl cellulose, hydroxyl ethyl cellulose
	Gum derivatives	gum indalca
Synthetic	Acrylic	polyacrylic acid, polyacrylamide
	Vinyl	polyvinyl alcohol

Source: Compiled by the authors.

To achieve an even print, there are certain essential qualities of thickeners, and they are:

- Compatibility with other ingredients of the print paste.

- Satisfactory adhesion to the fabric after printing and drying.
- No affinity towards the dye and other chemicals in the print paste.
- During fixation the thicker film should easily leave the dye and allow its penetration into the fibre.
- Should be easily removed from the fabric during washing.
- Should function satisfactorily for the printing method chosen; for instance, blocking of screen should not occur.

3. *Hydrotropic/hygroscopic agents*: These are water-soluble compounds such as urea, diethylene glycol, glycerine, etc. They facilitate steam condensation at the printed portion, thus creating a mini dyebath, so that dye can enter the fibre structure. Hygroscopic agents also swell the fibre to help the dye move into the fibre structure for fixation.
4. *Auxiliary chemicals*: Other auxiliaries such as solvents in print paste improve dye solubility. They also facilitate dye diffusion and improve colour yield. In addition to these, other chemicals may be added depending on the dyes and fibres. For instance, an acidic agent such as citric acid or tartaric acid may be added for acid dyes; an alkali is required for printing with most of the reactive dyes; discharging agents are required for discharge printing.

Following preparation of fabric and printing paste, the next step is printing of the fabric. This step involves various methods and styles. The **method of printing** involves the means by which the pattern is produced and is distinct from the **style of printing** which refers to the manner in which a printed effect is produced.

METHODS OF PRINTING

These methods refer to the tools or machines used for printing. There are several methods for printing of textiles such as block, roller printing, screen printing and transfer printing. Screen printing, both hand and rotary method, is of significant commercial importance while block and heat transfer printing are not widely used in the commercial production of textiles.

Block Printing

Block printing is the most ancient of all techniques and dates back to about 2000 BC. It was the first form of textile printing method that was developed in the ancient world. Archaeologists have found wooden blocks carrying design motifs in tombs near the ancient town of Panopolis in Egypt. Block printing was highly developed in ancient and medieval India. Indian block printed cotton fabrics were in high demand in Europe in the early seventeenth century.

Block printing is the simplest of the printing techniques. The design is drawn on the flat side of the block that is usually made of wood. The design is then carved by cutting away the spaces between the areas that form the pattern, thus placing the design in a raised position. The colour is then applied to the surface of the block, and the block is pressed onto the cloth. This process is repeated as many times as required according to the design needs. Multicoloured patterns require separate blocks for each colour.

Centers of Block Printing in India

Block printing is a method that continues to be a major commercial activity in India. Within Gujarat, places like Dhamadka, Baroda, Ahmedabad, Bhavnagar, Vasna, Rajkot, Jamnagar, Jetpur, Pethapur, Kutch and Porbandar are famous for their distinctive styles of block printing. In Pethapur, they use intricate blocks to print textiles using mud-resist method. The popular Ajrakh fabric from Dhamadka is popular for red and blue geometric prints that are achieved using natural colours. Rajasthan is famous for its brightly-printed bed covers, quilts and saris. Jaipur, Bagru, Sanganer, Pali and Barmer are important centres in Rajasthan. Sanganer is famous for its Calico printing where the outlines are first printed, and then the colour is filled in. Bagru is famous for its Dabu prints in which portions are hidden from the dye by applying a mud-resist paste. In Andhra Pradesh, block printing method is applied in the creation of the exquisite Kalamkari paintings of Masulipatnam. Kalamkari work is mainly done on bed covers, curtains and fabric for garments, using a combination of wooden block printing and hand painting. Besides these famous centres, block printing is also practiced in the states of Punjab, West Bengal, Madhya Pradesh, Uttar Pradesh and Maharashtra. Each place has its own unique style in terms of patterns, technique and colours in use.

Block Making

Wooden blocks for textile printing are generally made up of hardwood such as box, lime, holly, sycamore, plane or pear wood. Several layers of wood are glued together with the grain of the wood running in different directions in order to provide the required thickness and strength. The block is then buffed using sand paper to make it smooth and flat. The next step is to transfer the design onto the block. The design is drawn on a transparent paper using a paste of lampblack and linseed oil. The paper is then laid flat on the block facing down and rubbed so that the design lines are transferred onto the block. The block-maker distinguishes the parts to be preserved from the parts to be cut away by tinting the former in pink colour. The parts which are not required are chiselled away to a depth of 1/4th–3/4th of an inch.

Image 13.1: Wooden block

Source: Authors.

Each colour has a corresponding block; so if a design is made up of three colours, it is made using three different blocks.

Pitch pins are then marked on the blocks; **pitch pins** guide the printer about the repeat of the design and helps in matching of design while printing. Pitch pins print small dots and allow the succeeding blocks to be correctly positioned by accurately positioning on the already printed dots. In order to print large areas of colour, special colour blocks are used. For making these, the block-maker traces the design on the block and chisels away the design leaving a narrow wall of wood around the design area. Then thick felt which is pre-soaked in water and gum is placed on the block; the outline of the design makes its impression on the felt. The felt is then cut according to the required shape and pressed into the centre of design motif so that it is in level with the wooden wall. These blocks are very effective in printing big coloured areas because felt holds up the dye easily and produces an even effect on the fabric.

During the latter half of eighteenth century, a new technique was developed which enabled more intricate designs to be made on the fabrics. This was known as picotage or 'pinning' where fine pieces of copper or brass in the form of strips or pins were attached to the wooden blocks. The nineteenth century saw more advances in the field of hand block printing. Coppered blocks were developed which were totally made up of metal strips placed into wooden base. For making these, the design is first traced onto the wooden base and the design outline is cut away to a depth of about 1/4th of an inch. Then the metal strips are hammered into the groove. This enables very fine line designs to be printed on the fabrics.

Process of Block Printing

Image 13.2: Printing with block

Source: Authors.

A successful block printing process requires a proper printing table. It is very important to have a table which is firm, strong and stable and which can withstand the pressure of printing. The top is generally made of metal and has a resilient cover. A firm yet resilient cover is a prerequisite for printing as it allows the colour to be transferred from the block onto the fabric. The resilient cover is provided by using a thick woollen blanket which is protected from getting soiled by a waterproof cover. Over the waterproof cover an unstarched and unbleached cotton fabric known as **back grey** is spread firmly and tightly. Back grey serves an important function by absorbing any extra colour which penetrates to the underside of the fabric thus preventing a blurred print. Apart from the printing table, another component which is very important for block printing is the **sieve** which holds the print paste. The colour sieve consists of a tub (known as the swimming tub) half filled with starch paste or old colour to provide resiliency. On the surface of this floats a frame with a tightly stretched piece of mackintosh (waterproof cloth) at its bottom. Over this a fine, evenly woven woollen cloth is stretched. This structure resembles an ink-pad which is used to transfer ink to a stamp. The print paste containing the dye is evenly spread onto this woollen fabric with the help of a wedge of wood.

The fabric to be printed is spread onto the printing table and pinned in place. The printer takes the block and presses it into the sieve two times in opposite directions to ensure that colour is evenly spread onto the block. Then (s)he places the block on the fabric and gives a slight blow with his hand on the back of the block so that the colour is evenly transferred onto the fabric giving a uniform print (Image 13.2). In order to match the design repeats, the printer uses the position of pitch pins as a guide. In the case of multicolour prints, the dark colours and outlines are first printed and then the lighter colours. The process is repeated to print the required length of fabric. Table 13.2 summarises the advantages and limitations of block printing process.

Table 13.2: Advantages and limitations of block printing

Advantages	*Limitations*
Simple	Expensive
Hand-made craft	Slow
Capable of yielding highly artistic results	Pattern alignment is difficult
	Irregular depth of colour
	Fine lines, subtle gradations and tonal effects not possible
	Carving of blocks is difficult and requires specialised skills
	Maintenance of wooden blocks is difficult. Wood tends to warp out of shape and crack.

Source: Compiled by the authors.

Roller Printing

This method of printing is comparable to newspaper printing. It is a high-speed process capable of producing over 6000 yards of printed fabric per hour. The method is also known as **cylinder printing** or machine printing and was invented by Thomas Bell of Scotland (patented in 1783) in an attempt to reduce the cost of the earlier copperplate printing. Copperplate printing, also known as engraved printing, was done using a polished copperplate on which a design was etched or engraved. This method was popularly used then for reproducing small monochrome patterns. Roller printing method is an intaglio process where the printing is done from lines cut into the metal. The design is transferred onto the fabric by engraved copper rollers/cylinders. There are various ways of engraving the roller, like hand engraving, machine or mill engraving, pentagraph engraving and photographic engraving. There are a series of rollers, each imprinting a different colour on the fabric. The design is repeated

In **intaglio printing** the design or pattern is cut/etched into the printing surface or plate and the print paste held within these recessed areas is transferred to the fabric. It is the reverse of **relief printing** (as in block printing) where protruding areas of the surface are inked.

after every rotation of the roller. The size of the engraved cylinders is governed by the design to be printed.

Machine Components

The main component is the engraved design roller which can be of solid or shell type. The former has a diameter of 5–12 inches and the latter has a diameter of 5–30 inches. The shell type design roller is made up of thick copper deposit on steel roller. The design is engraved onto this roller. The design of each colour is accurately transferred to the roller by one of the following methods: (*a*) by hand, with a graver which cuts the metal away; (*b*) by etching, in which the pattern is dissolved out in nitric acid; and (*c*) by machine, in which the pattern is simply indented. After engraving, the design roller is given a plating of chrome which provides extra strength and extends the life of the roller.

Figure 13.1: Roller printing machinery

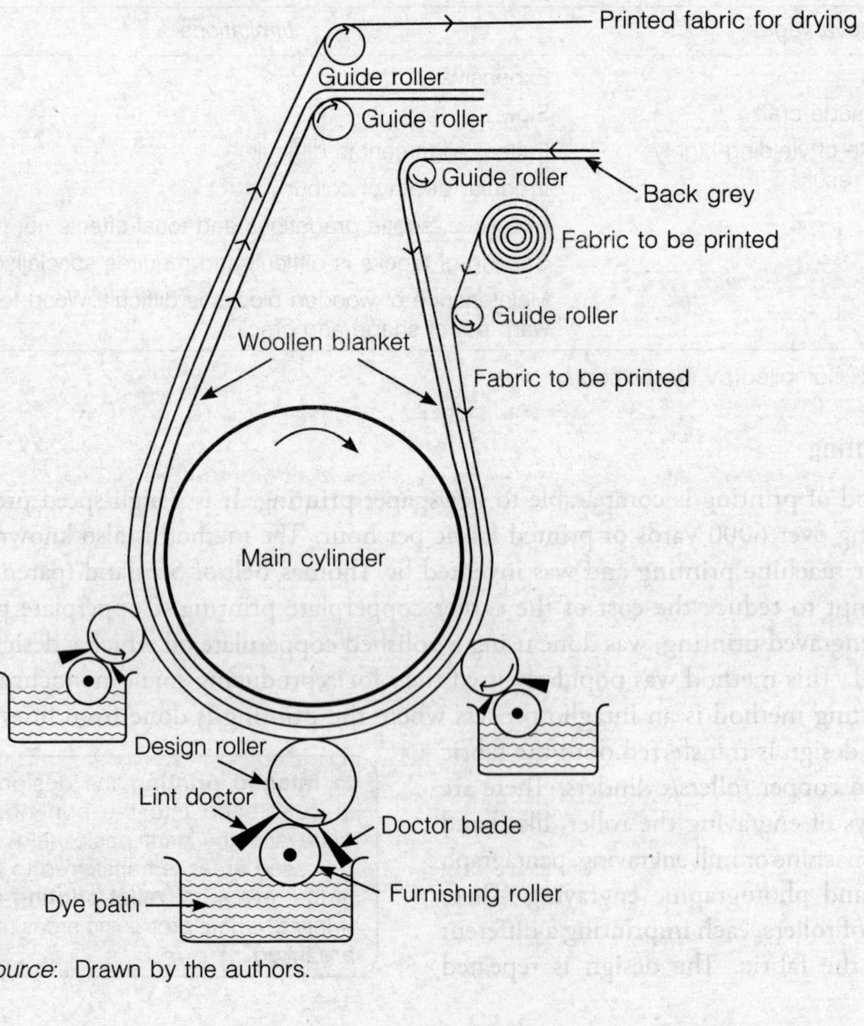

Printed fabric for drying
Guide roller
Guide roller
Guide roller
Back grey
Fabric to be printed
Guide roller
Woollen blanket
Fabric to be printed
Main cylinder
Design roller
Lint doctor
Doctor blade
Furnishing roller
Dye bath

Source: Drawn by the authors.

The central main cylinder/pressure bowl is made up of cast iron and acts as the printing table. This is covered with many layers of a special fabric known as lapping which provides resiliency. The lapping is made up of a fabric which has a warp of linen and weft of wool. This lapping is stretched carefully on the cylinder to avoid any creases. It is covered with a layer of woollen blanket which provides the perfect surface for printing. Nowadays, instead of woollen blanket mackintosh is used. The woollen blanket is covered with back grey which is an unbleached cotton cloth. The last layer is the fabric which is to be printed. Design rollers are arranged around the central cylinder. Up to 16 copper rollers can be used in a machine which implies that one can print a design with up to 16 colours. Below the design roller is a furnishing roller which is a small cylindrical roller covered with nylon bristles appearing as a rotating brush. The furnishing roller rotates in a small tub which contains the dye paste (dye bath). The furnishing roller picks up the colour from the dye bath and transfers it to the design roller.

A sharp steel blade known as doctor blade which has a transverse motion is placed in contact with the design roller just before it touches the fabric. The function of the doctor blade is to remove excess colour from the surface of the design roller which is not a part of the print design. Another blade called lint doctor clears the design roller of any impurities after it prints the fabric. Figure 13.1 depicts the roller printing machinery.

Process of Printing

Once the rollers have been prepared, they are installed in the exact position on the printing machine. The fabric to be printed moves over the rotating central cast iron cylinder, along with the back grey that will absorb excess dye and prevent it from being deposited on the drum. The design roller also rotates and takes up the colour from the furnishing roller. The doctor blade scrapes the excess dye from the roller, and the roller then rotates against the cloth and the design is imprinted. The fabric moves on to the second roller where the second colour is printed and further on in a continuous printing operation. As soon as the entire process is over, the printed cloth is dried and sent to a chamber where steam or heat sets the dye.

Roller printing is superior to other types of printing for fine and precise designs. However, it requires skilled labour and heavy manual work in the changing of the colour troughs and rollers. To keep the print in registration, the rollers must be aligned perfectly. If rollers are not correctly positioned, the resulting print will have one or more colours falling out of position, causing the print to be distorted. Also, production of small quantities of printed cloth is not economical as initial investment of time and money in the preparation of rollers and setting up of the machine is very high. As a result, production of roller-printed fabric has gone down significantly in the past few years. Table 13.3 summarises the advantages and disadvantages of the roller printing process.

> While printing an image that has more than one colour, it is necessary to ensure that each colour overlaps the others accurately. **Registration of the print** implies that each print exactly matches all other prints in the design. In case the prints do not exactly match other prints, the finished image will look fuzzy and blurred, or 'out of register'.

Duplex printing is a type of roller printing in which both the sides of the fabric are printed simultaneously with the same or different patterns. The design has the same clarity of design on both the sides. This type of printing is generally used for printing furnishing and drapery fabrics.

Table 13.3: Advantages and disadvantages of roller printing

Advantages	*Disadvantages*
Subtle gradations of colour possible	Number of colours limited by equipment
Very fine line designs can be produced	Creating engraved rollers is expensive
Less expensive method for large lengths of fabric	Out of register prints
Versatile in colours, pattern and scale	Scale limited by size of roller if rollers are not set properly
High rate of production	Lot of time and skill required in setting up the machine

Source: Compiled by the authors.

Stencil Printing

The art of stencilling is not new. It has been applied to the decoration of textile fabrics from time immemorial by the Japanese. Very fine and intricate patterns were achieved using stencils by the Japanese as early as the eighth century. This method involves cutting out the pattern from a sheet of stout paper or thin metal with a sharp-pointed knife, the uncut portions representing the part that is to be reserved or left uncoloured. The sheet is then laid on the material to be printed and the colour brushed through its interstices. The peculiarity of stenciled patterns is that they have to be held together by ties, that is certain parts of them have to be left uncut, so as to connect them with each other, and prevent them from falling apart in separate pieces. For instance, a complete circle cannot be cut without its centre dropping out, and therefore its outline has to be interrupted at convenient points by ties or uncut portions.

Screen Printing

Screen printing is a stencil process where the design is cut through the surface. The only difference between screen printing and stencil printing is that there are no ties (to hold the stencil together) in screens as in the case of stencils. Screen printing is a method whereby a closely-meshed woven fabric of silk or nylon is mounted on a wooden or metal frame. The design is created by painting out or making non-design portions of the screen opaque, thus preventing the print paste from passing through. Those areas where the print paste passes through will register as the printed pattern. Earlier, screens were made of fine and strong silk fabric and the technique was originally called **silk-screen printing**. Today synthetic fibres are being used. The screen is placed in contact with the fabric to be printed and the print paste forced through the screen by a squeegee. Squeegee is an essential tool used for spreading colour. Also known as a fill blade, squeegee is moved across the screen, forcing or pumping the

> The printing tool **squeegee** has a wooden handle with a flat, smooth rubber blade; it is used to spread the print paste evenly across the back of a stencil or screen.

> **Types of screen printing**
> - Flat-bed screen printing
>   - Hand screen printing
>   - Automatic flat-bed screen printing
> - Rotary screen printing or rotary printing

print paste through the mesh openings during the squeegee stroke. It helps in making a clean image on the printed surface. A screen is prepared for each colour of the design. There are two types of screen printing, each of which embodies the same principle: flat-bed screen printing and rotary screen printing.

Flat-bed Screen Printing

Preparation of Screen

Historically, the method of screen making was simple. A wooden frame was taken and fine silk cloth called 'bolting silk' or organdie fabric was stretched tightly over the wooden frame. Nowadays, to make screens that are much stronger and last longer, the frames are made up of metal rods and the mesh is made up of nylon and polyester. As it is very important to have a tightly stretched mesh, a special machine called stretcher is used. It stretches the warp and weft of the mesh tightly so that it is parallel to the metal frame sides. The mesh is fixed onto the frame using a strong adhesive which is resistant to heat and solvents. Earlier varnish was applied by hand on the background of the design; i.e. the design was drawn on the stretched screen and the background of the screen was covered with varnish, the area to be printed was left unpainted so that the print paste could pass through the fabric from this area.

Nowadays, photo-mechanical techniques are used to make the design on the screen. After the mesh is attached to the frame, it is covered with a light-sensitive emulsion using a flat and soft brush. It is done in diffused light as daylight may damage this photo-sensitive layer. After applying the emulsion the screen is left to dry properly. The same process is repeated for the other side of the screen so that both the sides are coated with the light-sensitive emulsion. Now the design is transferred onto the screen using 'tracings'. **Tracings** are separations of the design according to the colours in the pattern. Each colour has its own tracing. These are made on a transparent film wherein the design areas are covered in black ink and the background is left blank. After the tracing is prepared, it is placed on the emulsion-coated screen and exposed to light. For this purpose special tables which have tube-lights under a glass surface are used. The tracing is placed on this table and the screen is placed over it and then it is exposed for two to three minutes. After exposing, the screen is washed in warm water to remove the unfixed layer of the emulsion. The area which was exposed to light, i.e. the background of the design, is fixed onto the screen and the design area is left uncovered. After washing and drying, the screens are checked for any faults or clogged design areas. These faults are rectified and the screens are ready to be used for printing.

Process of Printing

Hand screen printing is done commercially on long tables that are up to 60 yards in length. The roll of fabric to be printed is spread smoothly onto the table, whose surface has first been coated with a light tack adhesive. The print operators then move the screen frames by hand along the whole table, printing one frame at a time, until the entire fabric is printed (Image 13.3).

Each frame will contain one colour of the print. A three colour print, for example, will require three screens and three applications to the fabric. The rate of production ranges from 50 to 90 yards per hour by this method.

Image 13.3: Screen printing

Source: Authors.

Automatic flat-bed screen printing is like hand screen printing except that the process is automated and therefore faster. Instead of the long table on which the fabric to be printed is spread, the fabric moves on a wide rubberised belt. The screens are mounted above the belt. Like hand screen printing, it is an intermittent process rather than a continuous one. In this process, the fabric moves on the belt. When the area of the fabric to be printed comes in position under the screens, the fabric stops for the lowering of the screens and squeegee action (which is done automatically). After the squeegee action, the screens are automatically raised and the fabric moves again to the next screen frame. The rate of production is about 500 yards per hour. Automatic screen printing is utilised for complete rolls of fabric only.

Rotary Screen Printing

Rotary printing is a continuous process like roller printing. The fabric being printed is moved on a wide rubber belt under the rotary screen cylinders which are in continuous movement. Rotary screen printing is the fastest method of screen printing, with a production of 2500 to more than 3500 yards per hour. Seamless, perforated metal or plastic screens are used. The largest rotary screens have a circumference of about 40 inches and the maximum repeat size of patterns is, therefore, about 40 inches. Rotary screen printing has become extremely popular and has almost completely taken over roller printing. The following are advantages of rotary screen printing over engraved roller printing:

1. It is a faster method of printing
2. Quick changeover of patterns
3. Continuous patterns
4. Does not require highly-skilled workers
5. Economical even for short lengths of fabrics

NEW DEVELOPMENTS IN PRINTING

Apart from the methods described above, there are many new techniques developed to impart designs to the fabric in an interesting and faster way. Some of the popular methods are discussed here.

Pigment Printing

Pigment printing has gained much importance today and for some fibres, such as cellulose fibres, it is by far the most commonly applied technique. Due to increased performance of modern auxiliaries, it is possible to obtain high-quality printing using this technique. Pigments can be used on almost all types of textile substrates. They do not penetrate the fibre but are affixed to the surface of the fabric by means of synthetic resins which are cured after application to make them insoluble. The pigments are insoluble, and application is in the form of water-in-oil or oil-in-water emulsions of pigment pastes and resins.

Pigment printing pastes contain a thickening agent, a binder and, if necessary, other auxiliaries such as fixing agents, plasticisers, defoamers, etc. After applying the printing paste, the fabric is dried and then the pigment is normally fixed with hot air (depending on the type of binder in the formulation, fixation can also be achieved by storage at 20°C for a few days). The advantage of pigment printing is that the process can be done without subsequent washing (which is needed for most of the other printing techniques). The colours produced are bright and generally fast except to crocking.

Heat Transfer Printing/Transfer Printing

Transfer printing literally means moving or transferring a design from one surface to another. In this method, the design is first printed on paper with printing inks containing suitable dyes. The printed paper (called transfer paper) is then stored until ready for use by the textile printer or converter. The dyes suitable for transfer printing should have the following characteristics:

- Excellent colour value
- Excellent heat transfer
- Dyeability
- Excellent colour fastness
- Excellent sublimation fastness
- Excellent printing ink suitability
- The dyes should be transferred at 195–210°C in 20 seconds on polyester fabrics.

Disperse dyes are the only dyes which can be sublimated and are thus the only ones which will respond in a way that permits heat transfer printing. The process is therefore limited to fabrics which are composed of fibres having affinity to this class of dyestuff. This includes acetate, acrylics, polyamides (nylon) and polyesters. The fabric to be printed is passed through a heat transfer printing machine which brings paper and fabric together face to face and passes them through the machine at about 204°C. Under this high temperature, the dye on the printed paper sublimates and is transferred onto the fabric. The process is relatively simple and does not require the skill required when producing roller or rotary screen prints. Table 13.4 lists the advantages and disadvantages of heat transfer printing.

Table 13.4: Advantages and disadvantages of heat transfer printing

Advantages	*Disadvantages*
Low production cost	Expensive short runs
Steaming and ageing not required	Cost of printed paper is high
Better quality prints	Selection of dye needs attention
Well-defined lines	Colourfastness may be poor if dye affinity is absent
No limitation of colours	Not suitable for all types of fibres
Can print both woven and knitted fabrics	
Can be done on fully-fashioned garments also	

Source: Compiled by the authors.

Blotch Printing

Blotch printing is a direct printing method where the background as well as the pattern design are printed onto a white fabric. The background colour is obtained by printing, and not dyeing, process. Blotch-printed fabrics generally have very less colour showing on the reverse of the fabric and thus can be differentiated from dyed ones where the background is visible on the reverse. Sometimes, blotch printing may be done to imitate the more expensive discharge-printed and resist-printed fabrics. On larger background areas there may be a problem of the print colour not covering the ground completely or may do so with uneven depth, thus leading to variations in shade from one point to another.

Digital Printing

Digital fabric printing is a relatively new technology with many applications. The main advantage of this type of printing is that one can produce smaller lengths of fabric of one design which is generally not the case with the other methods. This is possible as no screens or rollers are to be prepared for this method. In other conventional methods longer fabric lengths are produced in order to recover the cost of preparation of screens and rollers. In this method, wide-format inkjet printers—which have been developed to print on various substrates like paper, canvas, vinyl and fabric—are used.

The printing inks used are formulated specially for each type of fibre. The fabric to be printed is fed through the printer using rollers and the printing ink is applied to the fabric in form of several tiny droplets. The design to be printed is developed using any design software like Photoshop or any artwork or photographs can be digitally scanned and printed on the fabric.

After printing, the fabric is steamed in order to fix the colour. This method can be used to create customised fabrics like cushion covers with your photographs. The production is fast and therefore latest designs can be printed. The main limitation of this method is its high initial cost. Once the equipment is setup the running costs reduces with time.

Jet Spray Printing and Polychromatic Printing

Jet spray printing is a method of imparting colours to the fabrics by spraying colours in a controlled manner through nozzles. Polychromatic printing is a technique of jet spray printing which gives multicoloured effects. This method of printing does not require any special equipment like etched rollers or screens to transfer colours to the fabric. The dye is applied to the fabric in the form of streams through jets which are placed above the fabric. The fabric moves continuously through the machine in open-width form. The design is controlled through movement and position of jets and also the action and speed of fabric. Affinity for dyestuff depends on the fibre content and fabric structure as well as the type of dyestuff used. The fabric is then passed through heavy rollers which force the dye into the fabric. This method allows for machine-made tie-dye and batik hand printing effects. The pattern created is visible on both sides of the fabric. The process is fast and versatile.

Electrostatic Printing

Electrostatic printing is a printing technique done without any plate, ink or type form. This process of printing on fabric was patented by the Swiss company Heberlin & Co. They mixed the powdered dye with a carrier such as a resin that has high dielectric properties required for electrostatic printing. The dye-carrier mix is spread on a design screen and the fabric to be printed is passed into an electrostatic field under the screen which is held about ½ inch above the fabric. The dye mixture is pulled through the electrostatic field and passed onto the material in the form of the desired pattern. The fabric is dried with infrared heat followed by a suitable method of fixation depending on the dye-class. Electrostatic printers are widely used for short run printing.

Photo Printing

It is a method of printing in which fabric is coated with chemicals that are sensitive to light. The photographic prints are transferred onto fabric utilising photoengraving photography techniques of colour printing on paper. The results are similar to printing photographs on paper. This is a direct style of printing which does not require a roller or screen and produces permanent designs.

STYLES OF PRINTING

Traditional textile printing techniques may be broadly categorised into four styles of printing.

Styles of printing

1. Direct style
2. Discharge style
3. Resist or reserve style
4. Mordant style

Direct Style of Printing

It is one of the most common techniques of colour design application where the dye is directly applied onto the fabric. Dye is imprinted on fabric in paste form and any desired pattern may be formed using any of the printing methods—block printing, screen printing or roller printing. It is the simplest and oldest style of printing. The characteristic feature of direct style printing is that usually darker coloured prints emerge on lighter ground. It involves transfer of paste-containing dyes to the appropriate areas of the fabric. After drying the required localised dyeing of the fibres occurs during steaming or any other fixation process. Washing follows to remove the paste residue. In the case of pigment printing the pigments adhere to fabric surface with the cured binder film. No additional treatment is needed. In India the block printing of Rajasthan and Khari printing that have been done since ancient times are examples of this technique. Screen printing method is extensively used for direct style. The advantages of this style are:

- It is the easiest printing style to operate.
- Least expensive.
- Suitable for the printing of both simple and complicated designs.

Discharge Style of Printing

In this method the fabric is dyed and then printed with a chemical that will destroy the colour in design areas. This discharging of colour from previously dyed ground is carried out by a discharging agent which is actually an oxidising or a reducing agent capable of destroying colour by oxidation and reduction. Examples of these discharging agents are: potassium chlorate and sodium chlorate (oxidising agents), and Rongalite-C (sodium formaldehyde sulphoxylate) and stannous chloride (reducing agents). The effects produced are very striking as they have the characteristic white colour pattern on a darker background. However, due to the use of discharging agents the strength of the fabric may be affected.

In **colour-discharge printing**, one or more colours may be applied producing a coloured design instead of white on the dyed ground. This is achieved by adding a dye that is resistant to the bleaching action. The printing paste therefore contains the resistant dyestuff and the discharging agent. On application, while the discharging agent of the print paste discharges the ground colour at the printed area, the dyestuff deposits and fixes itself on the ground. As a result, colour discharge effect is obtained.

Resist Style of Printing

In resist printing the fabric is first printed with an agent that resists either dye penetration or dye fixation. During subsequent dyeing, only the areas free of the resist agent are coloured. The resisting agents employed function either mechanically or chemically or, sometimes, in both ways. The mechanical resisting agents include waxes, fats, resins, thickeners and pigments—such as china clay, the oxides of zinc and titanium, and sulphates of lead and barium. Such mechanical resisting agents simply form a physical barrier between the fabric and the colorant. They are mainly used for more decorative styles in which breadth of effect and variety of tone in the resisted areas are of more importance than sharp definition of the pattern. Chemical-resisting agents include a wide variety of chemical compounds, such as acids, alkalis, various salts, and oxidising and reducing agents. They prevent fixation or development of the ground colour by chemically reacting with the dye or with the reagents necessary for its fixation or formation. After the fabric has passed through the subsequent dyeing process, the resist paste is removed, leaving a white or light coloured pattern on a dark background. Some variations of resist printing are batik, tie and dye, dabu printing and ikat.

Batik printing is a hand printing cottage-industry that originated in the island of Java. This resist method of printing designs using rice starch and wax was known to them for about two millennia. It is believed that the art of batik printing has been practised since ancient times in some parts of southern India as well, especially in the Cholamandal village near Chennai. In this style, designs are made using wax as resist (Image 13.4) on the fabric which is subsequently immersed in the dye bath to absorb the colour on the unwaxed portions. The application of wax is done using various tools, such as brushes, tjap and tjanting. **Tjap** is a copper stamp (Image 13.5) and **tjanting** is a spouted tool which are used to draw dots and lines of the resist on the fabric (Image 13.6).

Batik gives an artistic effect with fine lines running irregularly across the fabric. And, it is a cheap method of printing. However, there are certain limitations. It is a very laborious and time-consuming process. Also, the dye has to be applied at a temperature lower than the melting point of wax.

Image 13.4: Application of wax resist before batik printing

Source: 'Batik-coloracio' by Jolle~commonswiki (CC by 3.0 SA; Wikimedia Commons).

Image 13.5: Tjap

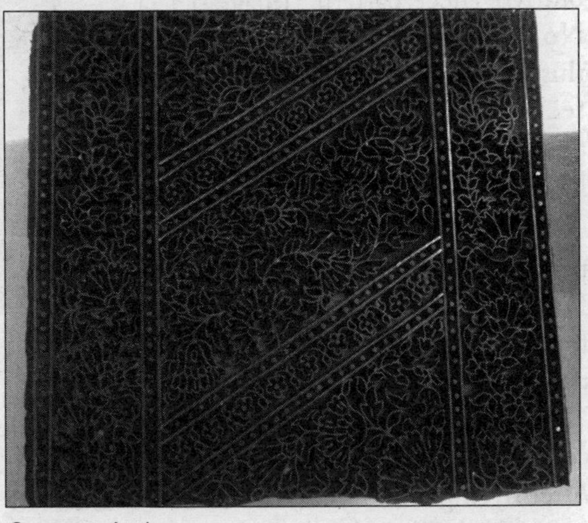

Source: Authors.

Image 13.6: Tjanting

Source: 'Batik 1' by Nalasa files (CC by SA 4.0; Wikimedia Commons).

Tie and dye, another type of resist style, is similar to batik printing. But here the dye is resisted by tying the fabric using various techniques before it is put in the dye bath. The untied portions take the colour but the tied areas do not allow penetration of the dye. This produces the characteristic blurred effect pattern (Image 13.7). No machine cost is involved in this method. Though it is a simple technique, it is time-consuming and laborious, and involves skilled labour.

Image 13.7: Tie and dye fabric

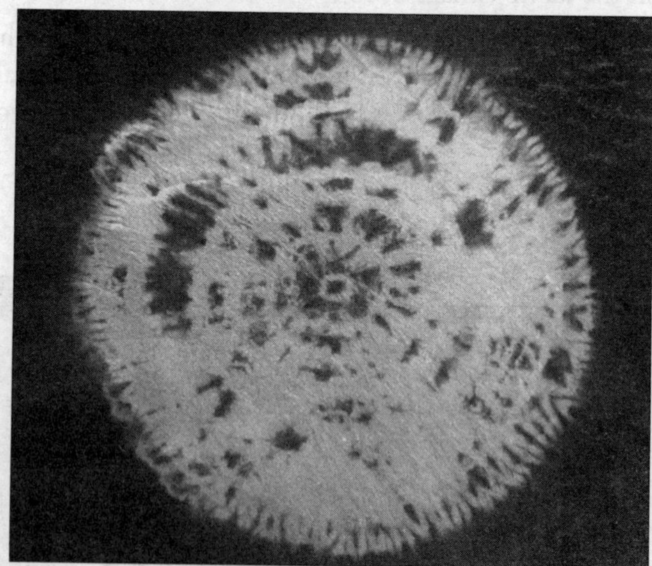

Source: Authors.

Dabu is a mud-resist hand-block printing style practised in Rajasthan. Dabu printing is also a unique art form. In this, a design is drawn onto the background cloth. This sketched design is covered with clay on which sawdust is sprinkled. The saw dust sticks to the cloth as the clay dries. Thereafter, the entire cloth is dyed in various colours. The area where clay and sawdust mixture is present resists the dye and remains colourless. After dyeing and drying, the cloth is washed to remove the clay and the mixture. The final print resembles batik style.

Ikat is another variation of resist style of printing. In this method, woven fabric is produced by using tie-dyed warp yarns or weft yarns or both. The effects are very beautiful and resemble prints though the design is actually achieved only after weaving (Image 13.8). A lot of planning goes into production of ikat textiles. These are produced in Gujarat, Andhra Pradesh and Odisha.

Image 13.8: Ikat fabric

Source: Authors.

Mordant Style

This style is also known as the 'dyed style of printing'. **Mordants** are chemical or natural substances used in printing and dyeing with natural dyes which serve as bridge links between the dye and the fibre. The mordant is printed in the desired pattern prior to dyeing the cloth; and during dyeing the colour adheres only where the mordant has been printed. The choice of mordant depends on the type of dye, fibre and the colours desired. Many natural mordants—such as alum, myrobalan, pomegranate and gallnuts—are high in tannins; they are eco-friendly and help to give very rich colours. Natural dyes produce interesting colours when they are combined with mordants; for example, alizarine dye with mordants such as alum, potassium dichromate and stannous chloride produces red, brown and pink colour, respectively.

DYE FIXATION

After printing, to retain the printed mark, drying is an essential operation. It is essential to remove the moisture applied from the print paste after printing in order to avoid marking off and staining of the unprinted areas. Drying prevents print paste run off from the patterned areas. Further print fixing treatments are given to transfer the dyestuff from the dry thickener film to the cloth. There are many ways of dye fixation depending on the dyestuff class and the substrate. Some of them are listed below

Dye fixing methods

1. Baking
2. Wet development
3. Pad-batch method
4. Normal steaming
5. Pressure steaming
6. High temperature steaming
6. Ageing

- *Baking*: This method is carried out in a gas or an electric baking machine that utilises dry heat. It is used for fixing pigment prints and reactive dyes. The dye is fixed in ventilated chambers filled with air heated at 160–180°C. Fabric moves continuously for 2–5 minutes on the rollers in the machine. The temperature and time may vary for different dyes and fabrics. On smaller printed areas baking treatment may be reproduced by ironing on the back side of the print at a temperature suitable to the fabric. Larger printed pieces may be folded and baked in an electric oven but that is not a very safe process.

- *Wet development*: In this process, the printed and dried fabric is passed in open width through a rectangular tank containing an alkaline solution and large quantities of electrolyte to suppress dyestuff bleeding. It has limited scope as it is a suitable method for fixing only certain reactive and azoic dyes on cellulosic fabrics. The treatment is carried out for 10–20 seconds at 100–110°C and is immediately followed by washing. The composition of the wet development bath, time and temperature vary with the type of dyestuff to be fixed.
- *Pad-batch method*: This process is recommended for cotton fabrics printed with reactive dyestuff. The printed fabric is first padded with sodium silicate on a padding mangle and then the fabric is immediately passed on to a perforated beam where it remains for anywhere between 10 minutes and 3 hours, depending on the type of reactive dye used. Then it is washed thoroughly in cold water. Pad-batch method can also be used for fixation of reactive dyes on wool in the presence of urea or after plasma treatment. This fixation technique should not be used for viscose as sodium silicate reacts with the fibres and affects the handle of the fabric.

Plasma treatment of textiles is done using a controlled stream of ionised gas which selectively modifies the substrate. This treatment improves the wettability, and thereby the dyeing, printing and finishing properties.

- *Normal steaming*: This method is carried out in industrial steamers at or slightly above atmospheric pressure and a temperature of about 100°C. It requires abundant supply of air-free steam. During normal steaming the printed areas absorb moisture and form a localised dye bath which is like a gel. The dye gets dissolved in this mini dye bath and diffuses inside the fibres which have also swollen in the meanwhile. The thickening agent prevents the dyestuff from spreading outside the printed area, i.e. it prevents 'bleeding'. If the print paste contains more quantity of the hygroscopic agent or if the steam is too moist, then the thickening agent becomes diluted and bleeding will occur. Also, if the steam is too dry, then the moisture absorbed by thickening agent is not sufficient and the dyestuff will not fix suitably.
- *Pressure steaming*: This method is done using steam under pressure which hastens the fixation of disperse dyes on polyester and acetates. This system produces full colour yield and bright prints. It is a batch process and needs more labour. The cost of production is high as it requires grey cloth for supporting the printed material. It is carried out in star steamers. The star steamer is a simple equipment with radiating arms for carrying the cloth. During the process, dry steam is fed at the bottom and the inside is kept completely moisture-free and under pressure after allowing all the air and condensate to go out.
- *High temperature steaming*: This rapid dye fixation method involves the usage of superheated steam. Superheated steam is high-temperature vapour that is generated by heating

saturated steam which is in turn obtained by boiling water. Superheated steam is dry and can therefore lose heat without condensing, i.e. changing state from gas to liquid. In this process live steam is raised to 180°C at atmospheric pressure by bringing it in contact with radiators which are at a temperature of over 200°C and passed over dyed products. High temperature steaming method can be carried out on many classes of dyes and fibres. Disperse dyes can be fixed in these steamers at 180°C in 1–2 minutes. It is a continuous process and gives full colour yield and bright prints. Use of superheated steam shows advantages of faster heating, shorter fixation time and less colour spread if the print has not been dried and also in the pad–steam situation, where there is usually more than sufficient water in the fabric.

- *Ageing*: Ageing is similar to steaming, with the only difference being the duration of treatment. Ageing is carried out in steamers for shorter duration ranging from 1–14 minutes in a flash ager, while steaming is a relatively longer treatment that takes anywhere between 45 minutes and 2 hours to finish. Ageing is generally used for vat dyes on cellulosic fabrics. The dye is printed on the fabric and padded with thickened solution of sodium hydrosulphite and alkali, followed by steaming for a very short period, generally 20–40 seconds, at 125–110°C.

Washing Off

The final treatment given to all printed textiles is washing off. Washing off is a physio-chemical process which removes the dirt and other substances from the textile and improves the texture. It involves rinsing the fabric in water to remove the print paste chemicals and additional unfixed dyestuffs, followed by a soaping treatment which is suitably selected according to the type of dyestuff and fabric. The steps involved are:

- Preparing the washing liquor using suitable detergents
- Achieving the required temperature and wetting the textile
- Emulsifying impurities after separating them from the textile
- Removing the liquor from the textile by rinsing
- Drying

Industrially, rinsing and soaping treatments are given either in open-width form or in rope form. In the case of delicate fabrics, the open-width method is preferred it avoids creases and wrinkles and permits processing of huge lots. However, rope-form is more effective as it provides a stronger mechanical action and therefore ensures better cleaning and relaxation of the fabric structure.

SUMMARY

- Printing can be described as localised dyeing. Here a concentrated dye paste is applied to the fabric along with a few with other additives.

- The various steps in printing are preparation of fabric, preparation of print paste, printing of fabric, drying after printing, fixing of prints, and washing off.
- The methods of printing refer to the tools or machines used for printing. Block printing, roller printing, screen printing and transfer printing are some methods of printing.
- Block printing is the simplest and oldest of the printing techniques. Here wooden blocks are used to print the fabric.
- Roller printing method is an intaglio process where printing is done using copper-engraved rollers/cylinders.
- Screen printing is a stencil process where the design is cut through the surface.
- Transfer printing involves transferring of a design from paper to fabric.
- Different styles of printing are direct style, discharge style, resist style and dyed/mordant style.

KEY WORDS

Print paste: A thickened solution of dyes/pigments and auxiliaries that is applied to the fabric using various techniques of printing.

Thickener: It is the main component of the print paste responsible for the viscosity and consistency of the print paste.

Sublimation: It is the transition directly from the solid to the gas phase without passing through the intermediate liquid phase.

Mordant: Mordants are chemicals or natural substances used in printing and dyeing with natural dyes. They serve as bridge links between the dye and the fibre.

Engraving: The cutting or etching of design into copper rollers is known as engraving.

Steaming: It is a fixation step carried out after printing of goods to fix the dye permanently onto the fabric.

EXERCISES

1. Define the following terms:
   a. Thickener
   b. Hygroscopic agent
   c. Resist style
   d. Ageing
   e. Discharging agent
2. Differentiate between:
   a. Styles of printing and methods of printing
   b. Resist style and dyed style
   c. Block printing and roller printing
3. What is the general composition of the print paste? Explain the importance of a thickener in a print paste.
4. Explain the process of block making. What are the requirements for a successful block print?

5. Give details of a roller printing machine with illustration.
6. Compare rotary screen printing and automatic flat-bed screen printing.
7. Explain the process of discharge style of printing.
8. What are the advantages and disadvantages of heat transfer printing?
9. Why is after-treatment of printed fabrics important? List the various after-treatment methods?
10. Enumerate recent developments in textile printing technology.

REFERENCES

Corbman, B. P. 1983. *Textiles: Fiber to Fabrics*. Sixth edition. Singapore: McGraw-Hill Book Co.

Joseph, L. Marjory. 1988. *Essentials of Textiles*. Fourth edition. New York: Holt, Rinehart and Winston Inc.

Kadolph J. Sara. 2009. *Textiles*. Tenth edition, Indian edition. New Delhi: Dorling Kindersley India Pvt. Ltd.

Miles, L. W. C. 1994. *Textile Printing*. Bradford, UK: Society of Dyers and Colourists.

Tortora, P. G., 1987, *Understanding Textiles*. New York: Macmillan.

ONLINE SOURCES (all accessed June 2016)

http://textilelearner.blogspot.in/2012/06/resist-printing-process-methods-of.html

http://library.iitd.ac.in/cd/1470-1482.pdf

http://uctemtn.blogspot.in/2012/12/dyeing-and-printing-overview.html

http://www.teonline.com/knowledge-centre/printing-textile-services.html

http://textilelearner.blogspot.in/2012/03/textile-washing-treatment-sequence-of.html

14

LABELLING AND STANDARDISATION IN TEXTILES

HIGHLIGHTS

- Labelling and types of labels
  - Content label
  - Manufacturer's label
  - Size label
  - Brand labels
  - Trademark or certification label
  - Grade labelling
  - Care labels
- Textile standardisation and organisations involved
- Consumer awareness and protection

Throughout this book we have seen many varieties of fibres and blends, and various manufacturing and processing techniques. These fibres and processing techniques are responsible for the characteristics of every textile product, i.e. its appearance, durability and suitability. Therefore, it is important that consumers are accurately informed about the product. The consumer has the right to know the composition, quality, care, price, etc. of the product (s)he intends to buy. In the absence of this information, the consumer may make the wrong choice and hence will not be able to get their money's worth. To avoid this, various governmental agencies, product manufacturers, retailers and consumer groups work together to establish labelling conventions and performance standards for textile products.

LABELLING

Labels describe the product and communicate factual information about its service qualities. This information is based on standard laboratory tests conducted by recognised institutions or organisations. Manufacturers should ensure that labels are attached permanently and securely and are legible during the useful life of the product; they should also be easily found by consumers at the point of sale. Moreover, additional information must appear on the outside of the package. Labelling is important because of the following reasons:

- Product specifications help the consumer in making an informed choice of product for a specific use.
- Provision of compliance to standards ensures that the product is reliable and will satisfy the consumer in terms of performance, safety, durability, etc.
- Care directions ensure that the products are maintained in a good condition for a longer duration.

Types of labels

- Content label
- Manufacturer's label
- Size label
- Brand label
- Trademark or certification label
- Grade label
- Care label
- Smart label

Different types of labels are attached to various types of garments. These are either square or rectangular in shape, attached inside the garment at the back neck or side seam in upper garments and at the back waistband in lower garments.

Content Label

Content label includes information of the percentage weight of each fibre present in the textile product along with its generic name in the descending order. If the presence of any fibre is less than 5 per cent by weight, it would be clubbed under 'Other fibre(s)'. However, if that fibre significantly affects the performance of the product, it should be specified. For example:

45% Nylon	45% Nylon
45% Polyester	45% Polyester
5% Rayon	5% Rayon
5% Other fibres	4% spandex, for elasticity
	1% Other fibres

Along with these pieces of information, the label also specifies the country of origin for imported fabrics and the country where the product was sewn in the case of readymade garments. For example, a label may read: 'Made in USA of fabric imported from Great Britain'.

Manufacturer's Label

Manufacturer's label provides information about the name and address of the manufacturer or the company producing the product. It also includes the Registered Identification Number (RN), a code given by USA's FTC (Federal Trade Commission), which helps identify the manufacturer.

Size Label

Size label designates specific set of body measurements which guide the consumers to select the best-fitting and comfortable garment. These sizes can be either numeric (8, 10, 12, 14, etc.) or letters (S [small], M [medium], L [large], XL [extra large], etc.). These size specifications vary from brand to brand depending upon the sizing systems followed by them. Besides, the size range for infants are in terms of months, as in 3 months, 6 months, 9 months, 12 months, etc. and for toddlers in terms of 1T, 2T, 3T or 4T.

Brand Labels

Brand label is one of the most important aspects of marketing which provides recognition to the product. Consumers identify with the name or the logo of the company which is in the form of a specific symbol, word or colour. The brand label offers assurance regarding the quality and performance of the product. For example, many popular company brands are associated with the suiting materials in India—Raymond, Grasim, Gwalior, Vimal, Belmonte, Reid & Taylor, Mayur, Siyaram, etc. Similarly, with readymade garments—Levis is associated with jeans, Tommy Hilfiger and Polo with shirts. Reputed brands have their logo on the right side of the product which grant these products their distintive identity and status symbol.

Sometimes, a small label called a **flag label** is attached outside at the side seam of the garment. It has a brand logo and is used as a design or distinguishing feature for various branded garments.

Trademark or Certification Label

Trademark or certification labels, also called special labels, can be a word, a symbol or a combination of both. Manufacturers use these marks or labels to certify its quality and add a distinct reputation to a particular product. These labels attract the attention of the consumer and help them to identify the product they wish to purchase within the product line of a particular company. Some trademarks are seal of cotton, woolmark, silk mark, handloom mark and ecolabels.

Seal of Cotton

The Seal of Cotton trademark is a trusted and recognised trademark for 100% cotton products made from cotton. This trademark is registered with Cotton Incorporated, a company representing US cotton producers and importers of cotton products, which was established in 1970. It aims to improve the profitabilty of cotton for the global cotton supply chain, and to promote interest in cotton products among consumers. As a brand, the Seal of Cotton trademark has recognition value and appeal to consumers, and enables them to readily identify apparel and other products made with cotton. Variations of the Seal of Cotton trademark were later developed to denote different percentages of cotton in products and, in some cases, product categories (Table 14.1).

Woolmark

Woolmark is owned by The Woolmark Company, the world's leading wool textile organisation. This mark guarantees consumers of the fibre content of the product and is an assurance of excellent quality. Each of the three logos developed by the company, i.e. woolmark, woolmark blend and wool blend (Table 14.2), is licensed separately.

Table 14.1: Seal of cotton trademarks

Name of trademark	*Seal of cotton trademark*	*Natural blend™ trademark*	*Cotton enhanced™ trademark*
Logo of trademark	cotton ®	cotton natural blend ™	cotton enhanced ™
Description	Fabrics and end products of 100% cotton	Fabrics and end products that contain a minimum of 60% cotton and feature performance characteristics	Specific to non-woven products like baby wipes containing a minimum of 15% cotton

Source: Cotton Incorporated.

Table 14.2: Woolmark, woolmark blend and wool blend trademarks

Name of trademark	*Woolmark*	*Woolmark blend*	*Wool blend*
Logo of trademark	® PURE NEW WOOL	® WOOL RICH BLEND	® WOOL BLEND PERFORMANCE
Description	This indicates that the product contains 100% pure new wool and meets and range of performance measures.	This indicates that the product contains 50%–99.9% new wool and meets and range of performance measures.	This indicates that the product contains 30%–49.9% new wool and meets and range of performance measures.

Source: Courtesy of The Woolmark Company; The Woolmark logo is a Certification mark in many countries.

The Woolmark Licensing Program, that awards these trademarks to products from various companies, is based on the exacting demands of today's customers and ensures that any product bearing its logo meets wool content, quality and performance criteria strictly.

Silk Mark

Silk Mark (Image 14.1) was launched on 17 June 2004 by Silk Mark Organisation of India (SMOI), a joint effort of the Central Silk Board, Ministry of Textiles and the Government of India. When a piece of textile is marked with this trademark, it is certified as being made from 100 per cent natural silk. This trademark helps promote fair practices in the sales of pure silk.

It also safeguards the interest of genuine traders and consumers of pure silk.

Image 14.1: Silk mark logo

Source: Silk Mark Organisation of India (SMOI). Reproduced by permission.

Handloom Mark

Handloom mark has been devised to promote genuine handloom products and assure consumers of the quality of the product in domestic and international markets. It has been implemented by the Textile Committee of the Government of India. There are two forms of handloom marks as shown in Image 14.2. A logo with the word 'handloom' written beneath it is placed on handloom products for domestic use, whereas, the same logo with the word 'hand woven IN INDIA' written beneath it is used for internationally-marketed handloom products.

Image 14.2: Handloom logo

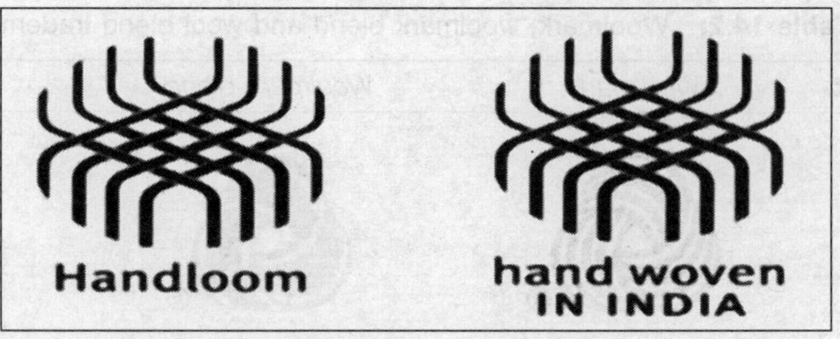

Source: Ministry of Textiles, Government of India.

Ecolabels (Environment Friendly Certification)

Any product which is made, used or disposed of in a way that is significantly less harmful to the environment could be considered as being an environment-friendly product. Increasing awareness to produce environmentally-safe products has resulted in the launch of environmental labelling programmes in 15 countries (including India) across the world. These trademarks for easy identification of environment-friendly products include 'Blue Angel' in Germany, 'Environmental Choice' in Canada, 'Green Seal' in United States and 'Austrian Eco-label' in Austria. In India, textiles are one of 16 product categories for which the notification of final criteria for environment-friendly certification has been laid down so far; and the eco-certification scheme launched in 1991 in this regard is known as 'Ecomark' (Image 14.3).

Image 14.3: Logo of 'Ecomark'

Source: The Ministry of Environment and Forests, Government of India.

Grade Labelling

A **grade label** is a grading mark on the commodity itself, or on a tag or label attached to the commodity that guides the consumer in selecting from a range of products. **Grading** implies a qualitative judgement of relative inferiority or superiority of a product. For example, wool fibres are graded as fine, medium, long and carpet wool on the basis of their length and fineness. Similarly, a fabric that has a high thread count is graded as better quality as it will be strong and durable. These grades are established on the basis of some basic quality or performance parameters. For example, 'pima cotton, 2×2 mercerised broadcloth, sanforised' means that the fabric is made of 2-ply yarn having extra-long cotton fibres twisted together for durability. The process of mercerisation has added strength and lustre to it. Moreover, it has minimum shrinkage (less than 1 per cent residual shrinkage) as it has been sanforised.

The Global Organic Textile Standard (GOTS) defines requirements that ensure that textiles are organic right from the cultivation of raw materials to the final product in order to provide credible assurance to the end consumer. Only those textiles which are produced and certified according to the provisions of the standards of GOTS can carry the GOTS grade label. The standards provide for a subdivision into two label-grades:

Label-grade 1: Organic

≥ 95% certified organic fibres; ≤ 5% non-organic natural or synthetic fibres

Label-grade 2: Made with X% organic

≥ 70% certified organic fibres; ≤ 30% non organic fibres, but a maximum of 10% synthetic fibres (up to 25% for socks, leggings and sportswear) (GOTS n.d.)

Care Labels

Another important part of informative labelling is the care label which describes the treatment procedures that would not damage the garment. The Federal Trade Commission (FTC) defines care label as 'a permanent label or tag, containing regular care information and instruction attached or affixed in some manner that will not become separated from the product and will remain legible during the useful life of the product' (ASTM 2014). These instructions should be applicable to the whole garment, including trimmings, zippers, linings, buttons, etc. Care labels are often a deciding factor when consumers shop for clothes. While some consumers look for the convenience of dry cleaning, others prefer the economy of washable garments.

The FTC established Permanent Care Labelling Rule on 3 July 1972, which required manufacturers and importers of textile apparel and of piece goods used for making apparel, to attach care instructions to garments. Updates to the Rule became effective on 1 September 2000. The items that are covered under the rule include, either wearing apparel and the piece goods made completely of textile product or items made primarily of textile product and some portions trimmed with genuine fur, leather or suede (e.g. textile coats with fur or leather patches on textile jackets). The following items are exempt under the Permanent Care Labelling Rule:

- Those items which are composed primarily of genuine fur, leather or suede and only partially of textile products (e.g. leather coats with textile linings).
- Textile accessories like hand wear (e.g. gloves), headwear (e.g. hats) and footwear (e.g. sneakers).
- Accessories like handkerchiefs, belts, suspenders and neckties which are not used to cover or protect parts of body.
- Products which do not require laundering and dry cleaning.
- Non-woven garments made for one-time use only.
- Piece goods sold for making apparel at home.
- Marked manufacturers' remnants up to 10 yards when the fibre content is not known and cannot be determined easily.
- Trim up to 5 inches wide.
- Articles intended to be sold at retail for $3.00 or less and which are completely washable under all normal circumstances.

Besides, few items don't need permanent care labels, but must have conspicuous temporary labels at the point of sale. These include totally reversible clothing without pockets, see-through garments and those garments which can be washed, bleached, dried, ironed or dry cleaned by the harshest procedures available. There are different standards for care labels for the different countries/regions of the world, such as the International Care Labelling System (GINETEX), ASTM Care Labelling System, British Care Labelling System, Canadian Care Labelling System, Indian Care Labelling System, etc. ASTM Care Labelling System is widely used and allows participation in a global marketplace.

ASTM Care Labelling System

The care labelling system developed by ASTM is an internationally-accepted system which provides a uniform system of symbols for the disclosure of care instructions on textile products. The best or the most economical care instructions are conveyed in easily understood, space-saving pictorial format that is not dependent on language. In this system there are five basic symbols: washtub with a water wave, triangle, square, iron and circle indicating the process of washing, bleaching, drying, ironing or pressing and dry-cleaning respectively. The consumer guide for ASTM Care Labelling System has been summarised in Table 14.3. Any additional symbols inside or outside the basic symbol represent the specific action required for that particular textile product. Temperature settings are indicated as numerical or as dots if regular use of high temperatures would harm the product.

If any of these five care procedures adversely affect the quality or performance of the textile product, then that process should be superimposed with the prohibitive symbol 'X'. For example, if ironing would harm a garment, the label should state 'Do not iron'. Also if the product is not colour fast, the label must say, 'wash with like colours' or 'wash separately' to avoid causing damage to other products while washing.

Table 14.3: Consumer guide for ASTM Care Labelling System

Washing	*Bleaching*	*Hand drying*	*Tumble drying*	*Ironing*	*Dry cleaning*
Machine wash, permanent press cycle (reduce mechanical action)	Any commercially-available bleach product may be used in the laundering process.	Drip dry, hang soaking wet	Tumble dry, normal (dried by machine)	Iron, any temperature, steam or dry	A Dry clean in any solvent
Machine wash, gentle or delicate cycle (greatly reduce mechanical action)	Cl Chlorine bleach may be used	Hang to dry after removing excess water	Tumble dry, normal, high heat	Hot (210°C/410°F)	P Dry clean in any solvent except trichloroethylene
95°C Max. temperature 95°C/203°F	Only non-chlorine bleach when needed	Dry flat after removing excess water	Tumble dry, normal, medium heat	Warm (160°C/320°F)	F Use fluorocarbon or petroleum solvent only
70°C Max. temperature 70°C/160°F	Do not use chlorine bleach	Dry in shade	Tumble dry, normal, low heat	Cool (120°C/248°F)	Dry clean, short cycle
60°C Max. temperature 60°C/140°F			Tumble dry, normal, no heat	Do not steam while ironing	Dry clean, reduced moisture

(*Contd.*)

Table 14.3: (*Contd.*)

Washing	*Bleaching*	*Hand drying*	*Tumble drying*	*Ironing*	*Dry cleaning*
50°C Max. temperature 50°C/122°F			Tumble dry, permanent press	Do not iron	Dry clean, low heat
40°C Max. temperature 40°C/104°F			Tumble dry, gentle		Dry clean, no steam
30°C Max. temperature 30°C/85°F			Do not tumble dry		Do not dry clean
Hand wash			Do not dry		
Do not wash			Do not wring		

Source: Compiled by the authors.

Indian Care Labelling System

The Indian Care Labelling System uses basic symbols like wash tub for washing, triangle for bleaching, iron for ironing, circle for dry-cleaning and circle inside a square for tumble drying. Any of the above symbols with a cross marked on it indicates that the process is not permitted on that particular garment. In case of delicate garments, a bar (for mild treatment) or a broken bar (for very mild treatment) is depicted under the symbol of wash tub or circle (Chatterjee et al. 2006).

Smart Label

Smart label is a creative bridge between fabrics and electronic media. Each smart label contains a unique Quick Response Code (QR Code) that can be scanned using any smartphone, iPad or tablet that is equipped with a QR reader. Once the code is scanned, the phone's browser will direct the user to a webpage, which gives information about the product. Smart labels give designers and brands greater flexibility over a standard woven label because it gives the opportunity to offer the information in great detail.

TEXTILE STANDARDS

A **standard** is a defined level of performance that a product must achieve in order to be considered as acceptable for use (Tortora 1978: 35). The prevalence of textile standards is beneficial for the consumers and makes it easier for them to compare the products before buying. **Company standards** are the set specifications applicable to that particular company in terms of product development, production, purchasing and quality assurance. **Industry standards** include common specifications followed by all the member companies in that particular industry. **Voluntary standards** include those standards which may be adopted by an individual or a company without any outside force; ASTM standards are examples of this type. **Mandatory standards** are imposed by law and should be adopted by all the producers of that particular product; these include standards related to fire safety and health issues.

International Organisations Engaged in Developing Standards

Individuals and companies come together with a common motive to develop and assess the quality standards of textile materials. There are several organisations working at the international level, and a few of them are:

- *ASTM International*: Established in 1898, American Society for Testing and Materials (ASTM) came to be known as ASTM International in 2001. It is a globally recognised non-governmental body which works towards development and delivery of international voluntary standards. It has 143 technical standard writing committees serving diverse industries. **ASTM International Committee D-13** takes care of the various test methods, specifications, guides and practices that support industries and governments in the field of textiles. The focus is only on identifying and assessing performance characteristics related to physical-mechanical procedures.

- *Society of Dyers and Colourists (SDC)*: SDC was established in 1884 as a professional, chartered society with its head office in Bradford, United Kingdom. It strives to develop eco-friendly, cost-saving solutions for dyeing and finishing. It is also responsible for evolving standard test methods for colour fastness and colour measurement.
- *American Association of Textile Chemists and Colorists (AATCC)*: Founded in 1921, AATC is today the world's leading non-profit association serving textile professionals. The focus of the association is to develop standard methods for assessing characteristics and performance of textile products during wet processing, i.e. while dyeing and finishing using water or any other liquid solvents.
- *American Society for Quality (ASQ)*: ASQ facilitates continuous improvement in the concepts and technology used for promoting quality of manufactured goods and services. Among its 15 divisions, the Textile and Needle Trade Division takes care of textile products.
- *American National Standards Institute (ANSI)*: ANSI was founded in 1918 as a private, non-profit institute. It is involved in the creation of thousands of norms to ensure safety and health of consumers and the protection of the environment in nearly every sector. It has an accredited certification programme that assess conformance to various standards for textile products as well.
- *International Organization for Standardization (ISO)*: ISO was founded in 1947 as an independent, non-government organisation comprising members of the national standard bodies of 164 countries. It has published more than 19,500 international standards, which help in breaking down barriers to international trade, in various fields. Two major ISO programmes are ISO 9000 for quality management and ISO 14,000 for environment management.
- *Textile/Clothing Technology Corporation ([TC]^2)*: Established in 1981, [TC]^2 emphasises on making the sewn products industry more productive and competitive by providing 3D body scanning solutions. It also provides assistance and innovative approaches for product development. Various efforts are made by the corporation to offer customised educational programmes for the sewn product industry developed through integrated research.
- *American Apparel & Footwear Association (AAFA)*: AAFA was founded in 2000 through the merger of two highly regarded trade associations—American Apparel and Manufacturers Association, and Footwear Industries of America. This collaborative forum promotes best practices and innovation in apparel, footwear and other sewn product companies.

National Organisations

There are several organisations and associations which are working for development in the field of textiles throughout India. These can be categoried as advisory bodies, export promotion councils, textiles research associations, statutory bodies, public sector undertakings and autonomous bodies.

Few of these organisations, specifically involved in research and development in different areas are as follows:

- *Bureau of Indian Standards (BIS)*: Formerly known as Indian Standards Institution (ISI), BIS was set up in 1947 as a corporate body with 25 members drawn from the central and state governments, industry, scientific and research institutions and consumer organisations. More than 18,000 standards have been formulated by BIS, which are primarily voluntary in nature. Only 68 standards are subject to mandatory certification. It also offers third-party assurance and certification schemes based on well-defined scheme of testing and inspection for the products and systems.
- *Apparel Export Promotion Council (AEPC)*: Apparel Export Promotion Council (AEPC) was established in 1978 to provide assistance to the exporters and buyers of Indian garments to encourage trade. Statistical records maintained up to date by the APEC help exporters access information on the trade potential of international markets and buyers choose the best sourcing destinations. It also facilitates the participation of its members in international fairs and trade delegations held globally. Besides, AEPC is making efforts to provide trained manpower to the apparel industry for its overall development.
- *All India Handlooms Board*: All India Handlooms Board was set up as an advisory body under the chairmanship of the Minister of Textiles in 1945 with the Development Commissioner (Handlooms) as the Member-Secretary. The Board advises the Government on various aspects of development of the handloom sector. It studies the various aspects of development like technical, marketing, organisational, artistic, etc. and suggests measures for further improvement in the handloom sector.
- *Textile Industry Research Associations*: The following are a few well-equipped textile research associations that carry cutting-edge research and provide support services to the Indian textile industry:
  - Ahmedabad Textile Industry Research Association (ATIRA), an autonomous non-profit association for textile research
  - Bombay Textile Research Association (BTRA)
  - South India Textile Research Association (SITRA) headquartered in
  - Northern India Textile Research Association (NITRA) headquate'

All these organisations strive for the highest standards of excellence ir and management through application-oriented scientific studies. T activities include ensuring optimum utilisation of men/machines/ma and product development, testing and consultancy services, ene energy sources, etc.

- *Central Institute for Research on Cotton Technology (* as a unit under the Indian Council for Agricultur

It is engaged in developing new technologies for improving the quality of cotton and also developing test methods and standards for fair practices in trade.

- *Central Silk Board (CSB)*: CSB was established in 1948 as a statutory body that functions under the Ministry of Textiles. Its efforts are in research and development to improve the productivity of natural silk, strengthen efficiency levels and ensure quality products by transfer of technology at all stages.
- *Synthetic & Art Silk Mills' Research Association (SASMIRA)*: SASMIRA was established in 1950. It is involved in the testing of synthetic and artificial silk materials. It is also involved in improvement of manufacturing process and development of machinery and appliances for greater productivity.
- *Indian Jute Industries' Research Association (IJIRA)*: IJIRA, Calcutta, is an autonomous co-operative research organisation. It has done pioneering work in improving the productivity and quality of traditional items in jute industry. It has been extensively involved in researching for quality upgradation, eco-friendly and cost-effective processing of jute fibres; producing an innovative product range; and devising new applications of technical textiles. IJIRA plays a lead role in keeping with the world trends and market today when jute is being used for enormous applications in conventional textile and non-textile fields,.
- *Wool Research Association (WRA)*: Established in 1963, WRA is the only national institute in the field of wool technology and is under the administrative control of the Ministry of Textiles. It is the most important national centre that deals with all aspects of wool and promotes research and development related to various applications on wool.
- *Man-made Textile Research Association (MANTRA)*: Founded in 1981, MANTRA is one of the national textile research laboratories catering to the multifarious needs of the man-made textiles industry at large. The major focus of the association is to carry out research and development activities as well as to provide testing and technical service facilities and hence impart better quality control.
- *Indian Technical Textile Association (ITTA)*: ITTA is the only association of the technical textiles industry in the country. It was registered in 2010. Since, technical textiles are one of the most recent and fastest growing segments it is essential to bring about active participation of the industry on issues and concerns. The association is working to promote, support, develop and increase productions, consumption and export of technical textiles so India can become a powerhouse of technical textiles in the days to come.

> **Technical textiles** are textile materials that are primarily manufactured for their technical performance and functional properties rather than for aesthetic and decorative purposes.

CONSUMER AWARENESS AND PROTECTION

It has been observed that the general public is not entirely aware of the beneficial schemes implemented by various organisations and agencies. Only those consumers who are aware of their rights and responsibilities can protect themselves effectively. Therefore, it is essential to educate the consumers so that they can fight back if they are exploited by the manufacturers and retailers.

Consumer Problems

Globalisation and liberalisation has increased the number of consumer-related issues. The following are major problems that consumers are facing today:

- The large and rapidly-changing variety of textile products that tend to confuse them.
- Due to lack of adequate information available for comparison, the consumer has to waste time looking for the right product.
- The manufacturers and the retailers are free to produce any type and number of goods or services and there is no regulation on the pricing.
- The manufacturers and the retailers use all the tricks of trade to mislead (through inaccurate advertising) and tempt consumers into buying low-grade products that are attractively packaged.
- The quality of goods manufactured locally are, at times, uncertain.
- In India, ignorance and illiteracy among people further leads to their exploitation as consumers.

Legislative Provisions

At the international level, various regulatory legislations have been passed to establish control over the manufacturing and marketing of various textile products. These are:

- Federal Trade Commission Ruling on the Weighting of Silk, 1938
- Wool Products Labelling Act, 1939
- Fur Products Act, 1951
- Flammable Fabrics Act, 1953
- Textile Fibre Product Identification Act, 1960
- Permanent Care Labelling Ruling of the Federal Trade Commission, 1972

In India, these legislations do not apply. Instead, consumer interests are safeguarded through Consumer Protection Act which was introduced in 1986. The act has laid down rights and responsibilities of a consumer and also makes provision for the establishment of consumer councils and consumer courts.

Consumer Rights and Responsibilities

Creating awareness regarding consumer rights and responsibilities is essential so that they can get the true worth of the money spent. According to the Consumer Protection Act, consumers can exercise the following rights:

- The right to safety against any hazardous or sub-standard products.

- The right to information on the quality and performance of the product.
- The right to choose from the variety of products without any outside unfair influence.
- The right to be heard if any case is filed against any manufacturer or retailer due to dissatisfaction from any product.
- The right to seek redressal to claim compensation if the quality of the product does not conform to the promise of the seller.
- The right to consumer education regarding their rights.

Therefore, it is the responsibility of each consumer to know their rights well. They should avoid buying in a hurry and without reading all the instructions carefully. Consumer should be aware of false advertising and never compromise on the quality of the product. They should always ask for the receipt of payment and also a guarantee card if applicable. In case of any dissatisfaction, a complaint should be filed for compensation.

Role of Government in Consumer Disputes Redressal

The Government of India has established Consumer Protection Councils at the central, state and district levels to promote and protect the rights of consumers. These councils function under the Department of Consumer Affairs and the Minister in-charge of the department is the chairman of the council. Similarly, consumer disputes redressal agencies or consumer courts have also been set up at the district, state and national levels. These are as follows:

- Consumer Disputes Redressal Forum or District Forum is established in each district by the state government. It is entitled to hear cases not exceeding Rs 20 lakh.
- Consumer Disputes Redressal Commission or State Commission, established in each state, can hear cases exceeding Rs 20 lakh but not above Rs 1 crore.
- National Consumer Disputes Redressal Commission or National Commission entertains complaints which involve compensation above Rs 1 crore.

The Department of Consumer Affairs under the Government of India has launched a consumer awareness programme named '*Jago Grahak Jago*' to safeguard the interests of the consumers. It has an official website (http://www.jagograhakjago.in/), where consumers can register their complaints. It also runs consumer awareness campaigns and provides funds to the voluntary consumer organisations for conducting comparative testing of various products available in market. Another organisation, Consumer Coordination Council (http://www.corecentre.co.in) was set up to bring all the consumer grievance redressal organisations together and empower the consumer groups.

Role of Non-governmental Organisations (NGOs) in Consumer Grievance Redressal

Various NGOs help consumers in lodging complaints and provide manpower and support to seek compensation against the loss. These organisations also work towards accelerating consumer awareness on their rights and responsibilities through the use of various media exposures so that they are not exploited by fraudulent practices. Some of the NGOs actively working in this context are: International Consumer Rights Protection Council, Consumer Guidance Society of India, Consumer Education and Research Centre, Consumer Grievance, Common Cause and Consumer Forum (www.consumer.org.in). Consumer Voice is a Delhi-based consumer advocacy group that offers suggestions for wise purchasing from a range of tested and verified products. A user-generated content and review site called MouthShut.com (www.mouthshut.com) provides feedback information on a wide range of new products or services.

SUMMARY

- Labels are an essential component of garments. They describe the product and communicate factual information about its service qualities.
- The different types of labels attached to the garments are: content label, manufacturer's label, size label, brand label, trademark or certification label, grade label and smart label.
- Smart label is a recent development that can be scanned using any smart phone, iPad or tablet to get information about the product from the internet.
- The ASTM Care Labelling System is a widely used and internationally-accepted system followed for care instructions on textile products.
- There are various mandatory or voluntary textile standards which define the performance level of the product. These standards are developed by international or national organisations either using test methods or through analytical assessment of the product.
- It is essential to make the consumer aware about the various legislative provisions, consumer rights and responsibilities and consumer redressal forums run by the government and non-government organisations.

KEY WORDS

Informative labelling: It is the information present on the label that describes the product and communicate factual information about its service qualities.

Trademark: It is certification label or a special label which provides a distinctive reputation to the product and certifies its quality.

Brand: It is the name or logo of the company which is in the form of a specific symbol, word or colour.

Textile standards: It is a measure that defines the level of performance of a textile and makes it easier for the consumers to compare the products before buying.

Legislative provisions: These are the various regulatory legislations which establish control over manufacturing and marketing of various textile products.

EXERCISES

1. What are the various types of labels on textile products? Discuss the importance of informative labelling.
2. What is brand labelling?
3. Briefly describe trademark and certification labels.

4. Write a note on woolmark logos?
5. Briefly discuss the care symbols used by ASTM Care Labelling System for washing.
6. Enumerate the various international organisations involved in development of textile standards and briefly discuss the functioning of any two.
7. Enlist the various problems consumers face while purchasing textile products.
8. What are the various regulatory legislations applicable to textile products?
9. Discuss the role of the government in grievance/disputes redressal of a consumer's dissatisfaction with a textile product.

REFERENCES

ASTM International. 2014. 'ASTM D5489-14: Standard Guide for Care Symbols for Care Instructions on Textile Products'. West Conshohocken, PA: www.astm.org (accessed July 2016).

Chatterjee, K. N., R. K. Nayak, S. Bhattacharya and N. R. Kansal. 2006. 'Care labelling of apparels'. *The Indian Textile Journal* 116 (12): 79–83.

Corbman, Bernard P. 1983. *Textiles: Fibre to Fabric*. New York: McGraw-Hill Book Company.

Global Organic Textile Standard (GOTS). n.d. 'Label-Grades'. Available at http://www.global-standard.org/the-standard/general-description.html (accessed February 2017).

Hess, Katharine Paddock. 1954. *Textile Fibres and Their Use*. New Delhi: Oxford and IBH Publishing Co.

Joseph, Marjory L. 1988. *Essentials of Textiles*. New York: Holt, Rinehart and Winston, Inc.

Kadolph, Sara J. 1998. *Quality Assurance for Textiles and Apparel*. New York: Fairchild Publications.

Sekhri, Seema. 2011. *Textbook of Fabric Science: Fundamentals of Finishing*. New Delhi: PHI Learning Private Limited.

Tortora, Phyllis G. 1978. *Understanding Textiles*. New York: Macmillan.

Vidyasagar, P. V. 1998. *Handbook of Textiles*. New Delhi: Mittal Publications.

Wingate, Isabel B. 1976. *Textile Fabrics and Their Selection*. New Jersey: Prentice-Hall.

ONLINE SOURCES

http://suchetadalal.com/?id=f2fe6bec-28fd-7230-4a49fbaa38d3&base=sections&f&t=NGOs+and+Consumer+ (accessed 6 July 2016)

http://www.aatcc.org (accessed 6 July 2016)

http://www.ansi.org (accessed 5 July 2016)

http://www.astm.org (accessed 6 July 2016)

http://www.atira.in (accessed 6 July 2016)

http://www.btraindia.com (accessed 5 July 2016)

http://www.ecolabelindex.com/ecolabel/ecomark-india (accessed 5 July 2016)

http://www.global-standard.org/the-standard/general-description.html (accessed 6 July 2016)

http://www.ijira.org (accessed 5 July 2016)

http://www.india-crafts.com/business-reports/indian-textile-industry/organization-association.htm (accessed 6 July 2016)

http://www.indiantextilejournal.com/articles/FAdetails.asp?id=390 (accessed 5 July 2016)

http://www.iso.org (accessed 6 July 2016)

http://www.ittaindia.org (accessed 6 July 2016)

http://www.mantrasurat.org (accessed 5 July 2016)

http://www.nitratextile.org (accessed 5 July 2016)

http://www.onlineclothingstudy.com/2013/04/7-common-garment-labels-and-information.html (accessed 5 July 2016)

http://www.sasmira.org (accessed 6 July 2016)

http://www.sitra.org.in (accessed 6 July 2016)

http://www.woolmark.com (accessed 5 July 2016)

http://www.wraindia.com (accessed 5 July 2016)

https://www.wewear.org (accessed 5 July 2016)

www.bis.org.in/ (accessed 6 July 2016)

www.cottoninc.com (accessed 5 July 2016)

http://www.cottoninc.com/product/NonWovens/Nonwoven-Resources/The-Seal-of-Cotton/What-is-the-Seal-of-Cotton.cfm (accessed 7 February 2017)

http://www.handloommark.gov.in/ (accessed 7 February 2017)

http://www.woolmark.com/brands/ (accessed 7 February 2017)

http://www.textileaffairs.com/c-common.htm#dry (accessed 7 February 2017)

ABOUT THE EDITORS AND CONTRIBUTORS

THE EDITORS

Deepali Rastogi is Associate Professor at the Department of Fabric and Apparel Science, Lady Irwin College, New Delhi. She has over 20 years of experience in research and teaching. Her research interests include fibre science, chemical processing of textiles, textile finishing, dyeing and printing, and research methodology. Besides publishing extensively in numerous international and national peer reviewed journals, she has a patent on 'Salt-free Dyeing of silk with Bifunctional Reactive Dyes'.

Sheetal Chopra is Assistant Professor at the Department of Fabric and Apparel Science, Lady Irwin College. She has over 20 years of research and teaching experience, and has published in numerous international and national peer reviewed journals. She was also the course writer for a diploma programme at the School of Continuing Education, IGNOU. Her research interests include dyeing and printing, textile finishing, development of sustainable wet processing technologies for textiles, statistics and research methods, and quality assurance, assessment and CSR in the textile and apparel industry.

THE CONTRIBUTORS

Sareekah Agarwaal teaches at the Department of Fabric and Apparel Science, Lady Irwin College. Her research interests include design intervention for persistence of traditional textile crafts and sustainability issues pertaining to this sector.

Ruchira Agarwal is Assistant Professor at the Department of Fabric and Apparel Science, Lady Irwin College. Her areas of interest include textile colouration, fibre science and fabric construction. She is currently pursuing her PhD in silk dyeing with the University of Delhi.

Anjali Agrawal is Assistant Professor at the School of Fashion and Design, G. D. Goenka University, Gurugram. She holds a PhD from the University of Delhi and was a post-doctoral fellow at IIT, Delhi in 2014. Her research interests include sustainable textiles, natural dyes and functional finishing.

Ashima Anand teaches at the Department of Fabric and Apparel Science, Lady Irwin College. Her areas of interest are pattern making, apparel construction, textile design development and computer applications.

Simmi Bhagat is Associate Professor at the Department of Fabric and Apparel Science, Lady Irwin College. Her research interests include Indian textiles, textile documentation and conservation, historic costumes, world textiles, fashion illustration and carbon footprint.

Manpreet Chahal is Assistant Professor at the Department of Fabric and Apparel Science, Lady Irwin College. Her areas of interest include apparel design, pattern making and construction, garment draping and apparel production. She has won awards for her research papers and posters at various international and national conferences.

Bhawana Chanana is Associate Professor at the Department of Fabric and Apparel Science, Lady Irwin College. She is currently on lien and working as Professor and Director, School of Fashion Design and Technology, Amity University, Mumbai. She is a principal member of MHD 16 committee and chairperson of the panel for re-drafting 'Specifications for Sanitary Napkins, IS: 5405–1980' of the Bureau of Indian Standards.

Nidhi Goyal is an academician and her expertise is in textiles, fashion and management. Her areas of interests include processing technologies, nanomaterials for textiles and product development. She holds a PhD in nanofinishing from the University of Delhi in collaboration with IIT, Delhi.

Kanika Kakkar teaches at the Department of Fabric and Apparel Science, Lady Irwin College. Her teaching interests are fabric science, apparel design and construction, fashion studies and extension. She is also a freelance apparel designer for 'Kikli'.

Neha Mehra is Assistant Professor at VJTI Engineering College, Mumbai. She holds a PhD from the Department of Fibres and Textile Processing Technology, Institute of Chemical Technology, Mumbai. Her research interests are colouration and finishing of textiles, medical textiles and nanotechnology.

Sabina Sethi is Associate Professor at the Department of Fabric and Apparel Science, Lady Irwin College. She holds a PhD in ultrasonic cleaning of textiles. Her research interests include consumer and marketing research, fashion retail and merchandising, life cycle assessment of textile items and traditional textiles.

Namita Sharma has taught at the Department of Fabric and Apparel Science, Lady Irwin College. Her interests are draping, pattern making, fibre science, textile documentation and conservation studies.

Divya Singhal has taught at the Department of Fabric and Apparel Science, Lady Irwin College. She is currently pursuing her PhD in textile conservation at the University of Delhi.

Vibha Yadav is Assistant Professor at the Department of Fabric and Apparel Science, Lady Irwin College. Her interests include textile design, Indian textiles, apparel design and construction, and fashion marketing and retail.